Results and Problems in Cell Differentiation

Series Editors:
W. Hennig, L. Nover, U. Scheer

33

Springer
Berlin
Heidelberg
New York
Barcelona
Hong Kong
London
Milan
Paris
Singapore
Tokyo

Paul R. Crocker (Ed.)

Mammalian Carbohydrate Recognition Systems

With 52 Figures and 10 Tables

Springer

Dr. PAUL R. CROCKER
The Wellcome Trust Building
Dept. of Biochemistry
University of Dundee
Dundee DD1 4HN
Scotland

ISBN 3-540-67335-0 Springer-Verlag Berlin Heidelberg New York

Library of Congress Cataloging-in-Publication Data

Mammalian carbohydrate recognition systems / Paul R. Crocker (ed.).
p. cm. – (Results and problems in cell differentiation ; 33)
Includes bibliographical references (p.).
ISBN 3540673350
1. Lectins. I. Crocker, Paul R., 1959– II. Series.

Springer-Verlag Berlin Heidelberg New York
a member of Bertelsmann Springer Science+Business Media GmbH

© Springer-Verlag Berlin Heidelberg 2001
Printed in Germany

Cover design: Meta Design, Berlin
Typesetting: Best-set Typesetter Ltd., Hong Kong
Printed on acid-free paper SPIN 10718859   39/313Oas – 5 4 3 2 1 0

Preface

Over the last decade, the number of known mammalian carbohydrate binding proteins (CBPs) has increased considerably. With the recent completion of the human genome, this number is likely to increase further still. Fortunately, most of the known CBPs can be placed into a small number of distinct families based on the primary amino acid sequence. Currently, there are five recognised categories, namely C-type lectins, Galectins, I-type lectins, P-type lectins and legume-related lectins. In addition, there are 'stand-alone' CBPs such as the hyaluronic acid-binding protein, CD44, although here, too, recent evidence suggests that this CBP may be part of a larger gene family. Each class of CBP has a characteristic protein fold required for carbohydrate recognition and, in most cases, the three-dimensional structure of at least one member has been determined in the presence of a bound oligosaccharide ligand. Given the enormous complexity of carbohydrates themselves and the huge array of glycoproteins and glycolipids that present them, it is not surprising that CBPs are involved in remarkably diverse functions. This is exemplified by the Galectins which mediate specific functions in the nucleus, cytoplasm, cell surface and in the extracellular milieu. Other CBPs are found in discrete cellular and extracellular compartments in a way that is intrinsically linked to their known functions. The 12 chapters of this book are focused on structure-function relationships of these CBP families, including CD44, and are written by leading experts in the field. The order of chapters is based on the cellular localisation of CBPs, beginning with a review on the endoplasmic reticulum chaperones, calreticulin and calnexin, and ending with the Collectins, a group of remarkable molecules that function in host defence to pathogens. The intervening chapters cover Galectins and a range of membrane proteins (ERGIC-53, mannose-6-phosphate receptors, CD44, the mannose/GalNAc-SO4 receptor, Siglecs and Selectins). Collectively, they discuss the molecular basis for carbohydrate recognition and how this triggers various biological responses, ranging from control of glycoprotein folding to initiation of leukocyte recruitment in inflammation. The new millennium offers exciting opportunities not only to identify and molecularly characterise new CBPs but also to deepen our understanding of how carbohydrate recognition by this diverse group of proteins leads to 'glycosignalling' and control of cellular behaviour. I would like to take this opportunity to thank all the authors for their excellent contributions.

Paul R. Crocker, Dundee July 2000

Contents

Galectins Structure and Function – A Synopsis
Hakon Leffler

Structure and Function of CD44: Characteristic Molecular Features and Analysis of the Hyaluronan Binding Site
Jürgen Bajorath

Structure and Function of the Macrophage Mannose Receptor
Maureen E. Taylor

The Man/GalNAc-4-SO₄-Receptor has Multiple Specificities and Functions

Alison Woodworth, Jacques U. Baenziger

Sialoadhesin Structure

Andrew P. May, E. Yvonne Jones

Ligands for Siglecs
Soerge Kelm

Functions of Selectins
Klaus Ley

Carbohydrate Ligands for the Leukocyte-Endothelium Adhesion Molecules, Selectins
Ten Feizi

Structures and Functions of Mammalian Collectins
Uday Kishore, Kenneth B.M. Reid

Addresses of Senior Authors

Dr Klaus Ley
University of Virginia
Health Science Center
School of Medicine
Dept of Biomedical Engineering
BOX 377
Charlottesville, VA 22908
USA

Dr Hakon Leffler
Department of Medical Microbiology
Solvegatan 23
S 22362 Lund
Sweden

Dr Maureen Taylor
Glycobiology Institute
Biochemistry Department
University of Oxford
South Parks Road,
Oxford OX1 3QU
UK

Dr Nancy M. Dahms
Department of Biochemistry
Medical College of Wisconsin
Milwaukee
Wisconsin 53226
USA

Professor
Dr Ten Feizi
The Glycosciences Laboratory
Northwick Park Hospital, Watford Road, Harrow, Middx.
HA1 3UJ
UK

Professor
Dr Ken Reid
MRC Immunochemistry Unit
Department of Biochemistry
University of Oxford
South Parks Road
Oxford OX1 3QU
UK

Dr Soerge Kelm
NW2, University of Bremen
Postfach 33 04 40
D-28334 Bremen
Germany

Dr Jürgen Bajorath
Senior Director
Computer-Aided Drug Discovery
New Chemical Entities, Inc.
18804 North Creek Pkwy. South
Bothell, WA 98011
USA
Tel.: 001-425-424-7297
Fax: 001-425-424-7299
jbajorath@nce-mail.com

Professor
Dr Jacques U. Baenziger
Washington University Medical School
Department of Pathology
660 S. Euclid Ave.
St. Louis, MO 63110
USA

Dr John J.M. Bergeron
Department of Anatomy and Cell Biology
McGill University
Montreal, Quebec H3A 2B2
Canada

Dr Andy May
Department of Structural Biology
Fairchild Building, Room D-139
Stanford University School of Medicine
Stanford, CA 94305-5126
USA

Dr A-C. Roche
Glycobiologie
Centre de Biophysique Moleculaire
Bat. B, rue Charles Sadron
45071 Orleans, cedex 2
France

Lectins of the ER Quality Control Machinery

C. A. Jakob, E. Chevet[1,2], D. Y. Thomas, J. J. M. Bergeron[1]

Abbreviations. CNX: calnexin   CRT: calreticulin   CPY: carboxypeptidase Y   CK2: casein kinase 2   ER: endoplasmic reticulum   UGGT:UDP-glucose glycoprotein glucosyltransferase

1 Introduction

Calnexin was first identified as a type I transmembrane phosphoprotein located in the endoplasmic reticulum (ER; Wada et al. 1991). In subsequent work, this protein has been shown to belong to a new family of chaperones or proteins with lectin-like properties with orthologs present in all eukaryotes so far studied (Wada et al. 1991; Hebert et al. 1995). Calnexin and calreticulin, an ER luminal homologue, share extensive sequence similarity and both bind calcium.

Calreticulin was originally isolated as a calcium storage protein from the sarcoplasmic reticulum and the ER. Only later was it discovered that calreticulin also binds oligosaccharides and hence belongs to the calnexin-like lectins. Two calnexin homologues, calmegin (Watanabe et al. 1994) and/or calnexin-t (Ohsako et al. 1994), have also been described in testis and display the same 'lectin' properties as calnexin in their luminal domains and also cytoplasmic C-terminal regions. Regions of these proteins share amino acid sequence identity ranging from 42–78% (Wada et al. 1991).

2 Structural Aspects

Common features in the luminal domain of all calnexin and calreticulin-like proteins are the high affinity Ca^{2+}-binding site and the proline-rich repeat motifs 1 (I-DPD/EA-KPEDWDD/E) and 2 (G-W-P-I-NP-Y; Fig. 1A). The proline-rich motif is also known as the P-domain in calreticulin (Trombetta and Helenius 1998). Calnexin, calreticulin and calmegin all contain multiple

[1] Department of Anatomy and Cell Biology, McGill University, Montreal, QC, H3A 2B2, Canada
[2] Genetics Group, Biotechnology Research Institute, National Research Council of Canada, 6100 Royalmount Avenue, Montreal, QC, H4P 2R2, Canada

Results and Problems in Cell Differentiation, Vol. 33
Paul R. Crocker (Ed.): Mammalian Carbohydrate Recognition Systems
© Springer-Verlag Berlin Heidelberg 2001

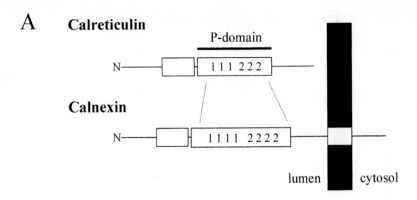

A **Calreticulin**

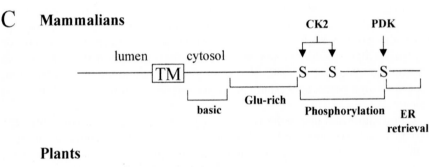

C **Mammalians**

Plants

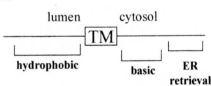

Fig. 1A–C. Structural features of calnexins and calreticulins. **A** Structural domains of calnexin and calreticulin. Both proteins contain the Ca^{2+} and the oligosaccharide binding domains. The oligosaccharide-binding domain contains two structural motifs (1 and 2, for details see text). **B** Alignments of all available calnexin amino acid sequences (see Table 1). The *upper half* represents the amino acid sequences immediately after the signal peptide. The *lower half* shows an alignment of amino acids of the luminal juxtamembrane domain. The beginning of the transmembrane domain is indicated (*TM*). The sequences were aligned by using the PileUp algorithm of the GCG Software package (gap creation penalty 12, gap extension penalty 4). Signal sequence cleavage sites and transmembrane domains were manually aligned afterwards, according to the published or predicted sites. The regions of similarity were shaded with Genedoc V2.5 (*black box* 93% identity, *dark grey* 67% identity, *light grey* 30% identity). **C.** Structural elements of C-terminal cytoplasmic regions of mammalian and plant calnexins (for details, see text)

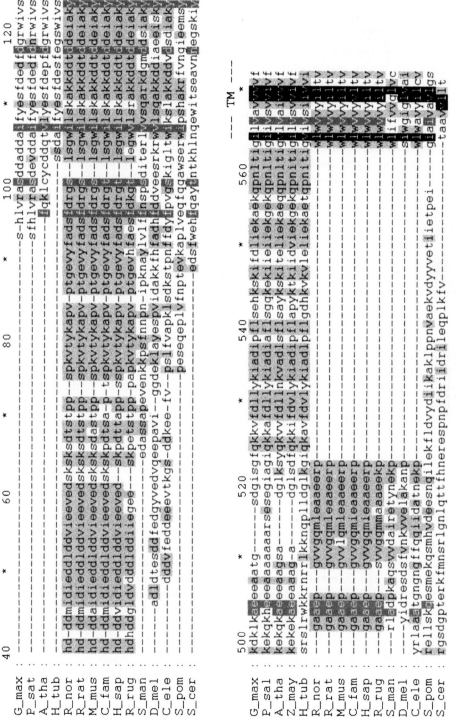

Fig. 1A–C. Continued

repeats of this motif; three in calreticulin, and four in calnexin. The binding of oligosaccharides occurs in this region (Trombetta and Helenius 1998).

We have aligned all the available amino acid sequences of calnexins and calreticulins. The amino acid sequences of calreticulin (30 sequences, see Table 1 for details) are very similar. All sequences contain the P-domain and a putative ER-retrieval sequence K/HDEL. None of the aligned proteins contain any other regions of similarity (data not shown).

The situation was more interesting when comparing the amino acid sequences of calnexin (16 sequences, see Table 1 for details about individual sequences). Mammals and the frog *Rana rugosa* possess a region rich in aspartate and glutamate immediately after the signal sequence. This motif, however less conserved, is also present in *Drosophila melanogaster, Caenorhabditis elegans* and *Schistosoma mansoni*, but is not found in plants (Fig. 1B). This motif is then followed by a proline-rich hydrophobic stretch and then another highly charged lysine-aspartate-rich region.

Other regions of divergence between phyla are located in the luminal side of the transmembrane domain. Mammals (*Rattus norvegicus, Rattus rattus, Mus musculus, Canis familiaris, Homo sapiens*) and amphibian *R. rugosa* calnexins contain a hydrophobic and glutamic acid-charged motif of approximately 20 amino acids (Fig. 1B). Plant (*Glycine max, Pisum sativum, Arabidobsis thaliana, Zea mays, Helianthus tuberosus*) calnexins have a different highly conserved juxtamembrane motif of 40 amino acids (Fig. 1B), although this region is also rich in hydrophobic and different charges of amino acids. No similar motifs have been found in any other proteins in the genome databases (C.A. Jakob, unpubl. data). The function of these regions is unclear, but they may be involved in recruiting or binding ER luminal proteins to calnexin and/or participate in modulating oligosaccharide binding, mediate folding of substrate proteins or be involved in other as yet unknown processes.

The cytoplasmic tails of calnexin of different species vary in size, from a single amino acid (*S. cerevisiae*) to 50 amino acids (plants) to approximately 100 amino acids in mammalian and amphibian calnexins. The sequence of the mammalian cytosolic regions is conserved with four defined regions (Wong et al. 1998): (1) a lysine-rich basic region, (2) an acidic (glutamic acid-rich) motif, (3) a phosphorylation domain, containing three phosphorylation sites (CK2 and PKC/proline-directed kinase sites), and (4) a putative ER retrieval motif (Fig. 1C). The C-terminal cytoplasmic tails of plant calnexins are shorter and contain only the basic region (Fig. 1 C), as well as potential serine and/or threonine phosphorylation sites. However, these potential sites of phosphorylation are not conserved in their position as found for the mammalian phosphorylation sites.

In summary, the luminal domains of calnexin are highly conserved with those of calreticulin in all eukaryotic cells. The calnexin of different phyla have their own characteristics. These include serine phosphorylation sites in the cytosolic tails of mammalian and plant calnexin and a stretch of hydrophobic

Table 1. Amino acid sequences of calnexin and calreticulin

Group	Calnexin (PID no.)	Calreticulin (PID no.)
Mammals	*R. norvegicus* (g543922) *R. rattus* (g310085) *M. musculus* (g543921) *H. sapiens* (g543920) *C. familiaris* (g3123183)	*R. norvegicus* (g1845572) *M. musculus* (g117502) *O. cuniculus* (g117504) *H. sapiens* (g87015) *B. taurus* (g348694) *B. taurus isoform* (g631545)
Plants	*G. max* (g3334138) *P. sativum* (g3702620) *A. thaliana* (g231683) *Z. mays* (g1181331) *H. tuberosus* (g510907)	*A. thaliana* (g2052383) *Z. mays* (g577612) *H. vulgare* (g439588) *N. plumbaginifolia* (g1419088) *N. tabacum* (g732893) *R. communis* (g1763297) *D. bioculata* (NID g1653976)[a] *B. napus* (g2429343) *B. vulgaris* (g3288109) *C. annuum* (NID g984112)[a]
Amphibians	*R. rugosa* (g1514959)	*R. rugosa* (g1514957) *X. laevis* (g258665)
Molluscs		*A. californica* (g345400)
Arthropods	*D. melanogaster* (g2213427)	*D. melanogaster* (g7686) *A. americanum* (g3924593)
Fungi	*S. cerevisiae* (g595528) *S. pombe* (g543923)	
Nematodes	*C. elegans* (g461686)	*C. elegans* (g6694) *O. volvulus* (g309630) *N. americanus* (g687326) *L. sigmodontis* (g2597972) *D. immitis* (g4115903)
Platyhelminths	*S. mansoni* (g422335)	*S. mansoni* (g345835) *S. japonicum* (g2829281)
Dictyostelium		*D. discoideum* (g3913371)
Euglenozoa		*E. gracilis* (g3970687) *L. donovani* (g1545977)

[a] only DNA sequence available

amino acids on the luminal side, which is in close proximity with the trans-membrane domain. These conserved elements may be important for associa-tion with other proteins or for the regulation of protein-protein interactions (see below).

3 Calnexin and Calreticulin in Quality Control of Glycoprotein Folding

N-linked glycosylation is highly conserved from yeast to mammalian cells. Preassembled oligosaccharides (Glc3Man9GlcNAc2) are transferred from a dolichyl phosphate precursor and are attached to the asparagine residue in the sequon Asn-X-Ser/Thr of newly synthesized secretory proteins (Kornfeld and Kornfeld 1985; Varki 1993; Roth 1995). The observation that calnexin is a mol-ecular chaperone with apparent specificity for N-linked glycoproteins was key in the elucidation of what has come to be termed the calnexin cycle of glyco-protein folding (Ou et al. 1993). For N-linked glycoprotein biosynthesis, the outermost glucose is quickly removed in the ER by glucosidase I, followed by the sequential removal of the two remaining glucoses by glucosidase II. Further processing occurs by mannosidases in the ER (yeast and mammals) and in the Golgi apparatus (in mammals only; Kim and Arvan 1995). During this glycan trimming, glycoproteins attain their correct tertiary and quaternary struc-tures. Folding is assisted by chaperones and mediated by isomerases (prolyl peptidyl isomerases, protein disulfide isomerases, including ERp57, etc.). Cor-rectly folded proteins are then able to leave the ER and are transported to other cellular compartments. Non-folded, incompletely folded or misfolded proteins have been found to remain in the lumen of the ER. Various lines of evidence have indicated that both calnexin and calreticulin are involved in mediating the folding process and also in retaining proteins in the ER lumen and through this function form a part of a quality control system (Fig. 2; Ou et al. 1993; Hammond et al. 1994).

The component of the calnexin cycle which apparently acts as a sensor for the folding state of glycoproteins and can discriminate between folded and non-correctly folded substrates is the enzyme UDP-glucose: glycoprotein glu-cosyltransferase (UGGT; Sousa et al. 1992). This enzyme can add a single glucose residue to glucosylate misfolded glycoproteins. By adding this single glucose residue, UGGT tags the incorrectly folded glycoprotein, which can bind to calnexin and calreticulin. Thus the misfolded glycoprotein is retained in the folding environment of the ER lumen. Glucosidase II removes the glucose residue from this oligosaccharide and thus releases the bound glycoprotein. Unfolded glycoproteins are postulated to undergo repeated cycles of binding to and release from calnexin and calreticulin. Glucosidase II does not recog-nize the conformation of the polypeptide but acts to remove the glucose residues (Zapun et al. 1997) and promote the detachment of the oligosaccha-ride from calnexin and calreticulin. If the glycoprotein is folded, it is not reglu-

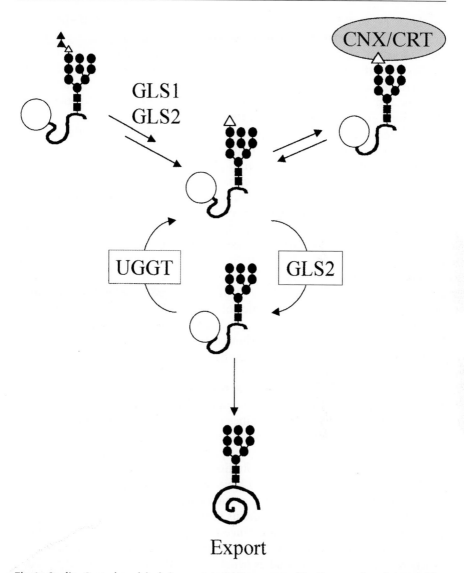

Fig. 2. Quality Control model of glycoprotein folding proposed by Hammond et al. (1994). The oligosaccharide of a nascent glycoprotein is trimmed by glucosidase I and II to a monoglucosylated form. If the glycoprotein has obtained its correct tertiary and quaternary structure, the glycoprotein will exit the ER lumen to subsequent compartments. If not correctly folded, the glycoprotein is subject to numerous cycles of de-/reglucosylation and association with calnexin/calreticulin until the correct fold is achieved (see text for details). Symbols: *triangle* glucose, *circle* mannose, *square* GlcNAc, *large open circle* chaperone. Abbreviations: *CNX* calnexin, *CRT* calreticulin, *GLS1* glucosidase I, *GLS2* glucosidase II, *UGGT* UDP-glucose glycoprotein glucosyltransferase

cosylated by UGGT and does not re-enter the calnexin cycle. If the glyco-protein is incompletely folded, it is reglucosylated by UGGT and re-enters the cycle (Sousa et al. 1992). This model for glycoprotein folding is supported by studies with the well-characterized model viral protein influenza hemagglu-tinin (Hebert et al. 1996),the VSV G protein (Hammond et al. 1994), and more recently and definitively by in vitro experiments with purified components (Hammond et al. 1994; Zapun et al. 1997).

The oligosaccharide-binding properties of calnexin and calreticulin have been characterized in detail and it appears that they bind glycoproteins in a lectin-type manner. The association of glycoprotein substrates with calnexin can be prevented experimentally either by glucosidase inhibitors, such as castanospermine and deoxynojirimycin, by the glycosylation inhibitor tuni-camycin or by mutation of the glycosylation sites (Zapun et al. 1997). From the literature, the number of oligosaccharides which are required for a stable calnexin/calreticulin interaction appear to vary. Using mutants of glycosy-lation sites, Hebert et al. (1997) found that two oligosaccharides were the minimal requirement for the co-immunopreciptiation of VSV G protein and calnexin. Similarly, Rodan et al. (1996) engineered a second glycosylation sequon into the amino acid sequence of RNase B for their studies of interac-tion in isolated microsomes. However, Zapun et al. (1997) using an in vitro system with defined components, showed that a single N-linked oligosaccha-ride chain is sufficient for interaction between the calnexin domain and G1 glycosylated RNAse B.

The idea of calnexin, a type I membrane protein, as a component of a quality control model where misfolded monoglucosylated glycoproteins are retained in the ER, is especially attractive. Since calnexin is anchored in the ER mem-brane, it might prevent bound, misfolded proteins from prematurely exiting the ER and being transported to subsequent cellular compartments. Calreti-culin has the same oligosaccharide binding properties as calnexin but is an ER luminal protein. Wada et al. (1995) have shown that calreticulin can be con-verted into a membrane-bound form, and hence perform a similar function to calnexin, i.e. retaining G1 glycoproteins in the ER. Moreover, calnexin and cal-reticulin both promote refolding, since both proteins have been reported to associate with the disulfide isomerase ERp57 (see below).

There have been several studies which seek to determine the glycoproteins, which bind to calnexin and calreticulin. It was found that calnexin and cal-reticulin show a partial overlap in the glycoproteins that associate with them (summarized in Helenius et al. 1997). For some glycoproteins, it was found that calnexin and calreticulin work in a concerted manner. For the folding and assembly of influenza HA, it was shown that calnexin first associates with the oligosaccharides during protein translocation, and then HA is transferred to calreticulin later, in protein folding. In this case it is believed that the binding of HA to calnexin favors the formation of the correct disulfide bonds between two cysteine residues, one located in the N-terminal portion of the protein, the other at the C-terminus. A conceptually similar scenario of a molecular chap-eron relay of calnexin and calreticulin in association with newly synthesized

MHC class I allotypes and TAP transporters has also been elucidated (Williams and Watts 1995; van Leeuwen and Kearse 1996; Pamer and Cresswell 1998).

Does calnexin recognize the peptide moiety of proteins? The reports in the literature which claim that calnexin can bind glycosylated polypeptides independently of glycosylation may be grouped into two classes. In one class of experiments, association of substrate with calnexin (revealed by co-immunoprecipitation) was not abrogated when glycosylation of a substrate was prevented either by inhibitory drugs or by mutation of the glycosylation sites (Carreno et al. 1995; Kim and Arvan 1995; Loo and Clarke 1995). However, in a report where the size of the complex was examined by sucrose density gradient, it appeared that a large aggregate was formed rather than the usual 1:1 calnexin to substrate ratio (Cannon et al. 1996). In the second group of studies, enzymatic cleavage of the oligosaccharides from a substrate which has been co-immunoprecipitated with calnexin did not result in its release from calnexin (Arunachalam and Dreswell 1995; Ware et al. 1995; Zhang et al. 1995). The release from calnexin was assessed by the presence of the substrate in the supernatant after treatment and recentrifugation of the immunoprecipitate. However, the substrates selected by in vivo immunoprecipitation are, by definition, unfolded proteins and these are likely to aggregate and are thus unlikely to be found in a soluble fraction after their release from calnexin. Thus, the current evidence points to a function of calnexin and calreticulin as lectins. However, it cannot be excluded that calnexin or calreticulin might associate with some unglycosylated proteins, and a definitive answer to this question will probably come from in vitro experiments with defined components. Indeed such in vitro studies now provide evidence for a direct influence of calnexin (and calreticulin) confor motional changes induced by ATP binding on the dissagregation of unglycosylated proteins (Ihara et al. 1999; Saito et al. 1999).

4 Calnexin and Calreticulin in Glycoprotein Degradation

The original quality control model proposed that misfolded glycoproteins be retained in the ER until the protein folds correctly. For proteins to permanently exit the calnexin cycle it has been proposed that in addition to their correct folding, a mannose residue is removed, making the glycoprotein a less-favored substrate of UGGT and perhaps permitting recognition by an ER mannose lectin (such as ERGIC-53; Arar et al. 1995; Itin et al. 1996).

Short-lived and misfolded ER or secretory proteins are degraded in the cytosol in a proteasome-dependent pathway. A possible link between quality control and the degradation of glycoproteins has been shown by work performed using mannosidase inhibitors. The degradation of the secretion-incompetent truncated variant α1-antitrypsin null Hong Kong in mouse hepatoma cells was blocked by deoxymannojirimycin, a specific mannosidase inhibitor (Liu et al. 1997). Moreover, the proteasome-dependent degradation of the T-cell receptor subunit CD3-δ in lymphocytes was abrogated with the same mannosidase inhibitor (Yang et al. 1998). Furthermore, recent work in S. cere-

visiae has shown the importance of a specific oligosaccharide structure for efficient degradation of a mutant allele of carboxypeptidase Y (Knop et al. 1996; Jakob et al. 1998). These results suggest that mannose residues of the N-linked oligosaccharides may play an important role in the quality control of glycoprotein folding (Fig. 3). Glycoproteins that fold correctly are able to exit the ER lumen and be transported to the Golgi apparatus, even if their oligosaccharides have not been trimmed correctly. Cell lines deficient for glucosidase II or incubated in the presence of glucosidase inhibitors can normally secrete proteins, indicating that extensive oligosaccharide trimming is not essential for export (Ora and Helenius 1995; Hebert et al. 1995). However, glycoproteins retained in the ER by the UGGT-calnexin-glucosidase II cycle (Bergeron et al. 1998), due to misfolding, are subject to complete trimming by mannosidases. Biochemical evidence suggests that a specific ER-resident $\alpha1,2$-mannosidase exists that shows a similar activity to the Mns1p of *S. cerevisiae* (Moremen et al. 1994), where removal of the middle $\alpha1,2$-linked mannose is essential for efficient degradation. This Man8GlcNAc structure might be recognized by a lectin-like receptor targeting the misfolded protein to retro-translocation and finally to degradation. Hence, the removal of specific mannose residues terminates the calnexin-mediated protein folding cycle.

Liu et al. (1999) have proposed that null Hongkong variant $\alpha1$-antitrypsin is tightly associated with calnexin and degraded in association with calnexin, indicating that calnexin might be involved in mediating degradation of misfolded glycoproteins. On the other hand, it has been previously shown that the affinity of calnexin towards monoglucosylated oligosaccharides markedly decreases with the number of mannoses present on the oligosaccharide. Further work has to prove the involvement of calnexin in degradation of misfolded ER protein.

In *S. cerevisiae*, the calnexin homologue Cne1p has only been shown to be involved in degradation of unglycosylated α-factor (McCracken and Brodsky 1996), but not in the degradation of misfolded glycoprotein CPY* or others (Knop et al. 1996). In *S. cerevisiae*, it is not known what the function of Cne1p is. In *S. pombe*, however, Cnx1p – the calnexin ortholog – is a key member of the calnexin cycle as described for higher eukaryotic cells (for review, see Parodi 1999)

We postulate a 'degradation-lectin' that specifically recognizes misfolded glycoproteins with the oligosaccharide structure GlcxMan8GlcNAc2 associated with an ER chaperone, such as BiP/Kar2p. This interaction would in turn target the bound protein for retro-translocation and for subsequent cytosolic degradation by the proteasomal pathway. This mechanism is proposed to work for soluble as well as membrane glycoproteins. We suggest that such a degradation-lectin exists not only in yeast *S. cerevisiae* but also in other eukaryotic cells. This hypothesis incorporates the mannose-timer/exit hypothesis (Su et al. 1993; Helenius et al. 1997) with the experimental findings that calnexin and calreticulin, but also UGGT preferably bind oligosaccharides of the structure Glc1Man9GlcNAc2 (Spiro et al. 1996; Zapun et al. 1997).

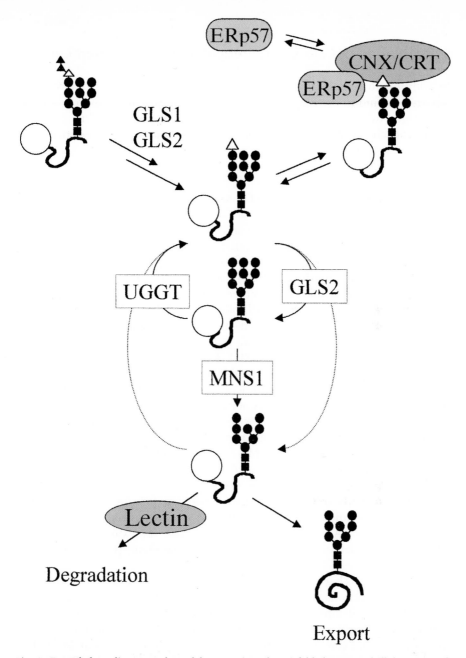

Fig. 3. Extended quality control model. A persistently misfolded protein will be trimmed by ER mannosidase I (mannose timer) and hence reduce the affinity for UGGT and calnexin/calreticulin, respectively. This will also allow for a lectin to target the misfolded protein for degradation. This extended model predicts a degradation of mutated, misfolded or slow-folding proteins, thus clearing the ER lumen. In the recent years, ERp57 has been shown to receive its monoglucose-specific activity by associating with calnexin/calreticulin. Symbols: *triangle* glucose, *circle* mannose, *square* GlcNAc, *large open circle* chaperone. Abbreviations: *CNX* calnexin, *CRT* calreticulin, *ERp57* monoglucose-dependent protein disulfide isomerase, *GLS1* glucosidase I, *GLS2* glucosidase II, *UGGT* UDP-glucose glycoprotein glucosyltransferase, *MNS1* ER mannosidase I

5 Regulation of Calnexin/Calreticulin-Substrate Interaction

In addition to oligosaccharides, a variety of other components have been reported to bind to the ER luminal domain of calnexin and also to its soluble homologue calreticulin. Moreover, modifications of the C-terminal cytoplasmic tail of calnexin appear to be important in modulating the properties of the complete protein.

5.1 Modifications of the Luminal Domain

Calnexin and calreticulin have been shown to bind to ERp57 in vitro and to confer specificity for monoglucosylated glycoproteins to the disulfide isomerase ERp57. Alone, ERp57 has oxidoreductase activity but little disulfide isomerase activity (Bourdi et al. 1995). This property contrasts with the other PDI family members, which reveal potent oxidoreductase and isomerase activities. ERp57 has also been shown (Elliott et al. 1997; Zapun et al. 1998; Hughes and Cresswell 1998) to participate in the disulfide bond formation of Glut-1 and MHC class I in vivo (Elliott et al. 1997; Hughes and Cresswell 1998) and RNAse B in vitro (Zapun et al. 1998). Moreover, calnexin and calreticulin behave in vitro as catalysts of the disulfide isomerase activity of ERp57 (Zapun et al. 1997). Thus, calnexin and calreticulin can both be regarded as a platform for ERp57 to perform its isomerase activity.

The ATP binding of calnexin was detected by its high absorbance at 260 nm and has been demonstrated by UV cross-linking experiments (Ou et al. 1995), However, the effect of Mg-ATP on the oligomerization of calnexin is controversial (Ou et al. 1995; Vassilakos et al. 1996) (Fig. 2). ATP binding to calnexin was abrogated by reduction (Ou et al. 1995). Association of ATP with calnexin apparently increases affinity of calnexin for oligosaccharides, but the specificity of this effect is questionable (Vassilakos et al. 1996). Interestingly, an effect of ATP on substrate release from calnexin and calreticulin has been also observed in vivo in co-immunoprecipitates (Otteken and Moss 1996).

Divalent cations have also been shown to play a crucial role in calnexin function. The interactions of calnexin with its substrates are highly dependent on divalent metal ions or polyamines, but are not influenced by detergent and sulfhydryl agents (Wiuff and Houen 1996). Moreover, calcium creates a protease-resistant core of the luminal domain of calnexin (Ou et al. 1995) (Fig. 2) and is necessary for oligosaccharide binding (Vassilakos et al. 1996) (Fig. 2). Furthermore, calcium has been shown to be required for ATP binding to calnexin as indicated by EGTA treatment which abolishes ATP binding (Ou et al. 1995; Vassilakos et al. 1996).

5.2 Alterations of the Cytosolic Tail

Mammalian calnexin is phosphorylated in vitro and in vivo on three serine residues within the cytosolic domain. Two of the phosphorylation sites are protein kinase CK2 motifs and the third was defined as a proline-directed kinase motif (Fig. 1 A) (Ou et al. 1992; Wong et al. 1998). In addition, calmegin has also been shown to be in vivo phosphorylated (Roderick et al. 1998 and 2000), but the responsible kinases have not been identified. Furthermore, phosphorylation of the calnexin cytoplasmic tail by CK2 has also been shown in the sarcoplasmic reticulum (Cala et al. 1993). In other species, calnexin homologues are also phosphorylated: i.e. Cnx1p in *S. pombe* (H. Wong and J.J.M. Bergeron, unpubl. data) and SmIrV1 in *S. mansoni* (Hawn and Strand 1994).

The role of phosphorylation has not yet determined. Nevertheless, *C. difficile* Cytotoxin B, which is retrotranslocated in the cytosol through the secretory pathway, promotes calnexin hyperphosphorylation (Schue et al. 1994). Moreover, treatment of cells with okadaic acid, a serine phosphatase inhibitor, leads to increased calnexin phosphorylation and to increased retention time of MHC class I molecules in the ER (Tector et al. 1994). Furthermore, MHC class I molecules have been shown to associate differentially with calnexin depending on their rate of egress from the ER (Capps and Zuniga 1994). Recent data show that phosphorylation may regulate the association of calnexin with ribosomes associated with the ER membrane (Chevet et al. 1999). The phosphorylation appears to be regulated through extracellular stimuli, indicating a response to the need for increased efficiency of protein synthesis and glycoprotein folding.

6 Conclusions

There is an increasing body of evidence which confirms the quality control model of Hammond and Helenius (Hammond et al. 1994). Many of its elements have been identified and characterized, amongst them the ER luminal lectins, calnexin and calreticulin. The lumen of the ER is a highly specialized compartment for a magnitude of processes such as protein synthesis, protein translocation through the ER membrane, protein folding, post-translational modification, protein sorting and transport. The integrated function of the folding process in the ER certainly results from the cooperation of its many components. The hypothesis that the ER resident proteins may interact loosely with each other to form a dynamic matrix (Sambrook 1990) is supported by the observations that several ER proteins can be co-immunoprecipitated or cross-linked (Ou et al. 1993; Ou et al. 1995; Cannon et al. 1996) under some conditions. As most ER-resident proteins are high capacity low affinity calcium binders, calcium ions may mediate their loose association. It has been proposed that this ER protein network may function as a chromatographic matrix for newly synthesized proteins which would remain adsorbed until losing their affinity by becoming folded (Tatu et al. 1995).

We are only starting to learn how protein interactions in the ER are formed, modulated and regulated. It has been a relatively short time since the initial identification of calnexin as a type I integral membrane protein of the ER and the discovery of its mechanism of action as a lectin via the calnexin cycle. Although less well understood, other lectins are clearly involved in sorting **of productive cargo to** exit from the ER (ERGIC-53; Agrar et al. 1995; Itin et al. 1996), or for retrotranslocation of misfolded glycoproteins to the proteasome (Jakob et al. 1998). The concentration of calcium, ATP, redox potential, protein phosphorylation and associated proteins play an important role in modulating the variety of functions occurring in the ER.

Acknowledgements. This work has been supported by the Swiss National Science Foundation (fellowship 81ZH-054340 to C.A.J), the Medical Research Council of Canada (to J.J.M.B. and D.Y.T).

References

Arar C, Carpentier V, Le-Caer JP, Monsigny M, Legrand A, Roche AC (1995) ERGIC-53, a membrane protein of the endoplasmic reticulum-Golgi intermediate compartment, is identical to MR60, an intracellular mannose-specific lectin of myelomonocytic cells. J Biol Chem 270:3551–3553

Arunachalam B, Cresswell P (1995) Molecular requirements for the interaction of class II major histocompatibility complex molecules and invariant chain with calnexin. J Biol Chem 270: 2784–2790

Bergeron JJM, Zapun A, Ou WJ, Hemming R, Parlati F, Cameron PH, Thomas DY (1998) The role of the lectin calnexin in conformation independent binding to N-linked glycoproteins and quality control. Adv Exp Med Biol 435:105–116

Bourdi M, Demady D, Martin JL, Jabbour SK, Martin BM, George JW, Pohl LR (1995) cDNA cloning and baculovirus expression of the human liver endoplasmic reticulum P58: characterization as a protein disulfide isomerase isoform, but not as a protease or a carnitine acyltransferase. Arch Biochem Biophys 323:397–403

Cala SE, Ulbright C, Kelley JS, Jones LR (1993) Purification of a 90 kDa protein (vol VII) from cardiac sarcoplasmic reticulum. Identification as calnexin and localization of casein kinase II phosphorylation sites. J Biol Chem 268:2969–2975

Cannon KS, Hebert DN, Helenius A (1996) Glycan-dependent and -independent association of vesicular stomatitis virus G protein with calnexin. J Biol Chem 271:14280–14284

Capps GG, Zuniga MC (1994) Class I histocompatibility molecule association with phosphorylated calnexin. Implications for rates of intracellular transport. J Biol Chem 269:11634–11639

Carreno BM, Solheim JC, Harris M, Stroynowski I, Connolly JM, Hansen TH (1995) TAP associates with a unique class I conformation, whereas calnexin associates with multiple class I forms in mouse and man. J Immunol 155:4726–4733

Chevet E, Wong HN, Gerber D, Cochet C, Fazel A, Cameron PH, Gushue JN, Thomas DY, Bergeron JJM (1999) Phosphorylation by CK2 and MAPK enhances calnexin association with ribosomes. EMBO J 18:3655–3666

Elliott JG, Oliver JD, High S (1997) The thiol-dependent reductase ERp57 interacts specifically with N-glycosylated integral membrane proteins. J Biol Chem 272:13849–13855

Hammond C, Braakman I, Helenius A (1994) Role of N-linked oligosaccharide recognition, glucose trimming, and calnexin in glycoprotein folding and quality control. Proc Natl Acad Sci USA 91:913–917

Hawn TR, Strand M (1994) Developmentally regulated localization and phosphorylation of SmIrV1, a *Schistosoma mansoni* antigen with similarity to calnexin. J Biol Chem 269: 20083–20089

Hebert DN, Foellmer B, Helenius A (1996) Calnexin and calreticulin promote folding, delay oligomerization and suppress degradation of influenza hemagglutinin in microsomes. EMBO J 15:2961–2968

Hebert DN, Simons JF, Peterson JR, Helenius A (1995) Calnexin, calreticulin, and Bip/Kar2p in protein folding. Cold Spring Harb Symp Quant Biol 60:405–415

Hebert DN, Zhang JX, Chen W, Foellmer B, Helenius A (1997) The number and location of glycans on influenza hemagglutinin determine folding and association with calnexin and calreticulin. J Cell Biol 139:613–623

Helenius A, Trombetta ES, Hebert DN, Simons JF (1997) Calnexin, calreticulin and the folding of glycoproteins. Trends Cell Biol 7:193–200

Hughes EA, Cresswell P (1998) The thiol oxidoreductase ERp57 is a component of the MHC class I peptide-loading complex. Curr Biol 8:709–712

Ihara Y, Cohen-Doyle MF, Saito Y, Williams DB (1999) Calnexin discriminates between protein conformational states and functions as a molecular chaperone in vitro. Mol Cell 4:331–341

Itin C, Roche AC, Monsigny M, Hauri HP (1996) ERGIC-53 is a functional mannose-selective and calcium-dependent human homologue of leguminous lectins. Mol Biol Cell 7:483–493

Jakob CA, Burda P, Roth J, Aebi M (1998) Degradation of misfolded endoplasmic reticulum glycoproteins in *Saccharomyces cerevisiae* is determined by a specific oligosaccharide structure. J Cell Biol 142:1223–1233

Kim PS, Arvan P (1995) Calnexin and BiP act as sequential molecular chaperones during thyroglobulin folding in the endoplasmic reticulum. J Cell Biol 128:29–38

Knop M, Hauser N, Wolf DH (1996) N-glycosylation affects endoplasmic reticulum degradation of a mutated derivative of carboxypeptidase Yscy in yeast. Yeast 12:1229–1238

Kornfeld R, Kornfeld S (1985) Assembly of asparagine-linked oligosaccharides. Annu Rev Biochem 54:631–664

Liu Y, Choudhury P, Cabral CM, Sifers RN (1997) Intracellular disposal of incompletely folded human alpha1-antitrypsin involves release from calnexin and post-translational trimming of asparagine-linked oligosaccharides. J Biol Chem 272:7946–7951

Liu Y, Choudhury P, Cabral CM, Sifers RN (1999) Oligosaccharide modification in the early secretory pathway directs the selection of a misfolded glycoprotein for degradation by the proteasome. J Biol Chem 274:5861–5867

Loo TW, Clarke DM (1995) P-glycoprotein. Associations between domains and between domains and molecular chaperones. J Biol Chem 270:21839–21844

McCracken AA, Brodsky JL (1996) Assembly of ER-associated protein degradation in vitro: dependence on cytosol, calnexin, and ATP. J Cell Biol 132:291–298

Moremen KW, Trimble RB, Herscovics A (1994) Glycosidases of the asparagine-linked oligosaccharide processing pathway. Glycobiology 4:113–125

Ohsako S, Hayashi Y, Bunick D (1994) Molecular cloning and sequencing of calnexin-t. An abundant male germ cell-specific calcium-binding protein of the endoplasmic reticulum. J Biol Chem 269:14140–14148

Ora A, Helenius A (1995) Calnexin fails to associate with substrate proteins in glucosidase II-deficient cell lines. J Biol Chem 270:26060–26062

Otteken A, Moss B (1996) Calreticulin interacts with newly synthesized human immunodeficiency virus type 1 envelope glycoprotein, suggesting a chaperone function similar to that of calnexin. J Biol Chem 271:97–103

Ou WJ, Bergeron JJM, Li Y, Kang CY, Thomas DY (1995) Conformational changes induced in the endoplasmic reticulum luminal domain of calnexin by Mg-ATP and Ca^{2+}. J Biol Chem 270: 18051–18059

Ou WJ, Cameron PH, Thomas DY, Bergeron JJM (1993) Association of folding intermediates of glycoproteins with calnexin during protein maturation. Nature 364:771–776

Ou WJ, Thomas DY, Bell AW, Bergeron JJ (1992) Casein kinase II phosphorylation of signal sequence receptor alpha and the associated membrane chaperone calnexin. J Biol Chem 267: 23789–23796

Pamer E, Cresswell P (1998) Mechanisms of MHC class I-restricted antigen processing. Annu Rev Immunol 16:323–358

Parodi AJ (1999) Reglucosylation of glycoproteins and quality control of glycoprotein folding in the endoplasmic reticulum of yeast cells. Biochim Biophys Acta 1426:287–295

Rodan AR, Simons JF, Trombetta ES, Helenius A (1996) N-linked oligosaccharides are necessary and sufficient for association of glycosylated forms of bovine RNase with calnexin and calreticulin. EMBO J 15:6921–6930

Roderick HL, Lechleiter JD, Camacho P (1998) Calnexin/Calmegin phosphorylation regulates Ca^{2+} wavee activity in Xenopus oocytes. Mol Biol Cell 9S:494–495

Roderick HL, Lechleiter JD, Camacho P (2000) Cytosolic phosphorylation of calnexin controls Intracellular Ca^{2+} oscillations via an interaction with SERCA2b. J Cell Biol 149:1235–1248

Roth J (1995) Biosynthesis: compartmentation of glycoprotein biosynthesis. In: Montreuil J, Schachter H, Vliegenthart JFG (eds) Glycoproteins. Elsevier, New York, pp 287–312

Saito Y, Ihara Y, Leach MR, Cohen-Doyle MF, Williams DB (1999) Galreticulin functions in vitro as a molecular chaperone for both glycosylated and non-glycosylated proteins. EMBO J 18:6718–6729

Sambrook JF (1990) The involvement of calcium in transport of secretory proteins from the endoplasmic reticulum. Cell 61:197–199

Schue V, Green GA, Girardot R, Monteil H (1994) Hyperphosphorylation of calnexin, a chaperone protein, induced by Clostridium difficile cytotoxin. Biochem Biophys Res Commun 203: 22–28

Sousa MC, Ferrero Garcia MA, Parodi AJ (1992) Recognition of the oligosaccharide and protein moieties of glycoproteins by the UDP-Glc:glycoptroteins glucosyltransferase. Biochemistry 31:97–105

Spiro RG, Zhu Q, Bhoyroo V, Soling HD (1996) Definition of the lectin-like properties of the molecular chaperone, calreticulin, and demonstration of its copurification with endomannosidase from rat liver Golgi. J Biol Chem 271:11588–11594

Su K, Stoller T, Rocco J, Zemsky J, Green R (1993) Pre-Golgi degradation of yeast prepro-alpha-factor expressed in a mammalian cell. Influence of cell type-specific oligosaccharide processing on intracellular fate. J Biol Chem 268:14301–14309

Tatu U, Hammond C, Helenius A (1995) Folding and oligomerization of influenza hemagglutinin in the ER and the intermediate compartment. EMBO J 14:1340–1348

Tector M, Zhang Q, Salter RD (1994) Phosphatase inhibitors block in vivo binding of peptides to class I major histocompatibility complex molecules. J Biol Chem 269:25816–25822

Trombetta ES, Helenius A (1998) Lectins as chaperones in glycoprotein folding. Curr Opin Struct Biol 8:587–592

van Leeuwen JE, Kearse KP (1996) Deglucosylation of N-linked glycans is an important step in the dissociation of calreticulin-class I-TAP complexes. Proc Natl Acad Sci USA 93:13997–14001

Varki A (1993) Biological roles of oligosaccharides: all of the theories are correct. Glycobiology 3:97–130

Vassilakos A, Cohen DM, Peterson PA, Jackson MR, Williams DB (1996) The molecular chaperone calnexin facilitates folding and assembly of class I histocompatibility molecules. EMBO J 15:1495–1506

Wada I, Imai S, Kai M, Sakane F, Kanoh H (1995) Chaperone function of calreticulin when expressed in the endoplasmic reticulum as the membrane-anchored and soluble forms. J Biol Chem 270:20298–20304

Wada I, Rindress D, Cameron PH, Ou WJ, Doherty JJd, Louvard D, Bell AW, Dignard D, Thomas DY, Bergeron JJ (1991) SSR alpha and associated calnexin are major calcium binding proteins of the endoplasmic reticulum membrane. J Biol Chem 266:19599–19610

Ware FE, Vassilakos A, Peterson PA, Jackson MR, Lehrman MA, Williams DB (1995) The molecular chaperone calnexin binds Glc1Man9GlcNAc2 oligosaccharide as an initial step in recognizing unfolded glycoproteins. J Biol Chem 270:4697–4704

Watanabe D, Yamada K, Nishina Y, Tajima Y, Koshimizu U, Nagata A, Nishimune Y (1994) Molecular cloning of a novel Ca^{2+}-binding protein (calmegin) specifically expressed during male meiotic germ cell development. J Biol Chem 269:7744–7749

Williams DB, Watts TH (1995) Molecular chaperones in antigen presentation. Curr Opin Immunol 7:77–84

Wiuff C, Houen G (1996) Cation-dependent interactions of calreticulin with denatured and native proteins. Acta Chem Scand 50:788–795

Wong HN, Ward MA, Bell AW, Chevet E, Bains S, Blackstock WP, Solari R, Thomas DY, Bergeron JJM (1998) Conserved in vivo phosphorylation of calnexin at casein kinase II sites as well as a protein kinase C/proline-directed kinase site. J Biol Chem 273:17227–17235

Yang M, Oemura S, Bonifacino JS, Weissman AM (1998) Novel aspects of degradation of T cell receptor subunits from the endoplasmic reticulum (ER) in T cells: importance of oligosaccharide processing, ubiquitination, and proteasome-dependent removal from ER membranes. J Exp Med 187:835–846

Zapun A, Darby NJ, Tessier DC, Michalak M, Bergeron JJ, Thomas DY (1998) Enhanced catalysis of ribonuclease B folding by the interaction of calnexin or calreticulin with ERp57. J Biol Chem 273:6009–6012

Zapun A, Petrescu SM, Rudd PM, Dwek RA, Thomas DY, Bergeron JJM (1997) Conformation-independent binding of monoglucosylated ribonuclease B to calnexin. Cell 88:29–38

Zhang Q, Tector M, Salter RD (1995) Calnexin recognizes carbohydrate and protein determinants of class I major histocompatibility complex molecules. J Biol Chem 270:3944–3948

MR60/ERGIC-53, a Mannose-Specific Shuttling Intracellular Membrane Lectin

Annie-Claude Roche[1] and Michel Monsigny[1]

1 Discovery

Monocyte-derived macrophages express a mannoside-specific membrane lectin (Man-R) at their surface, which efficiently binds and internalizes mannosylated proteins as well as mannans (for reviews, see Stahl et al. 1984, Stahl 1992, and Taylor, this Vol.); conversely, monocytes and their precursors do not express this lectin. Knowing that the majority (80%) of the Man-R is located inside the cells and that its surface expression occurs during the first few hours of differentiation into macrophages, coming from an intracellular precursor protein pool (Lennartz et al. 1987), we decided to look for the presence of an intracellular precursor in two immature cell lines, the human promyelocytic cells HL60 and the human promonocytic cells U937, by using fluorescent neoglycoproteins.

Neoglycoproteins were independently developed at the beginning of the 1970s by Lee and coworkers and ourselves (for reviews, see: Stowell and Lee 1980; Lee 1992; Lee and Lee 1994a,b; Monsigny et al. 1994) as tools for studying various lectin properties. In both cases, neoglycoproteins were made by adding activated sugars to serum albumin. The neoglycoproteins, we use, are obtained by substituting about half the lysine residues of bovine serum albumin by reaction with glycosyl-phenyl-isothiocyanate (Roche et al. 1983; Monsigny et al. 1984) Fluorescein-labeled neoglycoproteins usually contain about 23 ± 3 sugars and 2.5 ± 0.5 fluorescein residues. The apparent affinity of neoglycoproteins for lectins is more than 1000 times higher than that of the corresponding free sugar (the Kd values of single sugars are generally in the millimolar range, whereas those of neoglycoproteins are down to micromolar range and even to nanomolar range in some cases).

Using flow cytometry, it was confirmed that mature macrophages bind and take up F-Man-BSA, while HL60 and U937 cells, which are monocyte precursors, neither bind nor take up this mannosylated neoglycoprotein. Conversely, the binding of fluoresceinylated mannosylated neoglycoprotein in fixed and permeabilized cells was quite high in all tested cells, including promyelocytic (HL60) and promonocytic cells (U937) as well as in fully matured macrophages. These results suggested that immature cells which do not express cell surface

[1] Glycobiologie. Centre de Biophysique Moléculaire, CNRS and University of Orléans, Rue Charles Sadron 45071 Orléans cedex 02 France

Results and Problems in Cell Differentiation, Vol. 33
Paul R. Crocker (Ed.): Mammalian Carbohydrate Recognition Systems
© Springer-Verlag Berlin Heidelberg 2001

Man-R, express a putative intracellular mannose-specific lectin. The specificity was assessed by using a panel of neoglycoproteins. F-GlcNAc-BSA, which binds macrophage Man-R with a high affinity, led to a very dull labeling of the permeabilized immature cells while those cells were strongly labeled with F-Man-BSA; the binding of Man-BSA was concentration-dependent and saturable, as expected for receptor-specific labeling (Pimpaneau et al. 1991). Using affinity chromatography on a mannoside-substituted gel (agarose beads with amino groups, Affi-102 Bio-Rad, substituted by reaction with 4-isothiocyanatophenyl α-D-mannopyranoside), the Man-R (175,000 M_r) was easily isolated from macrophages. Using the same affinity gel, a mannose-specific membrane protein was also isolated from HL60 membrane proteins solubilized in Triton X-100. This sugar-binding protein which also required $CaCl_2$ to bind immobilized mannosides, was selectively eluted from the affinity gel by a solution of 0.2 M D-mannose. Using polyacrylamide gel electrophoresis in the presence of SDS and under reducing conditions, this mannose binding protein appeared as a single band corresponding to an M_r 60,000 protein and not to the expected M_r 175,000 Man-R (Pimpaneau et al. 1991). By omitting the reducing agent, this protein migrated as an M_r 120,000 protein. In addition, this lectin was shown to be insensitive to N-glycanase (Carpentier et al. 1994). This isolated protein was shown to aggregate beads substituted with Man-BSA but not those substituted with other neoglycoproteins, including Glc-BSA, indicating that this protein behaves as a lectin, according to the common lectin definition (Goldstein et al. 1980), and is specific for mannosides. We decided to call this intracellular mannoside-binding lectin MR60.

To elucidate the structure and function of MR60, the cloning of its cDNA was required. Purified MR60 was digested with trypsin and ten different peptides were isolated and sequenced. Based on these sequences, degenerated oligonucleotides were synthesized and used as primers in a reverse-transcriptase-polymerase chain reaction (RT-PCR): a short 550-base pair DNA fragment was obtained. This DNA was used as a specific probe to screen an HL60 cDNA library kindly provided by Dr M. Fukuda (San Diego, CA). The full-length cDNA was obtained and sequenced and the peptide sequence of MR60 was deduced (Arar et al. 1995). It was found that the sequence of MR60 was identical to that of ERGIC-53, except for one nucleotide in codon 153 leading to a seryl residue in MR60 and a threonyl residue in ERGIC-53 (Schindler et al. 1993). ERGIC-53 was isolated as an antigen defined as a marker of ERGIC, the endoplasmic reticulum (ER)-Golgi (G)-intermediate compartment (IC). The cDNA encoding ERGIC-53 had been obtained by screening a lambda-gt11 expression library with an affinity-purified antiserum. Later, Lahtinen et al. (1996) reported the cloning of a homologous protein from rat; this was called p58. The human MR60/ERGIC-53 gene was mapped by in situ hybridization on the q21.3-q22 region of the human chromosome 18 (Arar et al. 1996) and has recently been characterized by Nichols et al. (1999). It contains 13 exons, the first one including a part of the 5′UTR. Exon 12 encodes the putative transmembrane domain and exon 13 encodes the short cytosolic part of the protein and contains a long 3' untranslated sequence. The gene of MR60/ERGIC-53 corresponds to the gene responsible for combined factor V and factor VIII deficiency (Nichols et al. 1998). Using molecular analysis of the ERGIC-53 gene in 35 families with this deficiency (Neerman-Arbez et al. 1999), it was found that the gene presents several common polymorphisms with amino acid substitution not associated with the disease. In particular, there is a mutation in codon 153 (ACT to TCT) corresponding to a serine which replaces a threonine. This explains the unique difference in the cDNA obtained by Schindler et al. (1993) and Arar et al. (1995).

The finding that MR60/ERGIC-53 is a lectin was confirmed by expressing the MR60/ERGIC-53 cDNA in Cos cells (Itin et al. 1996). The recombinant protein binds to a mannose-substituted gel in a calcium-dependent manner, and it is specifically eluted with a D-mannose solution (100 mM, optimum at 150 mM); D-glucose and D-GlcNAc started to elute the MR60/ERGIC-53 bound to the mannose-substituted gel at a higher concentration, about 200 mM, optimum at 400 mM; D-galactose was unable to elute the protein. These results confirm that MR60/ERGIC-53 is a mannose-specific lectin; the elution with a high concentration of glucose or N-acetylglucosamine may be explained by the fact that these three sugars are identical except for their C-2 substituants. This behavior is reminiscent of the plant lectin concanavalin A which can be eluted from a gel containing immobilized mannose by those three sugars.

2 Structural Features

ERGIC-53 (Schweizer et al. 1988) and p58 (Saraste et al. 1987) were first described as antigens recognized by antibodies raised against purified Golgi membranes, the monoclonal antibody G1/93 and the p58 polyclonal antibody, respectively; both antigens were described as integral membrane proteins. The amino acid sequence of ERGIC-53, MR60 and p58, deduced from isolated cDNA is in agreement with a transmembrane topology (Schindler et al. 1993; Arar et al. 1995; Lahtinen et al. 1996). Based on biochemical data, on the amino acid sequence and on the hydropathy profile established according to Kyte and Doolittle (1982), the three teams have independently proposed that the protein is a type I transmembrane protein, with a large extracytoplasmic domain and a short cytosolic domain of 12 amino acids, and that the protein is devoid of any site for N-glycosylation. So, it appears that the 58-kDa cis-Golgi resident protein, purified by immunoaffinity from microsomes of rat pancreas and identified as a glycoprotein by Hendricks et al. (1991) is not p58.

The two Lys at positions 4 and 3 (before the C-terminus) correspond to an ER retention signal (Jackson et al. 1990; Shin et al. 1991) allowing the protein to recycle between cis-Golgi and ER. The rat p58 amino acid sequence is close to that of the human MR60/ ERGIC-53 protein (89% identity); the mature protein contains an eight amino acid insertion close to the N-terminus (Lahtinen et al. 1996). The cytoplasmic p58 tail is identical to that of MR60/

ERGIC-53, except for a conservative mutation at position 11 (a threonine instead of a serine). In addition, the sequence of a protein from frog was found to be close to that from mammalian cells, showing that this protein is highly conserved through evolution (Lahtinen et al. 1996). *Saccharomyces cerivisiae* EMP47p, a type I transmembrane protein with a di-lysine motif cycles between the Golgi apparatus and the ER (Schröder et al. 1995), and is also homologous to MR60/ERGIC-53. Although it was postulated that this protein should be a lectin, this protein does not have the characteristic features of a lectin, in that the key amino acids of the putative CRD are not present (see below).

The predicted secondary structure of MR60/ERGIC-53 showed that the amino-terminal moiety (1–250) contains mainly β-strands in addition to turns, while the COOH-terminal moiety contains mainly α-helices (Arar et al. 1995). Analyzing the sequence of p58, Lahtinen et al. (1999) pointed out the presence of four major α-helical regions, interrupted by proline- and glycine-rich sequences; the first one is heavily charged with alternating acidic and basic residues, the following three are amphipathic regions of decreasing lengths.

3 Intracellular Mannose-Specific Animal Lectins Are Homologous to Leguminous Plant Lectins

Fiedler and Simons (1994) proposed that the secretory pathway contains two proteins with homology to leguminous lectins. This proposal was based on sequence similarities between one domain of leguminous plant lectins and one domain of VIP36 and of ERGIC-53, two proteins from animal cells. VIP36 is a type I integral membrane protein localized in epithelial cells in the Golgi apparatus, in the apical and basolateral transport vesicles up to the cell surface (Fiedler et al. 1994; Fiedler and Simons 1996). The discovery that MR60, a mannose-specific lectin, and ERGIC-53, a marker specific for the intermediate compartment, were identical, strongly supports Fiedler and Simons hypothesis. On the basis of X-ray data, Lobsanov et al. (1993) reported a structural homology between galectin-2 and PsA, the mannose-specific lectin from pea seeds, although the primary structures were poorly related and the sugar specificities were different. The presence of short β-strands in the amino-terminal moiety of MR60/ERGIC-53 is reminiscent of those found in both the 3-D structure of galectins and of the pea seed lectin (Arar et al. 1995). Moreover, several amino acids of galectins which stabilize the side chains of the amino acids present in the binding site of the sugar ligand, according to Lobsanov et al. (1993), were also shown to be present in MR60/ERGIC-53 (Arar et al. 1995). Several leguminous lectins which have similar 3-D structures differ in their sugar specificity. LoL I (*Lathyrus ochrus*), for instance, binds mannose or glucose but not galactose and, conversely, EcorL (*Erythrina corallodendron*) binds galactose (see for reviews, Sharon and Lis 1990; Sharon 1993; Lis and

Sharon 1998). The amino acids that form the saccharide combining site (or carbohydrate recognition domain, CRD) of leguminous lectins are localized in four distinct regions of the polypeptide chains and belong to a loop structure. Replacements, by site-directed mutagenesis, of one aspartic acid and/or of one asparagine in the CRD by another amino acid (alanine) in several plant lectins results in loss of sugar-binding ability (EcorL, Adar and Sharon 1996; pea lectin, Van Eijsden et al. 1994; *Griffonia simplificia* lectin II, Zhu et al. 1996). These amino acids participate in coordinating the calcium ion present in all members of this family; this explains the requirement of calcium for carbohydrate binding. The key role of these amino acids has also been demonstrated for the MR60/ERGIC-53 mannose-specific lectin. In the CRD, two amino acids, Asp121 corresponding to Asp 89 and Asp 81 in ECorL and LOL I respectively, and Asn 156, corresponding to Asn 133 and Asn 125 in EcorL and LoL I, respectively, were mutated to alanine (Itin et al. 1996). The recombinant mutated MR60/ERGIC-53 proteins did not bind to immobilized mannoside, nor fluoresceinylated mannosylated neoglycoprotein under conditions where the recombinant wild type MR60/ERGIC-53 protein binds. These data strengthen the proposal that MR60/ERGIC-53 is an animal phytolectin-like protein, a new lectin family homologous to the family of leguminous lectins (for a review, see Roche and Monsigny 1996).

4 Oligomerization and Sugar Binding Activity

The newly synthesized human MR60/ERGIC-53 lectin (or rat p58 lectin) rapidly dimerizes and is partially transformed into hexamers (Schweizer et al. 1988; Pimpaneau et al. 1991; Lahtinen et al. 1992). It was shown by crosslinking experiments, that no additional protein was closely associated with this protein and that the oligomerization reached a steady state within 120 min (Schweizer et al. 1988). The high molecular mass proteins resulting from oligomerization are stabilized by disulfide bridges (Schweizer et al. 1988; Lahtinen et al. 1992); both forms are indeed sensitive to reducing agents. The structural requirements for this oligomerization and the relationship between the oligomerization state and the mannose binding activity have been investigated by studying the properties of various recombinant proteins. Using a c-*Myc* epitope-tagged p58 protein in which the tag was introduced downstream from the predicted N-terminus signal sequence cleavage site, Lahtinen et al. (1996) showed that the c-*Myc* epitope tag did not affect oligomerization, since homodimers, homo-hexamers as well as hetero-oligomers are formed. Various constructs have been used to map the structural determinants for the oligomerization of p58 and particularly the role of the cysteine residues (Lahtinen et al. 1999). Truncated MR60 constructs carrying a c-*Myc* epitope tag and a His$_6$ epitope tag at the C-terminal end have been developed independently(Carrière et al. 1999b). These recombinant proteins were overexpressed in various cells, BHK21 cells (Lahtinen et al. 1999), Cos-1 and HeLa cells (Carrière et al. 1999b), and their capacity to oligomerize was analyzed. The main conclusions are summarized below:

Neither the Cytoplasmic Tail Nor the Transmembrane Domain Are Required for the Oligomerization of p58 (Lahtinen et al. 1999) or MR60/ERGIC-53 Lectin (Carrière et al. 1999b).

The recombinant proteins lacking the transmembrane region and the cytosolic tail form dimers and hexamers.

The α-Helical Domains Do Not Participate in Oligomerization (Lahtinen et al. 1999).

To test whether electrostatic or amphipathic interactions between the α-helical regions contribute to oligomerization, the predicted α-helical domains were individually deleted and the mutants were expressed in BHK-21 cells.[35]S-methionine-labeled proteins were immunoprecipitated with c-*Myc*-specific antibodies and analyzed using SDS-PAGE under reducing and non-reducing conditions. None of these deletions affects oligomerization; all these deleted mutants give both the dimeric and the hexameric protein as in the case of the wild type lectin.

The Two Cysteines Close to the Membrane Are Required for the Formation of Dimers and of Hexamers (see Fig. 1).

The mature protein contains four cysteines: two (C190 and C230 in MR60; C198 and C238 in p58) are located in the N-terminal half part of the protein, and two others (C466 and C475 in MR60; C473 and C482 in p58) are close to the transmembrane domain. By replacing the two most N-terminal cysteines, present in the lectin domain of p58 by alanine, Lahtinen et al. (1999) have shown that those cysteines are not involved in the oligomerization of the protein. The role of the two membrane-proximal cysteines has been demonstrated by two approaches; either by using mutants in which one or two cysteines were replaced by alanine (Lahtinen et al. 1999) or by using truncated proteins lacking a part of the luminal domain: $MR60_{\Delta473}$, which still contains one cysteine (C 466) and $MR60_{\Delta378}$, which does not contain the two membrane-proximal cysteines (C466 and C 475; Carrière et al. 1999b). The recombinant proteins in which either the two cysteines C473A and C482A in p58 were replaced or were absent as in the shortest truncated protein, were detected only in a monomeric form. These results confirm that the two first cysteines (C190 and C230 in p58) are not involved in the stabilization of the dimers and indicate that the two other cysteines (473 and 482 in p58 and 466 and 475 in MR60/ERGIC-53) which are close to the putative transmembrane domain are involved in the stabilization of the oligomeric state. The cysteine 466 stabilizes the dimeric form since the $MR60_{\Delta473}$ protein appears as a dimer in the absence and as a monomer in the presence of a reducing agent. The cysteines 466 and 475 are involved in stabilizing the higher oligomeric forms since the absence of one of those cysteine residues in the $MR60_{\Delta473}$ protein is correlated with a total lack of hexamers, while both the full-length recombinant protein and the endogenous MR60/ERGIC-53 protein, which both contain the cysteines 466 and 475, appear in both the hexameric and dimeric forms. The single p58

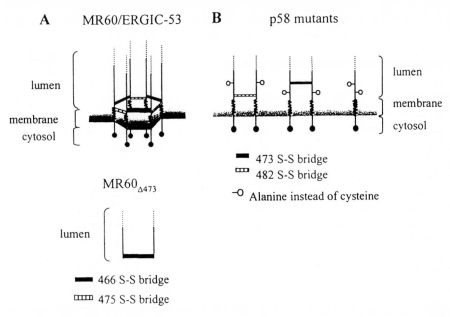

Fig. 1A,B. Schematic organization of the intracellular mannose-specific membrane lectin. **A** full length MR60/ERGIC-53 and the truncated counterpart MR60$_{\Delta473}$. **B** p58 mutants: two single mutants (C473 A, C482 A) and one double mutant (C473 A, C482 A)

mutants C473A and C482A behave identically: they are both able to form dimers but fail to form hexamers (Lahtinen et al. 1999); either one of these two membrane-proximal cysteines leads to a stable dimeric form and each one of these two cysteines must be present to allow the formation of hexamers.

A Dimeric State Is Required for Eliciting Sugar Binding Activity. A close relationship between oligomerization state and lectin activity has already been demonstrated for several animal and plant lectins. For instance, the human hepatic asialoglycoprotein receptor (ASGP-R) is a non-covalent hetero-oligomer (mainly trimer) composed of two closely related subunits, H1 and H2. These two subunits must be present to achieve a high affinity binding of asialoglycoprotein ligands at the cell surface, as shown by studies dealing with the expression of H1 or H2 and of both H1 and H2 proteins (McPhaul and Berg 1986; Shia and Lodish 1989); it was concluded that three galactoses borne by a given oligosaccharide interact simultaneously with the sites of the two H1 subunits and one H2 subunit. Recently, it has been suggested that the stalk segment of the receptor subunits oligomerizes, giving an α-helical coiled coil on top of which the carbohydrate recognition domains are exposed and available for ligand binding (Bider 1996). An other example is given by the native mannan-binding protein (MBP), a calcium-dependent mammalian serum lectin, with a collagen-like domain characterized by an NH$_2$-terminal cysteine-rich domain, a neck domain and a carbohydrate recognition domain; this protein forms several disulfide-dependent oligomeric (up to hexamer of trimers) structures (for a review, see Epstein et al. 1996). A truncated recombinant protein lacking a large part of the collagen domain is able to form trimers by non covalent association of the neck domain and binds sugar with an efficiency similar to that of the native form (Eda et al. 1998). Usually, lectins have one sugar binding site per subunit, but some lectins such as wheat germ agglutinin (WGA), which is a dimeric plant lectin, contains four carbohydrate binding sites (Privat et al. 1974); the binding sites are non-cooperative, spatially distinct and are formed by amino acid residues of both protomers (Wright 1984). Conversely, the high affinity of a lectin for its ligand derives from the presence of multiple carbohydrate binding recognition domains on one subunit, as in the case of the macrophage mannose-specific receptor (see Taylor, this Vol.). We have shown that the ability of MR60/ERGIC-53 to bind mannosides is correlated with its oligomerization state. Both the full-length protein, which properly oligomerizes in dimers and hexamers, and the truncated $MR60_{\Delta473}$ protein, which only forms dimers, bind to a mannoside-substituted gel and, to be eluted they require a high mannoside concentration (0.2 M mannose). Similar results were obtained with the p58 mutants able to form dimers and hexamers; recombinant proteins expressed in BHK cells bind immobilized mannose (Lahtinen et al. 1999). In contrast, the shorter recombinant protein which did not form (stable) dimers was neither retained nor even retarded on a gel containing immobilized mannoside (Carrière et al. 1999b). This result shows that the affinity of this short monomeric protein for a gel containing immobilized mannoside is too low to be efficient, i.e. lower than about $10^3 \, l \times mol^{-1}$; indeed, it was shown that proteins such as hydrolases with an affinity as low as $10^3 \, l \times mol^{-1}$ for their ligand (substrate or substrate analogues) are able to bind such an affinity gel (Cuatrecasas 1971).

In conclusion, the sugar binding activity of MR60/ERGIC-53/p58 requires an oligomeric state

5 Cytological Features

ERGIC (for reviews, see Hauri and Schweizer 1992; Farquhar and Hauri 1997) was first identified as a compartment in which, at 15 °C, the transport of newly synthesized Semliki Forest virus spike glycoprotein is blocked (Saraste and Kuismanen 1984). The intermediate compartment was defined as a compartment labeled with anti p53 antibody (Schweizer et al. 1990) and by subcellular fractionation it was shown that it exibits different features from rough ER and cis-Golgi (Schweizer et al. 1991). By using antibodies, either a monoclonal antibody G1/93 (Schweizer et al. 1988) or an anti rat p58 (Saraste et al. 1987) or an anti-luminal p58 peptide antiserum (Lahtinen et al. 1996), the endogenous human MR60/ERGIC-53 and rat p58 proteins were shown to be mainly localized in tubulo-vesicular clusters near the cis-side of the Golgi apparatus, although they were found to a lesser extent throughout the cell, including the cell periphery (Schweizer et al. 1988; Saraste and Svensson 1991; Carpentier et al. 1994). Recently, genetic immunization has been used to obtain highly specific antibodies directed against MR60/ERGIC-53. By intradermal injection of a plasmid encoding either the full length or the truncated MR60$_{\Delta473}$ protein (Carrière et al. 1999a), we obtained immune sera with a high specificity and high affinity for endogenous or recombinant MR60; these immune sera were efficient for in situ visualization of the lectin (Fig. 2) and for immunoprecipitation of the lectin from cell extracts. MR60/ERGIC-53 and p58 proteins are not exclusively restricted to ERGIC, they are also found in the first *cis*-cisternae of Golgi stacks as well as in the ER, suggesting a recycling pathway (Schweizer et al. 1988; Saraste and Svensson 1991; Saraste and Kuismanen 1992). Both the human and the rat lectins are also markers for the vesiculo-tubular clusters, VTCs (Bannykh et al. 1996 ; see for a review, Bannykh et al. 1998). VTCs have been shown to travel along microtubules and to flatten onto the *cis* face of the Golgi (Presley et al. 1997; Scales et al. 1997). The subcellular distribution of MR60/ERGIC-53 and p58 is temperature dependent; lowering the temperature from 37 to 15 °C led to their concentration in the Golgi area (Lippincott-Schwartz et al. 1990; Schweizer et al. 1990; Saraste and Svensson 1991) and to their co-localization with the KDEL receptor (Tang et al. 1995). The 15 °C compartment (immature VTC or ERGIC) is unable to fuse with the Golgi. Upon returning the cells to 37 °C, the distribution changed; MR60/ERGIC-53 (and p58) acquired a widespread distribution in the cytoplasm, suggesting a membrane recycling process. It was suggested that the ER was the compartment from which MR60/ERGIC-53 was recruited when it accumulated in the ERGIC at 15 °C, and to which MR60/ERGIC-53 recycles upon

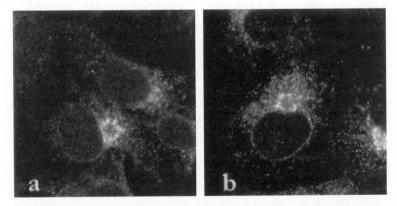

Fig. 2a,b. Immunofluorescence staining of the endogenous MR60/ERGIC-53. Cos cells were successively fixed with paraformaldehyde, permeabilized with saponin (1 mg/ml), and either incubated with anti-MR60 antiserum (1/500) obtained by DNA immunization (**a**, Carrière et al. 1999a) or G1/93, an anti-ERGIC-53 monoclonal antibody (a gift of H.P Hauri, Basel,**b**) followed by FTC-goat anti-mouse. Cell labeling was analyzed with a confocal microscope (Bio-Rad, MRC 1024)

rewarming (Klumperman et al. 1998). Different models of membrane traffic have been proposed in which vesicles and/or tubules play an important role. Recent data dealing with the Golgi structure in three dimensions (Ladinsky et al. 1999) show that the stacks of cisternae are localized between two types of ER, the *cis* and the *trans* ER. The ERGIC, which is structurally different from all the Golgi cisternae, is a collection of polymorphic vesicles. Tubular extensions from *cis*-Golgi reach ERGIC; in this model, ERGIC looks like VTCs elements. The clustering of ERGIC over compact regions of the underlying cisternae suggests an interaction with the *cis*-most Golgi cisternae.

Recombinant wild type proteins displayed an expression pattern close to that of constitutive endogenous proteins in the Golgi area and also in the ER. When MR60/ERGIC-53 was overexpressed in Cos cells, the intracellular retention system became saturated and MR60/ERGIC-53 appeared at the cell surface (Kappeler et al. 1994; Itin et al. 1995a); similar results were obtained in BHK cells when c-*Myc*-tagged p58 was overexpressed (Lahtinen et al. 1996). In all cases, these recombinant proteins were transmembrane proteins. Cell surface MR60/ERGIC-53 was efficiently endocytosed, showing a new type of endocytosis signal related to the ER-retrieval signal, but not identical, since arginine can replace lysine residues (Kappeler et al. 1994; Itin et al. 1995a). It is interesting that VIP36, a lectin which localized to both the plasma membrane and the Gogi (Fiedler et al. 1994), contains a C-terminal tetrapeptide KRFY, which can promote endocytosis. This peptide motif is related to the KKFF of MR60/ERGIC-53.

MR60/ERGIC-53 and p58 carry a dilysine ER-retrieval signal in their cytosolic domain, an RSQQE targeting determinant, adjacent to the membrane (Itin et al. 1995b) and two phenylalanines in a C-terminal position that reduce the retrieval efficiency of the dilysine signal and modulate the RSQQE targeting determinant. These two phenylalanines are important for precise ERGIC localization. In addition, Itin et al. (1995b) suggested that the RSQQE together with the luminal domain also plays a role in the precise targeting of MR60/ERGIC-53, since a construct containing the CD4 instead of the MR60/ERGIC-53 luminal domain showed an ER distribution with a strong perinuclear ring. Using site-directed mutagenesis, Kappeler et al. (1997) provided evidence that luminal and transmembrane domains mediate retention of MR60/ERGIC-53 in ER rather than in ERGIC. MR60/ERGIC-53 and p58 proteins, and other proteins (such as proteins of the p24 family (Dominguez et al. 1998) contain the sequences KKXX and KXKXX in their cytosolic tail, or internal FF residues which could be involved in the recruitment of COPs (coat proteins). Indeed, coatomers are essential for the retrieval of dilysine tagged protein (Letourneur et al. 1994); recently, Tisdale et al. (1997) have shown that MR60/ERGIC-53 binds COP I and is required for selective transport through the early secretory pathway (for review see Hong 1998). The two C-terminal phenylalanine residues in the cytosolic tail are the ER-exit determinants by interacting with the Sec23p of the Sec23p-Sec24p complex of COP II. Sequential coupling between COP II and COP I coat proteins coordinate and direct bi-directional vesicular traffic between the ER and pre-Golgi (ERGIC) (Aridor et al. 1995), see Aridor and Balch 1996; Schekman and Orci 1996).

In a recent paper, Nichols et al. (1999) showed that 10 out of 19 combined factor V and factor VIII deficiency mutations were identified in the ERGIC-53 gene. The mutations reported result in premature termination or truncation, removing either the C-terminal KKFF motif, or the whole transmembrane and cytosolic domains.

6 Deciphering the Role of MR60/ERGIC-53: Looking for Highly Specific Oligosaccharide and Natural Glycoprotein Ligands

In 1994, based on homology of VIP36 and ERGIC-53 with plant lectins, Fiedler and Simons speculated that these two proteins are members of a novel family of lectins involved in the sorting and/or the recycling of proteins, lipids, or both, in animal cells. In addition to the well-understood role of N-glycans in lysosomal protein sorting, N-glycans may have a role in the secretory pathway (see for a review, Fiedler and Simons 1995). As suggested by recent data (Scheiffele et al. 1995; Gut et al. 1998), N-glycans may play a role as apical sorting signals in polarized cells by binding to lectin-like molecules. The finding that ERGIC-53, – which constitutively recycles between ER, ERGIC, and cis-Golgi – is identical to MR60, an intracellular mannose-specific membrane lectin, supports the idea that this lectin is involved in the traffic of early processed glycoproteins. MR60/ERGIC-53 has all the features of a sorting receptor: oligomeric structure providing multiple CRDs, selectivity for glyco-conjugates containing mannose, and recycling between the different organelles in the early secretory pathway. It was proposed (Itin et al. 1996) that MR60/ERGIC-53 could recognize fully folded glycoproteins containing high-mannose oligosaccharides when glycoproteins are released from the quality control machinery (see Williams, this Vol.) or may carry incorrectly trimmed glycoproteins from ERGIC or cis-Golgi back to the ER. Until now, the natural ligands have not yet been identified, but several reports are in favor of an in vivo lectin function of MR60/ERGIC-53. Vollenweider et al. (1998), by overexpressing a mutant of MR60/ERGIC-53 unable to leave the ER (in the construct the two COOH-terminal phenylalanines were mutated to alanine, KKAA), and by using an inducible system for transient expression of the stably transfected protein, showed that the secretion of a glycoprotein – identified as procathepsin C – was reduced. The authors concluded that MR60/ERGIC-53 was required for efficient intracellular transport of a small subset of glycoproteins, but not for the majority of glycoproteins.

The lack of sugar binding capacity of the truncated $MR60_{\Delta 378}$ protein sheds new light on the molecular mechanism related to the mutations observed in ERGIC-53 in the case of the combined factors V and VIII deficiency patients

(Nichols et al. 1998). Indeed, in one patient the splice donor mutation after the codon 383 allowed the synthesis of a truncated protein slightly longer than MR60$_{\Delta378}$. By using either anti-p58 polyclonal antibodies or anti-Myc antibodies, we showed clearly that this short protein was synthesized and found in the early secretory pathway. In the paper by Nichols et al., the short protein was not visualized inside the cells because the authors were using a monoclonal antibody (G1/93) which was probably directed against an epitope not present in the short protein; this short protein was also not revealed on a western blot which was designed to show proteins with a molecular mass close to that of the full-length ERGIC-53 protein. Moreover, the short protein corresponding to the splice donor mutation lacks the cysteine residues close to the membrane and therefore can presumably neither form a dimer nor act as a lectin, i.e. a sugar-binding protein. Thus, in the case of the splice donor mutation, the truncated protein, even if it is present, cannot act as a sugar-dependent chaperone. In another patient, the frameshift mutation associated with a G insertion at the level of codons 29–30 led to a peptide containing only 101 amino acids, 71 of which do not belong to the normal MR60/ERGIC-53 protein; this protein does not contain the CRD. Neerman-Arbez (1999) confirmed that mutations in the MR60/ERGIC-53 gene are responsible for the deficiency in 74% of the families analyzed; these results are in agreement with those presented in Nichols et al. (1999). They pointed out that both factor V and factor VIII are subject to extensive post-translational modifications, including addition of multiple saccharide residues predominantly in the B domain of those factors and that MR60/ERGIC-53 probably interacts with this domain through a lectin recognition event. MR60/ERGIC-53 may also be required for the secretion of glycoproteins whose loss is not sufficient to cause a clinically recognizable phenotype. In addition, the presence of a non functional MR60/ERGIC-53 or the absence of MR60/ERGIC-53, deriving from a genetic defect may also be compensated by other pathways which either are not yet known or are specially used by cells suffering from the absence of an active MR60/ERGIC-53 protein.

MR60/ERGIC-53 has been discovered by using a fluoresceinylated and mannosylated glycoprotein (neoglycoprotein), serum albumin substituted with about 25 mannose residues, and MR60/ERGIC-53 has been isolated on a mannoside substituted gel (Pimpaneau et al. 1991). However, such a monosaccharide is not really specific for one single endogenous lectin; neoglycoproteins bearing mannose residues will be able to bind any mannose-specific lectin. Interestingly, in the case of oligosaccharides, the situation is quite different; the number of monosaccharides exposed on a complex oligosaccharide may change both the affinity and the specificity for a lectin. Lee et al. (1983) showed, for the asialoglycoprotein receptor of liver parenchymal cells (ASGP-R), that the affinity of an oligosaccharide containing a single antenna bearing a galactose residue in a terminal non-reducing position is about $10^3 \, l \times mol^{-1}$, with two antennae, i.e. with two galactose residues in a non-reducing terminal position, about $10^6 \, l \times mol^{-1}$, and with three or four antennae, i.e. with three or four terminal galactose residues, about $10^9 1 \times mol^{-1}$. Oligosaccharides isolated from living organisms, or released by hydrolysis from glycoconjugates, may be easily transformed into glycosynthons such as a β-oligosaccharyl-pyroglutamyl-amido-ethyl-dithio-2-pyridine, thanks to a new method developed in our laboratory (Sdiqui et al. 1995; Quétard et al. 1997, 1998; for a review on the preparation of glycosynthons, see Monsigny et al. 1998). This last compound is quantitatively reduced with tris(carboxyethyl)phosphine, and then transformed into a fluorescent conjugate by reacting, for example, with iodoacetamido-fluorescein. The conjugate β-oligosaccharyl-pyroglutamyl-amido-ethyl-thioacetamido-fluorescein, was finally readily purified by gel filtration and characterized by nuclear magnetic resonance and mass spectrometry. With these tools we are trying to identify a high affinity ligand amongst a panel of complex oligosaccharides containing mannose residues. Several different oligomannosides were purified from biological fluids, including urine and milk by J.C. Michalski and co-workers (Lille, France); oligosaccharides bearing from five to eight mannoses were selected and transformed into glycosynthons and then into fluorescein-labeled derivatives (Fig. 3).

The capacity of fluorescent oligomannoside derivatives to label the endogenous MR60/ERGIC-53 lectin was analyzed on fixed and permeabilized cells. Among the ten F-oligomannosides tested so far, only one gave a specific label-

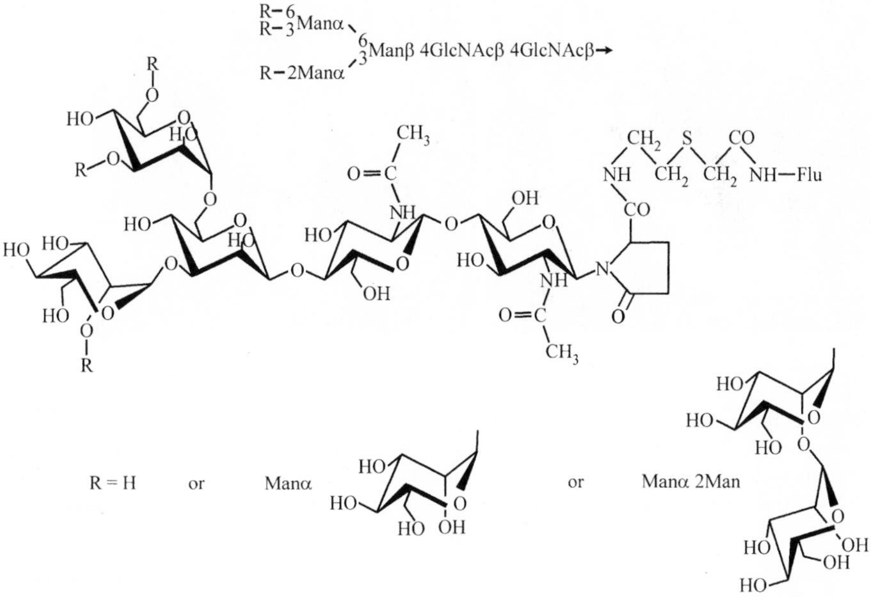

Fig. 3. Structure of fluorescent oligomannoside derivatives. Hydroxyl groups in position 2 of the Man α3 arm and in positions 3 and 6 of the Man α6 branch are either free or substituted with a monosaccharide Manα or with Man α-2Manα leading to 27 oligomannosides containing 3,4,5,6,7,8 or 9 mannose residues out of the 36 structures shown in Fig. 5

ing of the ERGIC region (Fig. 4a), the other F-oligomannosides either gave a nonspecific labeling of the cytosol or were concentrated in vesicles scattered throughout the cytoplasm (Fig. 4b).The precise structure of the oligomanno-side tested is being determined by NMR in collaboration with J.C. Michalski and will be reported elsewhere. Knowing that the absence of MR60/ERGIC-53 does not disturb the fate of many common glycoproteins, but only that of a few of them, it may be postulated that MR60/ERGIC-53 binds unusual oligoman-nosides. Mannose release in the Golgi following the removal of glucose by ER-glucosidase I and II leads to the $Man_5 GlcNAc_2$ through the main pathway, but it is also known that alternative early processing routes exist. Starting from the tetradecasaccharide added 'en bloc' on a protein, a large number of oligoman-nosides may theoretically be encountered (see Fig. 5 for all possible oligo-mannoside structures). The release of mannose residues can begin in the ER (Bischoff and Kornfeld 1983) with ER mannosidase I and II and in the Golgi with an endomannosidase that specifically cleaves the $\alpha 1 \rightarrow 2$ linkage between the glucose substituted mannose residue and the more internal portion of its polymannose branch leading to one $Man_8 GlcNAc_2$ isomer (for a review, see Moremen et al. 1994, Weng and Spiro 1996 and references therein); in addition,

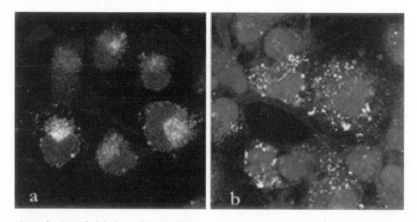

Fig. 4a,b. Specific labeling of intracellular mannose-specific lectins with F-oligomannosides. Cos cells fixed and permeabilized with saponin in the same conditions as in Fig. 2, were incubated 45 min at room temperature with 20 µg/ml of two different F-oligomannosides, **a** and **b** respec-tively. The ERGIC-like labeling (**a**) was obtained with a unique oligosaccharide while the vesicu-lar fluorescence throughout the cells (**b**) was obtained with various oligomannosides from Man_5 to Man_8. (All oligomannosides were a gift of J.C Michalski, Lille)

Fig. 5. Oligomannoside structures which can theoretically be obtained by partial truncation of the initial nona-mannopyranoside called M_9 found in *N*-glycans of glycoproteins in the ER. Struc-tures tagged with * are products deriving from the action of known ER α mannosidases, Golgi α mannosidases or endomannosidases acting on $Glc_2 Man_9$ or $Glc_1 Man_9$. All these mannosidases cleave the mannosyl α-2 mannose linkage

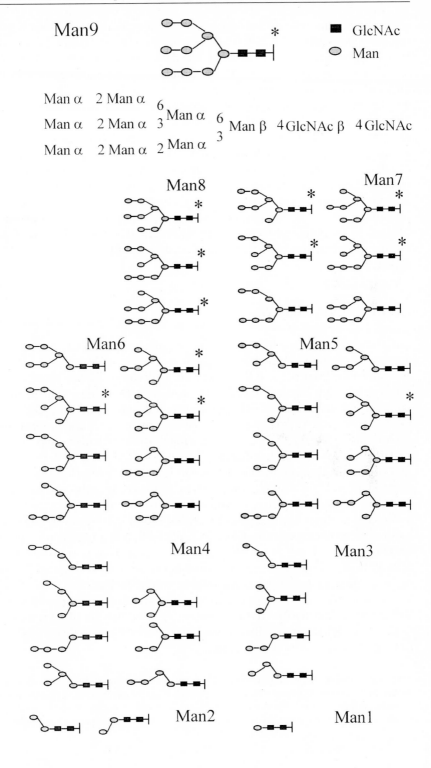

Man9

Man α 2 Man α 6
                  Man α 6
Man α 2 Man α 3   Man β 4 GlcNAc β 4 GlcNAc
                      3
Man α 2 Man α 2 Man α

Man8 Man7

Man6 Man5

Man4 Man3

Man2 Man1

some other pathways may have not yet been discovered. Among other hypotheses about the function of MR60/ERGIC-53, two must be taken into consideration because, on the basis of available data, they have not yet been ruled out. One is that MR60/ERGIC-53 could be a chaperone, recognizing both an oligosaccharide and a peptide. The other hypothesis is that glucose borne by an oligomannose such as in the premature N-glycan (Glcα 3Manα 2Manα 2Manα 3) could reach a putative buried subsite in the MR60/ERGIC-53 binding site, although MR60/ERGIC-53 does not bind immobilized glucose on either a bead or a protein (such as in the case of neoglycoprotein). This lack of interaction could, however, derive from an inappropriate presentation, for instance because the distance between the glucose residue and the matrix (bead or protein) is too short. Currently, efforts are being made to test these hypotheses and to isolate natural ligands, i.e., endogenous glycoproteins recognized by immobilized recombinant MR60/ERGIC-53 with the aim of determining the structure of the oligosaccharide naturally recognized by MR60/ERGIC-53.

7 Concluding Remarks

Endogenous membrane lectins such as the galactose-specific lectin (asialoglycoprotein receptor) and the mannose-6-phosphate lectin, the mannose specific-lectin (Man-R) and the newcomer (MR60/ERGIC-53) share common features: they are intrinsic membrane proteins with a single transmembrane domain, they act as a shuttle either between endosomes and the plasma membrane, between the Golgi apparatus (or plasma membrane) and pre-lysosomes, or, for MR60/ERGIC-53, between the endoplasmic reticulum and cis-Golgi, passing through ERGIC. In addition, MR60/ERGIC-53 has a structure close to that of the liver galactose-specific lectin: both have a short cytosolic tail, a trans-membrane spanning region, a stem and a single carbohydrate recognition domain. Both are active as oligomers; a trimer for the galactose-specific lectin, and a dimer and hexamer for MR60/ERGIC-53. However, there is one main difference between these two lectins: the first is a type II protein while MR60/ERGIC-53 is a type I protein. Another specific feature is that the CRD of MR60/ERGIC-53 has stretches of amino acids favoring a series of β-strands close to those found in galectins, soluble animal lectins, and in leguminous lectins. MR60/ERGIC-53 appears, therefore, as an intracellular phytolike lectin specific for some oligosaccharides containing mannose, shuttling from the ER to the cis-Golgi apparatus. Its precise sugar specificity as well as its biological function have yet to be determined.

Acknowledgements. We thank Jean-Claude Michalski for kindly providing oligomannosides and for helpful discussion, Véronique Piller and Friedrich Piller for discussion and critical reading of the manuscript, Hans-Peter Hauri and Ulla Lahtinen for their antibodies. We are especially grateful to Violaine Carrière and Christophe Quétard for their help in preparing the graphs in this manuscript, to Sylvain Bourgerie, Eric Duverger and Christophe Quétard who prepared all the fluorescent oligomannoside derivatives and to Marie-Thérèse Bousser for her valuable help in cell experiments. This work was supported by grants from the Agence Nationale de Recherche sur le SIDA and from the Association pour la Recherche sur le Cancer (ARC 6132). ACR is Research Director at the Institut National de la Santé et de la Recherche Médicale and MM is Professor at the University of Orléans.

References

Adar R, Sharon N (1996) Mutational studies of the amino acid residues in the combining site of *Erythrina corallodendron* lectin. Eur J Biochem 239:668–674

Arar C, Carpentier V, LeCaer JP, Monsigny M, Legrand A, Roche AC (1995) ERGIC-53, a membrane protein of the endoplasmic reticulum-Golgi intermediate compartment, is identical to MR60, an intracellular mannose-specific lectin of myelomonocytic cells. J Biol Chem 270: 3551–3553

Arar C, Mignon C, Mattei M, Monsigny M, Roche A, Legrand A (1996) Mapping of the MR60/ERGIC-53 gene to human chromosome 18q21.3–18q22 by in situ hybridization. Mamm Genome 7:791–792

Aridor M, Balch WE (1996) Principles of selective transport: coat complexes hold the key. Trends Cell Biol 6:315–320

Aridor M, Bannykh SI, Rowe T, Balch WE (1995) Sequential coupling between COPII and COPI vesicle coats in endoplasmic reticulum to Golgi transport. J Cell Biol 131:875–893

Bannykh SI, Rowe T, Balch WE (1996) The organization of endoplasmic reticulum export complexes. J Cell Biol 135:19–35

Bannykh SI, Nishimura N, Balch WE (1998) Getting into the Golgi. Trends Cell Biol 8:21–25

Bider MD (1996) The oligomerization domain of the asialoglycoprotein receptor preferentially forms 2:2 heterotetramers in vitro. J Biol Chem 271:31996–32001

Bischoff J, Kornfeld R (1983) Evidence for an alpha-mannosidase in endoplasmic reticulum of rat liver. J Biol Chem 258:7907–7910

Carpentier V, Vassard C, Plessis C, Motta G, Monsigny M, Roche AC (1994) Characterization and cellular localization by monoclonal antibodies of the 60 kDa mannose specific lectin of human promyelocytic cells, HL60. Glycoconj J 11:333–338

Carrière V, Landemarre L, Altemayer V, Motta G, Monsigny M, Roche AC (1999a) Intradermal DNA immunization: antisera specific for the membrane lectin MR60/ERGIC-53. Biosci Rep 6:559–570

Carrière V, Piller V, Legrand A, Monsigny M, Roche AC (1999b) The sugar binding activity of MR60, a mannose specific shuttling lectin, requires a dimeric state. Glycobiology 9:995–1002

Cuatrecasas P (1971) Affinity chromatography. Annu Rev Biochem 40:259–278

Dominguez M, Dejgaard K, Füllekrug J, Dahan S, Fazel A, Paccaud JP, Thomas DY, Bergeron JJM, Nilsson T (1998) gp25L/emp24/p24 protein family members of the cis-Golgi network bind both COP I and II coatomer. J Cell Biol 140:751–756

Eda S, Suzuki Y, Kawai T, Ohtani K, Kase T, Sakamoto T, Wakamiya N (1998) Characterization of truncated human mannan-binding protein (MBP) expressed in *Escherichia coli*. Biosci Biotechnol Biochem 62:1326–1331

Epstein J, Eichbaum Q, Sheriff S, Ezekowitz RAB (1996) The collectins in innate immunity. Curr Opin Immunol 8:29–35

Farquhar MG, Hauri HP (1997) Protein sorting and vesicular traffic in the Golgi apparatus. In: Berger EG, Roth J (eds) The Golgi apparatus. Birkhäuser-Verlag, Basel, Switzerland, pp 63–129

Fiedler K, Simons K (1994) A putative novel class of animal lectins in the secretory pathway homologous to leguminous lectins. Cell 77:625–626

Fiedler K, Simons K (1995) The role of N-glycans in the secretory pathway. Cell 81:309–312

Fiedler K, Simons K (1996) Characterization of VIP36, an animal lectin homologous to legumi-
    nous lectins. J Cell Sci 109:271–276
Fiedler K, Parton RG, Kellner R, Etzold T, Simons K (1994) VIP36, a novel component of glycol-
    ipid rafts and exocytic carrier vesicles in epithelial cells. EMBO J 13:1729–1740
Goldstein IJ, Hughes RC, Monsigny M, Osawa T, Sharon N (1980) What should be called a lectin?
    Nature 285:66
Gut A, Kappeler F, Hyka N, Balda MS, Hauri HP, Matter K (1998) Carbohydrate-mediated Golgi
    to cell surface transport and apical targeting of membrane proteins. EMBO J 17:1919–1929
Hauri HP, Schweizer A (1992) The endoplasmic reticulum-Golgi intermediate compartment. Curr
    Opin Cell Biol 4:600–608
Hendricks LC, Gabel CA, Suh K, Farquhar MG (1991) A 58-kDa resident protein of the cis-Golgi
    cisterna is not terminally glycosylated. J Biol Chem 266:17559–17565
Hong W (1998) Protein transport from the endoplasmic reticulum to the Golgi apparatus. J Cell
    Sci 111:2831–2839
Itin C, Kappeler F, Linstedt AD, Hauri HP (1995a) A novel endocytosis signal related to the KKXX
    ER-retrieval signal. EMBO J 14:2250–2256
Itin C, Schindler R, Hauri HP (1995b) Targeting of protein ERGIC-53 to the ER/ERGIC/cis-Golgi
    recycling pathway. J Cell Biol 131:57–67
Itin C, Roche AC, Monsigny M, HauriHP (1996) ERGIC-53 is a functional mannose-selective and
    calcium-dependent human homologue of leguminous lectins. Mol Biol Cell 7:483–493
Jackson MR, Nilsson T, Peterson PA (1990) Identification of a consensus motif for retention of
    transmembrane proteins in the endoplasmic reticulum. EMBO J 9:3153–3162
Kappeler F, Itin C, Schindler R, Hauri HP (1994) A dual role for COOH-terminal lysine residues
    in pre-Golgi retention and endocytosis of ERGIC-53. J Biol Chem 269:6279–6281
Kappeler F, Klopfenstein DR, Foguet M, Paccaud JP, Hauri HP (1997) The recycling of ERGIC-53
    in the early secretory pathway. ERGIC-53 carries a cytosolic endoplasmic reticulum-exit deter-
    minant interacting with COP II. J Biol Chem 272:31801–31808
Klumperman J, Schweizer A, Clausen H, Tang BL, Hong W, Oorsschot V, Hauri HP (1998) The
    recycling pathway of protein ERGIC-53 and dynamics of the ER-Golgi intermediate com-
    partment. J Cell Sci 111:3411–3425
Kyte J, Doolittle RF (1982) A simple method for displaying the hydropathic character of a protein.
    J Mol Biol 157:105–132
Ladinsky MS, Mastronarde DN, McIntosh JR, Howell KE, Staehelin LA (1999) Golgi structure in
    three dimensions: functional insights from the normal rat kidney cell. J Cell Biol 144:
    1135–1149
Lahtinen U, Dahllof B, Saraste J (1992) Characterization of a 58-kDa cis-Golgi protein in pan-
    creatic exocrine cells. J Cell Sci 103:321–333
Lahtinen U, Hellman U, Wernstedt C, Saraste J, Pettersson RF (1996) Molecular cloning and
    expression of a 58-kDa cis-Golgi and intermediate compartment protein. J Biol Chem 271:
    4031–4037
Lahtinen U, Svensson R, Pettersson RF (1999) Mapping of structural determinants for the
    oligomerization of p58, a lectin-like protein of the intermediate compartment and cis-Golgi.
    Eur J Biochem 260:392–397
Lee YC (1992) Biochemistry of carbohydrate-protein interaction. FASEB J 6:193–200
Lee YC, Lee RT (1994a) Neoglycoconjugates, part A. Synthesis. Methods Enzymol 242:1–238
Lee YC, Lee RT (1994b) Neoglycoconjugates, part B. Biochemical application. Methods Enzymol
    242:1–450
Lee YC, Townsend RR, Hardy MR, Lönngren J, Arnap J, Haraldsson M, Lönn H (1983) Binding of
    synthetic oligosaccharides to the hepatic Gal/GalNAc lectin. Dependence on fine structural
    features. J Biol Chem 258:199–202
Lennartz MR, Wileman TE, Stahl PD (1987) Isolation and characterization of a mannose-specific
    endocytosis receptor from rabbit alveolar macrophages. Biochem J 245:705–711
Letourneur F, Gaynor EC, Hennecke S, Demolliere C, Duden R, Emr SD, Riezman H, Cosson P
    (1994) Coatomer is essential for retrieval of dilysine-tagged proteins to the endoplasmic retic-
    ulum. Cell 79:1199–1207

Lippincott-Schwartz J, Donaldson JG, Schweizer A, Berger EG, Hauri HP, Yuan LC, Klausner RD (1990) Microtubule-dependent retrograde transport of proteins into the ER in the presence of brefeldin A suggests an ER recycling pathway. Cell 60:821–836

Lis H, Sharon N (1998) Lectins: carbohydrate-specific proteins that mediate cellular recognition. Chem Rev 98:637–674

Lobsanov YD, Gitt MA, Leffler H, Barondes SH, Rini JM (1993) X-ray crystal structure of the human dimeric S-Lac lectin, L-14-II in conylex with lactose at 2.9Å resolution. J Biol Chem 268:27034–27038

McPhaul M, Berg P (1986) Formation of functional asialoglycoprotein receptor after transfection with cDNAs encoding the receptor proteins. Proc Natl Acad Sci USA 83:8863–8867

Monsigny M, Roche AC, Midoux P (1984) Uptake of neoglycoproteins via membrane lectin(s) of L1210 cells evidenced by quantitative flow cytofluorometry and drug targeting. Biol Cell 51:187–196

Monsigny M, Roche AC, Midoux P, Mayer R (1994) Glycoconjugates as carriers for specific delivery of therapeutic drugs and genes. Adv Drug Deliv Rev 14:1–24

Monsigny M, Quétard C, Bourgerie S, Delay D, Pichon C, Midoux P, Mayer R, Roche AC (1998) Glycotargeting: the preparation of glyco-aminoacids and derivatives from unprotected reducing sugars. Biochimie 80:99–108

Moremen KW, Trimble RB, Herscovics A (1994) Glycosidases of the asparagine-linked oligosaccharide processing pathway. Glycobiology 4:113–125

Neerman-Arbez M, Johnson KM, Morris MA, McVey JH, Peyvandi F, Nichols WC, Ginsburg D, Rossier C, Antonarakis SE, Tuddenham EGD (1999) Molecular analysis of the ERGIC-53 gene in 35 families with combined factor V-factor VIII deficiency. Blood 93:2253–2260

Nichols WC, Seligsohn U, Zivelin A, Terry VH, Hertel CE, Wheatley MA, Moussali MJ, Hauri HP, Ciavarella N, Kaufman RJ, Ginsburg D (1998) Mutation in the ER-Golgi intermediate compartment protein ERGIC-53 cause combined deficiency of coagulation factor V and VIII. Cell 93:61–70

Nichols WC, Terry VH, Wheatley MA, Yang A, Zivelin A, Ciavarella N, Stefanile C, Matsushita T, Saito H, de Bosch NB, Ruiz-Saez A, Torres A, Thompson AR, Feinstein DI, White GC, Negrier C, Vinciguerra C, Aktan M, Kaufman RJ, Ginsburg D, Seligsohn U (1999) ERGIC-53 gene structure and mutation analysis in 19 combined factors V and VIII deficiency families. Blood 93:2261–2266

Pimpaneau V, Midoux P, Monsigny M, Roche AC (1991) Characterization and isolation of an intracellular D-mannose-specific receptor from human promyelocytic HL60 cells. Carbohydr Res 213:95–108

Presley JF, Cole NB, Schroer TA, Hirschberg K, Zaal KJ, Lippincott-Schwartz J (1997) ER-to-Golgi transport visualized in living cells. Nature 389:81–85

Privat JP, Delmotte F, Monsigny M (1974) Protein-sugar interactions. Association of β-(1–4) linked N-acetyl-D-glucosamine oligomer derivatives with wheat germ agglutinin (lectin). FEBS Lett 46:224–228

Quétard C, Normand-Sdiqui N, Mayer R, Roche AC, Monsigny M (1997) Simple synthesis of novel glycosynthons: oligosylpyroglutamyl derivatives. Carbohydr Lett 2:415–422

Quétard C, Bourgerie S, Normand-Sdiqui N, Mayer R, Strecker G, Midoux P, Roche AC, Monsigny M (1998) Novel glycosynthons for glycoconjugate preparation: oligosaccharylpyroglutamylanilide derivatives. Bioconj Chem 9:268–276

Roche AC, Monsigny M (1996) Trafficking of endogenous glycoproteins mediated by intracellular lectins: facts and hypothesis. Chemtracts Biochem Mol Biol 6:188–201

Roche AC, Barzilay M, Midoux P, Junqua S, Sharon N, Monsigny M (1983) Sugar-specific endocytosis of glycoproteins by Lewis lung carcinoma cells. J Cell Biochem 22: 131–140

Saraste J, Kuismanen E (1984) Pre- and post-Golgi vacuoles operate in the transport of Semliki Forest virus membrane glycoproteins to the cell surface. Cell 38:535–549

Saraste J, Kuismanen E (1992) Pathways of protein sorting and membrane traffic between the rough endoplasmic reticulum and the Golgi complex. Semin Cell Biol 3:343–355

Saraste J, Svensson K (1991) Distribution of the intermediate elements operating in ER to Golgi transport. J Cell Sci 100:415–430

38 A.-C. Roche and M. Monsigny

Saraste J, Palade GE, Farquhar MG (1987) Antibodies to rat pancreas Golgi subfractions: identification of a 58-kD cis-Golgi protein. J Cell Biol 105:2021–2029

Scales SJ, Pepperkok R, Kreis TE (1997) Visualization of ER-to-Golgi transport in living cells reveals a sequential mode of action of COP II and COP I. Cell 90:1137–1148

Scheiffele P, Peränen J, Simons K (1995) N-glycans as apical sorting signals in epithelial cells. Nature 378:96–98

Schekman R, Orci L (1996) Coat proteins and vesicle budding. Science 271:1526–1533

Schindler R, Itin C, Zerial M, Lottspeich F, Hauri HP (1993) ERGIC-53, a membrane protein of the ER-Golgi intermediate compartment, carries an ER retention motif. Eur J Cell Biol 61:1–9

Schröder S, Schimmöller F, Singer-Krüger B, Riezman H (1995) The Golgi-localization of yeast Emp47p depends on its di-lysine motif but is not affected by the ret-1-1 mutation in alpha-COP. J Cell Biol 131:895–912

Schweizer A, Fransen JA, Bachi T, Ginsel L, Hauri HP (1988) Identification, by a monoclonal antibody, of a 53-kD protein associated with a tubulo-vesicular compartment at the cis-side of the Golgi apparatus. J Cell Biol 107:1643–1653

Schweizer A, Fransen JA, Matter K, Kreis TE, Ginsel L, Hauri HP (1990) Identification of an intermediate compartment involved in protein transport from endoplasmic reticulum to Golgi apparatus. Eur J Cell Biol 53:185–196

Schweizer A, Matter K, Ketcham CM, Hauri HP (1991) The isolated ER-Golgi intermediate compartment exhibits properties that are different from ER and cis-Golgi. J Cell Biol 113:45–54

Sdiqui N, Roche AC, Mayer R, Monsigny M (1995) New synthesis of glycopeptides. Carbohydr Lett 1:269–275

Sharon N (1993) Lectin-carbohydrate complexes of plants and animals: an atomic view. Trends Biochem Sci 18:221–226

Sharon N, Lis H (1990) Legume lectins: a large family of homologous proteins. FASEB J 4:198–208

Shia MA, Lodish HF (1989) The two subunits of the human asialoglycoprotein receptor have different fates when expressed alone in fibroblasts. Proc Natl Acad Sci USA 86:1158–1162

Shin J, Dunbrack RL Jr, Lee S, Strominger JL (1991) Signals for retention of transmembrane proteins in the endoplasmic reticulum studied with CD4 truncation mutants. Proc Natl Acad Sci USA 88:1918–1922

Stahl PD (1992) The macrophage mannose receptor: current status. Curr Opin Immunol 4:49–52

Stahl PD, Wileman TE, Diment S, Shepherd VL (1984) Mannose-specific oligosaccharide recognition by mononuclear phagocytes. Biol Cell 51:215–218

Stowell CP, Lee YC (1980) Neoglycoproteins: the preparation and application of synthetic glycoproteins. Adv Carbohydr Chem Biochem 37:225–281

Tang BL, Low SH, Hong W (1995) Segregation of ERGIC53 and the mammalian KDEL receptor upon exit from the 15 °C compartment. Eur J Biochem 68:398–410

Tisdale EJ, Plutner H, Matteson J, Balch WE (1997) p53/58 binds COP I and is required for selective transport through the early secretory pathway. J Cell Biol 137:581–593

Van Eijsden RR, De Pater BS, Kijne JW (1994) Mutational analysis of the sugar-binding site of pea lectin. Glyccoonj J 11:375–380

Vollenweider F, Kappeler F, Itin C, Hauri HP (1998) Mistargeting of the lectin ERGIC-53 to the endoplasmic reticulum HeLa cells impairs the secretion of a lysosomal enzyme. J Cell Biol 142:377–389

Weng S, Spiro RG (1996) Evaluation of the early processing routes of N-linked oligosaccharides of glycoproteins through the characterization of Man$_8$GlcNAc$_2$ isomers: evidence that endomannosidase functions in vivo in the absence of glucosidase blockade. Glycobiology 6:861–868

Wright CS (1984) Structural comparison of the two distinct sugar binding sites in wheat germ agglutinin isolectin II. J Mol Biol 178:91–104

Zhu K, Bressan RA, Hasegawa PM, Murdock LL (1996) Identification of N-acetylglucosamine binding residues in Griffonia simplicifolia lectin II. FEBS Lett 390:271–274

The Cation-Dependent Mannose 6-Phosphate Receptor

Jung-Ja P. Kim[1] and Nancy M. Dahms[1]

1 Introduction

The 46kDa cation-dependent mannose 6-phosphate receptor (CD-MPR) and the 300kDa cation-independent mannose 6-phosphate receptor (CI-MPR) are the sole members of the P-type family of lectins, which derive their name from their ability to bind phosphorylated mannose residues. The MPRs are found ubiquitously in higher eukaryotes, and most cell types express both receptors. A key function mediated by the MPRs is the targeting of newly synthesized soluble acid hydrolases bearing mannose 6-phosphate (Man-6-P) residues to the lysosome, an acidified organelle that is responsible for the degradation of both internalized and endogenous macromolecules (de Duve 1963). The generation of functional lysosomes is essential for the survival of the organism as evidenced by the autosomal recessive human lysosomal storage disorder, mucolipidosis II (I-cell disease; Neufeld 1991), a disease characterized by severe psychomotor retardation, hepatomegaly, and cardiomegaly that results in death within the first decade of life (Kornfeld and Sly 1995). I-cell disease is caused by the absence of the enzyme (i.e., phosphotransferase) which generates the Man-6-P recognition marker on lysosomal enzymes, resulting in the mistargeting of these enzymes into the secretory pathway. As expected, patients affected with this disease exhibit elevated levels of lysosomal enzymes in their serum and extracellular fluids and their cells contain numerous dense vacuoles filled with storage material.

Numerous studies have been conducted in order to understand at the molecular level how the MPRs interact with lysosomal enzymes. The lack of obvious sequence homology between the MPRs and other known proteins left many questions about their structure unresolved, including whether they have a carbohydrate-recognition domain (CRD) that is similar to other animal or plant lectins. This review will summarize the information gained recently from the crystal structure of the CD-MPR which has provided the first view of the CRD of a P-type lectin.

---

[1] Department of Biochemistry, Medical College of Wisconsin, Milwaukee, WI 53226

Results and Problems in Cell Differentiation, Vol. 33
Paul R. Crocker (Ed.): Mammalian Carbohydrate Recognition Systems
© Springer-Verlag Berlin Heidelberg 2001

2 Intracellular Trafficking of the MPRs and Lysosomal Enzymes

2.1 Generation of the Mannose 6-Phosphate Recognition Marker

The biogenesis of lysosomes is accomplished, in part, by the selective delivery of soluble acid hydrolases from their site of synthesis in the endoplasmic reticulum (ER) to the lysosome. Soluble lysosomal enzymes, like secretory proteins, are synthesized as precursors that contain an N-terminal signal sequence and undergo co-translational glycosylation at selected Asn residues (Fig. 1).

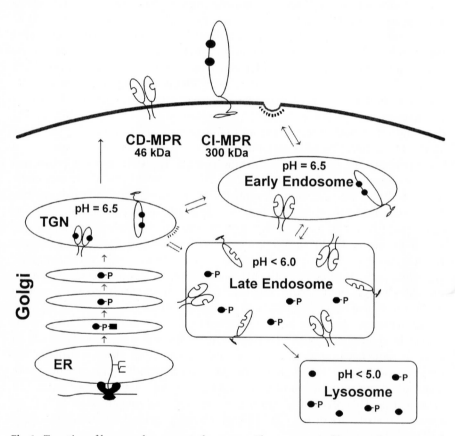

Fig. 1. Targeting of lysosomal enzymes to lysosomes. The movement of lysosomal enzymes and the MPRs between the various intracellular compartments and the cell surface are shown. Trafficking from the cell surface and TGN occurs via clathrin-coat pits (eyelashes). Lysosomal enzymes, which are released from the MPRs in the late endosomal compartment, become packaged into lysosomes where dephosphorylation of their N-linked oligosaccharides (branched tree) may occur. Occupied (oval with two filled circles) and unoccupied (oval with two circular indentations) Man-6-P binding sites of the CI-MPR and CD-MPR are shown. Lysosomal enzymes that are not phosphorylated (●), contain a phosphodiester (●-P-■) with N-acetylglucosamine (■), or contain a phosphomonoester (●-P) are shown

However, lysosomal enzymes become diverted from the secretory pathway by the acquisition of Man-6-P residues in early Golgi compartments which serve as a high affinity ligand for the MPRs. Two enzymes are responsible for the generation of this specific carbohydrate modification. The first enzyme, UDP-*N*-acetylglucosamine:lysosomal enzyme *N*-acetylglucosamine-1-phosphotransferase, transfers *N*-acetylglucosamine 1-phosphate to one or more mannose residues to give rise to a phosphodiester intermediate (von Figura and Hasilik 1986). This selective phosphorylation of N-linked high mannose-type oligosaccharides on lysosomal enzymes is achieved by the ability of the phosphotransferase to recognize a conserved three-dimensional polypeptide determinant which is enriched in one or more lysine residues (Cuozzo et al. 1998). The second enzyme, *N*-acetylglucosamine-1-phosphodiester a-*N*-acetylglucosaminidase, removes the *N*-acetylglucosamine residue to generate the phosphomonoester (von Figura and Hasilik 1986). Oligosaccharides isolated from lysosomal enzymes are quite heterogeneous as they can contain one or two phosphomannosyl residues that can be located at five different positions in the oligosaccharide chain (Varki and Kornfeld 1980).

2.2 Subcellular Distribution of the MPRs

The CD-MPR and CI-MPR traverse similar intracellular trafficking pathways (Fig. 1) Immunolocalization and biochemical analyses demonstrate that at steady state the majority of the MPRs are located in the endosomes and trans Golgi network (TGN), with ~5–10% of the receptor molecules on the cell surface and virtually none in the lysosome (von Figura and Hasilik 1986; Kornfeld 1992). The MPRs bind newly synthesized lysosomal enzymes bearing Man-6-P residues in the Golgi and deliver these ligands to a late endosomal compartment where the low pH of this compartment causes the receptor:enzyme complex to dissociate. The receptors return to the Golgi to repeat the process whereas the lysosomal enzymes are packaged into lysosomes. Measurements of the number and half-life of MPRs and the rate of ligand internalization indicate that the MPRs are reutilized and can undergo many rounds of ligand delivery. In addition, studies using antibodies to label receptors on the cell surface indicate that all MPRs in the cell are in rapid equilibrium (Sahagian 1984). This suggests there is only one pool of receptor and that a single MPR functions in both the endocytic and biosynthetic pathways. However, the CI-MPR, unlike the CD-MPR, is able to internalize extracellular ligands from the cell surface, resulting in their delivery to the lysosome via the endocytic pathway.

2.3 Targeting Signals in the Cytoplasmic Region of the MPRs

The identification of the components of the cellular machinery which specifically recognize the MPRs and mediate their intracellular movements between the TGN, plasma membrane, and endosomal compartments has been actively pursued in recent years. Both receptors have been shown to utilize clathrin-coated vesicles for export from the TGN and for endocytosis from the cell surface, and additional proteins, including the adaptor proteins (AP-1 and AP-2; Glickman et al. 1989), Rab9 GTPase (Riederer et al. 1994), and a novel protein called TIP47 (Diaz and Pfeffer 1998) have now been identified which function at various sites along the pathway. The cytoplasmic region of the MPRs contains the signals required for endocytosis and for efficient intracel-lular sorting of lysosomal enzymes (Fig. 2). Endocytosis of the CD-MPR is directed by three sequences, a phenylalanine-containing sequence (FPHLAF), a YRGV sequence, and a di-leucine motif located within the C-terminal seven residues (Denzer et al. 1997), whereas internalization of the CI-MPR from the cell surface is mediated by the sequence YKYSKV (Chen et al. 1997). Two signals in the cytoplasmic region of the CD-MPR, an acidic cluster followed by a di-leucine motif (HLLPM) located at the C-terminus and a di-aromatic signal (FW) which is modulated by palmitoylation (Schweizer et al. 1997), are required for the efficient targeting of lysosomal enzymes by the CD-MPR. A similar C-terminal di-leucine motif (LLHV) has been identified in the CI-MPR which is required for the efficient endosomal sorting of this receptor (Chen et al. 1997).

3 Primary Structure and Biosynthesis of the CD-MPR

3.1 Primary Structure

The CD-MPR is a type I transmembrane protein that has been cloned from bovine (Dahms et al. 1987), human (Pohlmann et al. 1987), and murine (Ma et al. 1991) sources. The bovine CD-MPR has an apparent molecular mass of approximately 46 kDa and consists of four structural/functional domains: a 28-residue amino terminal signal sequence, a 159-residue extracytoplasmic domain, a single 25-residue transmembrane region, and a 67-residue carboxy-terminal cytoplasmic domain (Dahms et al. 1987; Fig. 2). Sequence analysis of the bovine cDNA revealed that it contains five potential N-linked glycosylation sites (Asn-X-Ser/Thr) which are located within the extracytoplasmic domain at positions 31, 57, 68, 81, and 87. Alignment of the bovine, human, and mouse amino acid sequences demonstrated that the mature proteins are 93–95% identical and that the five glycosylation sites are conserved. Although there are no obvious primary sequence similarities between the two MPRs' signal sequences, transmembrane regions, or their cytoplasmic domains, the extra-cytoplasmic region of the CD-MPR shares significant, yet limited, sequence identity (14–28%) with each of the 15 repeating domains of the CI-MPR (Lobel et al. 1988; Fig. 2). In addition, the six cysteine residues in the extracytoplas-mic region of the CD-MPR, which are required for the folding and proper assembly of the CD-MPR during its biosynthesis (Wendland et al. 1991a), align

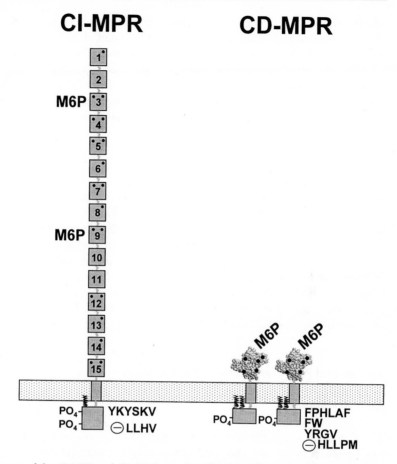

Fig. 2. Diagram of the CI-MPR and CD-MPR proteins. The three-dimensional structure of the extracytoplasmic domain of the CD-MPR (Roberts et al. 1998) is depicted using a space-filling model. The 15 repeating domains of the CI-MPR (numbering starts at the amino terminus) are depicted. The locations of the Man-6-P (M6P) binding sites and the potential N-linked glycosylation sites (●) are indicated. The CI-MPR binds two mol of Man-6-P per polypeptide chain and the two Man-6-P binding sites have been localized to domain 3 and domain 9. The serine phosphorylation (PO₄) and palmitoylation (coil) sites in the cytoplasmic region are indicated. The sequences in the cytoplasmic domain of the bovine MPRs, some of which contain a region of acidic residues (-), that are important for internalization and lysosomal enzyme trafficking are listed

with the conserved positions of the CI-MPR's cysteine residues, suggesting the two MPRs have a similar tertiary structure.

3.2 Genomic Structure

The genes for the two MPRs have been shown to map to different chromosomes: the CI-MPR has been localized to human chromosome 6 and contains

48 exons (Rao et al. 1994), whereas the CD-MPR has been mapped to human chromosome 12 and contains 7 exons (Pohlmann et al. 1987; Klier et al. 1991). Analysis of their genomic structure has revealed that the position of the intron/exon splice junctions is conserved between several repeating domains of the CI-MPR and the extracytoplasmic region of the CD-MPR (Szebenyi and Rotwein 1994). However, no correlation exists between the exon boundaries and the structural or functional protein domains of the receptors.

3.3 Oligomeric Structure

The predominant form of the CD-MPR in membranes is a dimer. However, trimeric and tetrameric species of the receptor have also been detected, both in membranes and in solution (Dahms and Kornfeld 1989; Li et al. 1990; Waheed et al. 1990). Further studies have shown that oligomerization does not require the transmembrane region since a truncated CD-MPR containing only the extracytoplasmic domain is dimeric when expressed in mammalian (Wendland et al. 1989) or insect cells (Marron-Terada et al. 1998a). Thus, the extracytoplasmic domain of the CD-MPR contains all of the information required for the formation of a dimeric structure.

3.4 Co- and Post-Translational Modifications

3.4.1 Acylation

The CD-MPR is synthesized on membrane-bound ribosomes in the ER and undergoes several co- and post-translational modifications (Fig. 2) as it traverses through the ER and Golgi compartments (Fig. 1). The two cysteine residues (Cys30 and Cys34) in the cytoplasmic tail of the CD-MPR are palmitoylated in a reversible manner via a thioester linkage. The addition of a fatty acid to the CD-MPR has been shown to be essential for its normal intracellular trafficking as substitution of Cys 34 results in accumulation of the mutant receptor in dense lysosomes (Schweizer et al. 1996). An implication of these studies is that palmitoylation functions to alter the conformation of the cytoplasmic region of the CD-MPR by serving as an attachment point to the lipid bilayer, thereby influencing the interaction of the receptor with other cellular proteins involved in the regulation of its intracellular trafficking.

3.4.2 Phosphorylation

The CD-MPR undergoes reversible phosphorylation at a single serine residue in its cytoplasmic region (Hemer et al. 1993). The function of this modification remains unclear. Although phosphorylation does not appear to influence the ability of the receptor to target lysosomal enzymes to the lysosome (Johnson and Kornfeld 1992; Hemer et al. 1993), recent studies have shown that this phosphorylation site is required for the expression of the receptor on the cell surface (Breuer et al. 1997).

3.4.3 Glycosylation

Four out of the five potential N-linked glycosylation sites of the CD-MPR are utilized (Dahms et al. 1987; Wendland et al. 1991b) and the receptor has been shown to contain both high mannose- and complex-type oligosaccharides (Hoflack and Kornfeld 1985b). Since the CD-MPR is a heavily glycosylated protein in which carbohydrates constitute approximately 20% of the total mass of the receptor, numerous studies have addressed the role of N-linked carbohydrates in the functioning of the receptor. Analysis of glycosylation-deficient mutants generated by site-directed mutagenesis indicate that oligosaccharides facilitate the proper folding of the protein and aid in the stabilization of the ligand binding conformation (Wendland et al. 1991b). Equilibrium binding experiments have shown that a truncated glycosylation-deficient form of the CD-MPR which contains only a single N-linked oligosaccharide chain retains a similar affinity for the lysosomal enzyme, ß-glucuronidase, as the fully glycosylated CD-MPR (Marron-Terada et al. 1998a). These results demonstrate that three out of the four N-linked oligosaccharides of the CD-MPR are not required for high affinity binding of ligand. However, studies by Li and Jourdian (1991) have provided evidence that the presence of sialic acid and polylactosamine residues on the CD-MPR can inhibit the ability of the receptor to bind phosphomannosyl-containing ligands.

4 Carbohydrate Recognition by the CD-MPR

4.1 Lysosomal Enzyme Recognition

Soluble acid hydrolases constitute a heterogeneous population of >40 enzymes that differ in size, oligomeric state, number of N-linked oligosaccharides, extent of phosphorylation, and the position of Man-6-P in the oligosaccharide chain. Since both MPRs are present in most cell types, several experimental approaches have been undertaken to evaluate the relative contribution of each MPR to the targeting of this diverse population of enzymes to the lysosome. Analysis of cell lines that are deficient in either the CD-MPR or the CI-MPR have demonstrated that both receptors are necessary for the efficient sorting of all lysosomal enzymes to the lysosome as neither MPR can fully compensate for the other (Munier-Lehmann et al. 1996). These results raise the possibility that the two receptors recognize distinct subsets of lysosomal enzymes. Analyses of the lysosomal enzymes secreted by cells expressing either the CD-MPR or the CI-MPR by two-dimensional gel electrophoresis was used to test this hypothesis. The results showed that three different subsets of lysosomal enzymes can be identified: the first interacts preferentially with the CD-MPR, the second interacts preferentially with the CI-MPR, and the third interacts equally with both MPRs (Munier-Lehmann et al. 1996). Thus, the two MPRs serve complementary roles in the biogenesis of lysosomes.

Numerous studies have been performed to characterize the carbohydrate binding properties of the MPRs. The CD-MPR binds Man-6-P with an affinity of 8×10^{-6} M (Tong and Kornfeld 1989; Tong et al. 1989). The specificity for binding is determined by the 2-hydroxyl group and the 6-phosphate monoester group based on the observation that fructose 1-phosphate, which is structurally similar to Man-6-P in its pyranose form, is a competitive inhibitor ($K_i = 1 \times 10^{-5}$ M) whereas mannose and glucose 6-phosphate ($K_i = 1\text{--}5 \times 10^{-2}$ M) are not (Tong and Kornfeld 1989). Inhibition studies using chemically synthesized oligomannosides or neoglycoproteins demonstrated that the presence of the phosphomonoester Man-6-P at a terminal position is the major determinant of receptor binding. In addition, linear mannose sequences which contained a terminal Man-6-P linked α1,2 to the penultimate mannose were shown to be the most potent inhibitors (Distler et al. 1991), suggesting that the CD-MPR binds an extended oligosaccharide structure which includes the Man-6-Pα1,2Man sequence.

The CD-MPR, like the CI-MPR, displays optimal ligand binding at ~pH 6.3 and no detectable binding below pH 5 (Tong and Kornfeld 1989; Tong et al. 1989; Distler et al. 1991), which is consistent with their function of releasing ligands in the acidic environment of the late endosomal compartment. The CI-MPR retains phosphomannosyl binding capabilities at neutral pH which explains the ability of this receptor to bind and internalize lysosomal enzymes at the cell surface. In contrast, ligand binding of the CD-MPR is dramatically reduced at a pH > 6.3. This loss of binding activity at neutral pH is the likely cause for the observed inability of the CD-MPR to bind and internalize ligands from the cell surface. The presence of cations has no effect on the binding affinity of the CI-MPR, whereas for the bovine CD-MPR, the presence of cations enhances its affinity towards phosphomannosyl residues fourfold (Hoflack et al. 1987; Junghans et al. 1988; Tong and Kornfeld 1989). The CD-MPR, unlike the CI-MPR, is unable to interact with phosphodiesters, such as the N-acetyl-glucosamine-1-phosphate-6-mannose moiety which is generated as an intermediate during the biosynthesis of the Man-6-P recognition marker (Hoflack et al. 1987; Tong and Kornfeld 1989). Additional studies have shown that diphosphorylated oligosaccharides bind the MPRs with an affinity greater than Man-6-P, with the CI-MPR exhibiting a 100-fold greater affinity for a diphosphorylated oligosaccharide than the CD-MPR (Tong and Kornfeld 1989). These *in vitro* binding studies are consistent with the observed glycosylation state of the lysosomal enzymes that are bound by each MPR *in vivo*: the CI-MPR preferentially binds acid hydrolases enriched in oligosaccharides containing two phosphomonoesters while the CD-MPR preferentially interacts with acid hydrolases containing oligosaccharides which bear only a single phosphomonoester (Munier-Lehmann et al. 1996).

4.2 Expression of Mutant Forms of the CD-MPR

The generation of various mutant forms of the CD-MPR has greatly facilitated the identification of structural features of the receptor that are required for carbohydrate recognition. Equilibrium dialysis experiments demonstrate that the CD-MPR binds 1 mol of Man-6-P (Tong and Kornfeld 1989; Distler et al. 1991) and 0.5 mol of a diphosphorylated high-mannose oligosaccharide per monomeric subunit (Tong and Kornfeld 1989). Since the CD-MPR exists predominantly as a dimer, the CD-MPR contains two Man-6-P binding sites in its functional form in the membrane. Expression of the extracytoplasmic domain alone, which was shown to retain the ability to bind Man-6-P (Dahms and Kornfeld 1989; Wendland et al. 1989), demonstrated that sequences within the transmembrane and cytoplasmic regions of the CD-MPR do not contribute to carbohydrate recognition. The observation that the extracytoplasmic domain, when expressed as a monomer in *Xenopus laevis* oocytes, was functional in binding Man-6-P, and indicates that each polypeptide of the CD-MPR homodimer is capable of folding into an independent CRD (Dahms and Kornfeld 1989). Although the CI-MPR contains 15 repeating units which are homologous to the CD-MPR (Fig. 2), equilibrium dialysis experiments have demonstrated that this receptor binds only 2 mol of Man-6-P (Tong et al. 1989; Distler et al. 1991). N-terminal sequencing of proteolytic fragments combined with expression of truncated forms of the CI-MPR have localized the CRDs of the CI-MPR to domains 1–3 and 7–9 (Westlund et al. 1991; Dahms et al. 1993) and sequences within domain 3 and domain 9 are required for carbohydrate recognition (Dahms et al. 1993).

Site-directed mutagenesis experiments have identified His105 and Arg111 of the CD-MPR as essential components of Man-6-P recognition (Wendland et al. 1991c; Dahms et al. 1993). Arg111 of the CD-MPR is conserved among bovine, human, and mouse species and is also conserved in domain 3 and domain 9 of the CI-MPR. Replacement of this conserved arginine in domain 3 (Arg435) and domain 9 (Arg1334) with alanine or lysine, in a construct encoding domains 1–3 or domains 7–9, respectively, resulted in the complete loss of ligand binding as assessed by affinity chromatography on pentamannosyl phosphate-agarose columns (Dahms et al. 1993). Taken together, these mutagenesis studies suggest that the CRDs of the CD-MPR and CI-MPR exhibit similarities in their structure and in the amino acids used for the recognition of phosphomannosyl residues.

4.3 Crystal Structure of the CD-MPR in the Presence of Bound Man-6-P

4.3.1 Polypeptide Fold

The first view of the CRD of the P-type lectins has been obtained by crystallizing the extracytoplasmic domain of the bovine CD-MPR in the presence of Man-6-P and solving its three-dimensional structure to 1.8 Å resolution (Roberts et al. 1998). Although a truncated, glycosylation-deficient form of the bovine CD-MPR (Asn81/STOP155) was used to generate crystals, this truncated version of the receptor has been shown to bind a lysosomal enzyme, β-glucuronidase, with an affinity identical to that of the full-length wild-type receptor (Marron-Terada et al. 1998a). Asn81/STOP155, which consists of residues 1–154 of the mature protein, has four out of its five potential N-glycosylation sites removed by replacing the asparagine residues at positions 31, 57, 68, and 87 with glutamine, and utilizes the remaining N-glycosylation site at position 81. The crystal structure reveals that the extracytoplasmic domain of the CD-MPR folds into a compact domain which contains one α helix located near the N-terminus followed by nine β strands. The predominantly anti-parallel β strands form two β sheets which are positioned orthogonal to each other (Fig. 3A). The interactions between the two β sheets are primarily hydrophobic in nature, involving residues Leu29, Phe34, Phe49, Val51, Leu64, Val77, Ile92, Leu94, Ala113, Val115, Ile117, Phe128, Phe145, Met147, and Leu151, and results in a hydrophobic core that collapses the interior of the protein to generate a flattened β barrel structure. All six cysteine residues of the extracytoplasmic domain of the CD-MPR are involved in disulfide bond formation (Cys6-Cys52, Cys106-Cys141, and Cys119-Cys153). The

A   B

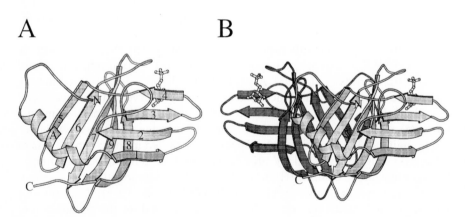

Fig. 3 A,B. Ribbon diagram of the bovine CD-MPR. **A** Monomer. The ß strands are numbered from N to C terminus, beginning with 1, and the α helix is labeled *A*. **B** Dimer. The location of Man-6-P is shown (ball-and-stick model)

disulfide bonds have been shown to be important for the functional integrity of the receptor, since a complete loss of binding activity occurs upon reduction of the CD-MPR with dithiothreitol (Li et al. 1990). The structure confirms that disulfide bonds are important for ligand binding since the linkage between Cys106 and Cys141 is critical for the stabilization and/or orientation of the loop involved in Man-6-P binding (see below).

4.3.2 Structural Similarity to Biotin-Binding Proteins

The primary sequence of the CD-MPR exhibits no significant sequence homology to other known proteins except to the CI-MPR. However, a comparison of its structure with those in the PDB database revealed that the polypeptide fold of the CD-MPR bears no resemblance to other lectins for which structural information is available, but, surprisingly, contains a fold topologically similar to that of the biotin-binding proteins, avidin and streptavidin (Roberts et al. 1998). Although avidin exhibits less than 10% sequence similarity to the CD-MPR, a root mean square deviation of only 1.9 Å exists between the residues that compose their two β sheets, excluding the variable loop regions. Figure 4 shows the nearly identical arrangement of the β strands between the CD-MPR and avidin. In addition, both proteins bind their ligand at a similar location in the structure. However, the biotin binding site is located deeper within the avidin molecule. The significance of this structural similarity is not readily

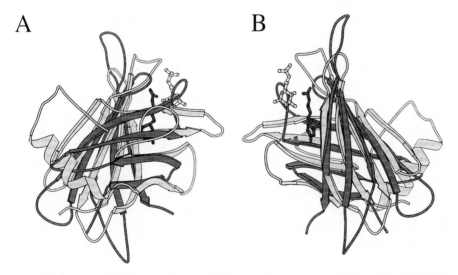

Fig. 4 A,B. Similarity between the CD-MPR and avidin folds. The CD-MPR structure (*light gray*) is overlaid with that of avidin (*dark gray*, PDB accession code 1AVE). The two views are oriented 180° to each other. The locations of Man-6-P (*light gray* ball-and-stick model) and biotin (*dark gray* ball-and-stick model) are shown

apparent, but the possibility exits that the CD-MPR utilizes the same polypeptide fold as avidin in order to facilitate high-affinity binding of its carbohydrate ligands.

4.3.3 Dimeric Structure

The CD-MPR crystallizes as a dimer, which is consistent with its dimeric state in membranes (Dahms and Kornfeld 1989; Li et al. 1990; Fig. 3B). Similar to that observed with the interactions between the β sheets of one monomer, the interactions between the two monomers are mediated by predominantly hydrophobic residues and includes residues Val9, Gly10, Phe86, Trp91, Met93, Ile95, Val114, Met116, Val131, Val138, Phe142, and Leu144. This hydrophobic contact surface area observed between the two monomers of the dimer constitutes approximately 20% of the entire surface area of a single monomer, indicating that the dimer exists as a stable complex. The two carbohydrate binding sites of the dimer are spaced 40 Å apart and are predicted to face outward from the lipid bilayer for interaction with Man-6-P-containing ligands found in the extracellular environment (Roberts et al. 1998).

4.3.4 Carbohydrate Binding Pocket

The structure of the CD-MPR was determined in the presence of bound Man-6-P. Each monomer contains a single Man-6-P molecule which interacts with a number of main chain and side chain atoms, including the two residues (His105 and Arg111) that were previously mutated and shown to be important for Man-6-P recognition (Wendland et al. 1991c; Dahms et al. 1993), located in loop regions as well as in β strands from both β sheets (Fig. 5). Unlike many other lectins (Weis and Drickamer 1996), the recognition of carbohydrate by the CD-MPR does not involve the stacking of aromatic residues of the polypeptide with the nonpolar face of the sugar ring. Biochemical studies using sugar analogs of Man-6-P have demonstrated the importance of the phosphate group and the axial 2-hydroxyl group of mannose for specificity of ligand binding (Tong and Kornfeld 1989), which is reflected in the number of interactions these moieties have with the polypeptide chain (Fig. 6). The phosphate group interacts with a number of amino acids, including the main chain atoms of Asp103, Asn104, and His105, the imine nitrogen of His105, as well as the divalent cation, Mn^{+2}. These residues are located within a loop between β strands 6 and 7, which is stabilized by a disulfide bond (Cys106-Cys141). Arg111, although not in close proximity for hydrogen bonding with the phosphate oxygens, may play a role in ligand stabilization by providing a positively charged environment for the phosphate moiety of Man-6-P. A number of residues are involved in forming hydrogen bonds with the hydroxyl groups of mannose, including Arg135 (C4 hydroxyl), Glu133 (C3 and C4 hydroxyls),

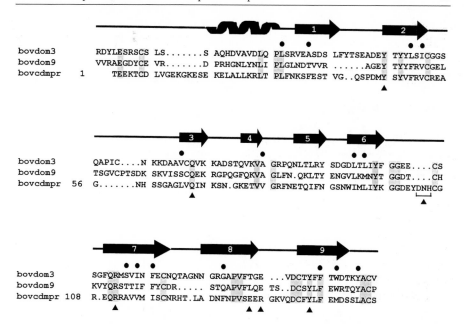

Fig. 5. Comparison of the bovine CD-MPR with domain 3 and domain 9 of the bovine CI-MPR. The secondary structure of the CD-MPR is shown *above* the aligned sequence. Residues that are conserved are shaded in *gray*. Residues of the CD-MPR that are involved in the association of the two β sheets are indicated (•). Residues involved in Man-6-P binding in the CD-MPR are indicated (▲)

Tyr143 (C2 and C3 hydroxyls), Arg111 (C2 hydroxyl), Gln66 (C2 and C3 hydroxyls), and Tyr45 (C1 hydroxyl; Fig. 6). These residues, along with Asp103, Asn104, and His105, are absolutely conserved among bovine, human, and murine CD-MPRs. Site-directed mutagenesis studies have shown that amino acid substitutions at position 66, 105, 111, 133, 135, or 143 exhibit a dramatic loss in the ability to bind to a pentamannosyl phosphate-agarose affinity column (Wendland et al. 1991c; Olson, unpublished data), thus confirming the importance of these residues in Man-6-P binding. The crystal structure also reveals that ~65% of the Man-6-P surface is buried by the receptor, with the phosphate moiety being almost completely concealed. This observation provides an explanation for the inability of the CD-MPR to bind oligosaccharides containing Man-6-P residues in a phosphodiester linkage with either *N*-acetylglucosamine or a methyl group (Hoflack et al. 1987; Tong and Kornfeld 1989; Distler et al. 1991) since the buried nature of the phosphate group leaves no room in the structure to accommodate a phosphodiester.

The bovine CD-MPR was originally identified and purified based on its ability to bind to phosphomannosyl affinity columns in the presence of divalent cations, particularly Mn^{+2} (Hoflack and Kornfeld 1985a,b). Analysis of the three dimensional structure reveals that the Mn^{+2} cation, which ligates one of the phosphate oxygen atoms of the Man-6-P ligand, is coordinated in the Man-

A

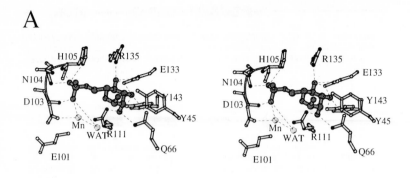

B

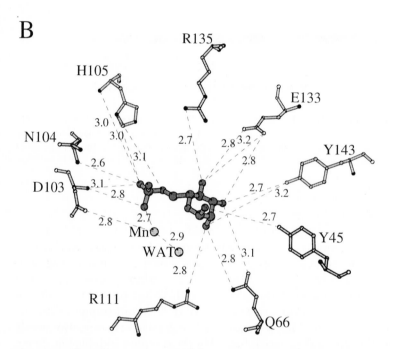

Fig. 6 A,B. Schematic representation of the potential interactions between Man-6-P (M6P) and amino acids of the CD-MPR. **A** A three-dimensional, stereo view of the binding site of the CD-MPR. **B** A flattened view of the binding site to show the potential hydrogen bond distances (in Angströms). The *black and white* balls of the amino acids represent nitrogen and oxygen atoms, respectively. The water molecule (*WAT*) and Mn^{2+} found in the binding pocket are also shown

6-P binding pocket by Asp103 and a water molecule (Fig. 6). It is predicted from the structure that the absence of a cation would result in Asp103 directly interacting with the phosphate oxygen. However, this negative charge interaction is likely to interfere with ligand binding. Thus, the presence of a positive ion may act as a 'shield' between two negative charges, Asp103 and a phosphate oxygen(s), resulting in an increased affinity for Man-6-P.

4.3.5 Comparison of the CRDs of the CD-MPR and CI-MPR

The extracytoplasmic domain of the CD-MPR shares approximately 25% sequence identity with the two domains (i.e., domain 3 and domain 9) of the CI-MPR known to be involved in lysosomal enzyme recognition (Lobel et al. 1988). In order to better evaluate the two CRDs of the CI-MPR, a structure-based sequence alignment was generated between the extracytoplasmic domain of the CD-MPR and domain 3 and domain 9 of the CI-MPR. In this alignment, the positions of the cysteine residues involved in disulfide bond formation were maintained and gaps and insertions were placed within the loop regions (Fig. 5). The observation that the hydrophobic residues involved in the association of the two β sheets in the CD-MPR are conserved in domain 3 and domain 9 supports the hypothesis that the two CRDs of the CI-MPR adopt a similar fold as the CD-MPR. Several of the residues involved in Man-6-P binding by the CD-MPR (i.e. Tyr45, Gln66, Arg111, and Tyr143) are also conserved in domain 3 and domain 9 of the CI-MPR. However, Glu133 and Arg135 are not conserved and have been changed to Thr and Glu, respectively, in domain 3 and Leu and Glu, respectively, in domain 9. Another significant difference between the CRDs of the two MPRs that is predicted from this alignment concerns residues in the loop between β strands 6 and 7. In the CD-MPR, this loop functions in the binding of the phosphate moiety, resulting in its rather buried position within the binding pocket. In both domain 3 and domain 9, this loop is four residues shorter and is predicted to result in the exposure of the phosphate group. An exposed region surrounding the phosphate portion of Man-6-P is consistent with the ability of the CI-MPR, but not the CD-MPR, to bind oligosaccharides containing phosphodiesters (Hoflack et al. 1987; Tong and Kornfeld 1989).

5 Concluding Remarks

The crystal structure of the CD-MPR has provided a wealth of information concerning the mechanism by which this receptor recognizes its phosphorylated ligands. Unlike the CD-MPR which contains two identical CRDs in the homodimer, recent studies have demonstrated that the two CRDs of the CI-MPR monomer are not functionally equivalent (Marron-Terada et al. 1998b). It is possible that during evolution the CI-MPR has optimized its ligand binding ability by incorporating two different Man-6-P binding sites in order to recognize a diverse population of lysosomal enzymes. The basis for the apparent functional differences between the CD-MPR and the CI-MPR as well as between the two CRDs of the CI-MPR must await the three dimensional structure of the CI-MPR.

Acknowledgements. This work was supported by National Institutes of Health Grant DK42667. This work was done during the tenure of an Established Investigatorship from the American Heart Association to N.M.D.

References

Breuer P, Korner C, Boker C, Herzog A, Pohlmann R, Braulke T (1997) Serine phosphorylation site of the 46-kDa mannose 6-phosphate receptor is required for transport to the plasma membrane in Madin-Darby canine kidney and mouse fibroblast cells. Mol Biol Cell 8:567–576

Chen HJ, Yuan J, Lobel P (1997) Systematic mutational analysis of the cation-independent mannose 6-phosphate/insulin-like growth factor II receptor cytoplasmic domain. An acidic cluster containing a key aspartate is important for function in lysosomal enzyme sorting. J Biol Chem 272:7003–7012

Cuozzo JW, Tao K, Cygler M, Mort JS, Sahagian GG (1998) Lysine-based structure responsible for selective mannose phosphorylation of cathepsin D and cathepsin L defines a common structural motif for lysosomal enzyme targeting. J Biol Chem 273:21067–21076

Dahms NM, Kornfeld S (1989) The cation-dependent mannose 6-phosphate receptor. Structural requirements for mannose 6-phosphate binding and oligomerization. J Biol Chem 264: 11458–11467

Dahms NM, Lobel P, Breitmeyer J, Chirgwin JM, Kornfeld S (1987) 46 kd mannose 6-phosphate receptor: cloning, expression, and homology to the 215 kd mannose 6-phosphate receptor. Cell 50:181–192

Dahms NM, Rose PA, Molkentin JD, Zhang Y, Brzycki MA (1993) The bovine mannose 6-phosphate/insulin-like growth factor II receptor. The role of arginine residues in mannose 6-phosphate binding. J Biol Chem 268:5457–5463

de Duve C (1963) The lysosome concept. Churchill, London

Denzer K, Weber B, Hille-Rehfeld A, Figura KV, Pohlmann R (1997) Identification of three internalization sequences in the cytoplasmic tail of the 46 kDa mannose 6-phosphate receptor. Biochem J 326:497–505

Diaz E, Pfeffer SR (1998) TIP47: a cargo selection device for mannose 6-phosphate receptor trafficking. Cell 93:433–443

Distler JJ, Guo JF, Jourdian GW, Srivastava OP, Hindsgaul O (1991) The binding specificity of high and low molecular weight phosphomannosyl receptors from bovine testes. Inhibition studies with chemically synthesized 6-O-phosphorylated oligomannosides. J Biol Chem 266:21687–21692

Glickman JN, Conibear E, Pearse BM (1989) Specificity of binding of clathrin adaptors to signals on the mannose-6-phosphate/insulin-like growth factor II receptor. EMBO J 8:1041–1047

Hemer F, Korner C, Braulke T (1993) Phosphorylation of the human 46-kDa mannose 6-phosphate receptor in the cytoplasmic domain at serine 56. J Biol Chem 268:17108–17113

Hoflack B, Kornfeld S (1985a) Lysosomal enzyme binding to mouse P388D1 macrophage membranes lacking the 215-kDa mannose 6-phosphate receptor: evidence for the existence of a second mannose 6-phosphate receptor. Proc Natl Acad Sci USA 82:4428–4432

Hoflack B, Kornfeld S (1985b) Purification and characterization of a cation-dependent mannose 6-phosphate receptor from murine P388D1 macrophages and bovine liver. J Biol Chem 260:12008–12014

Hoflack B, Fujimoto K, Kornfeld S (1987) The interaction of phosphorylated oligosaccharides and lysosomal enzymes with bovine liver cation-dependent mannose 6-phosphate receptor. J Biol Chem 262:123–129

Johnson KF, Kornfeld S (1992) A His-Leu-Leu sequence near the carboxyl terminus of the cytoplasmic domain of the cation-dependent mannose 6-phosphate receptor is necessary for the lysosomal enzyme sorting function. J Biol Chem 267:17110–17115

Junghans U, Waheed A, von Figura K (1988) The 'cation-dependent' mannose 6-phosphate receptor binds ligands in the absence of divalent cations. FEBS Lett 237:81–84

Klier HJ, von Figura K, Pohlmann R (1991) Isolation and analysis of the human 46-kDa mannose 6-phosphate receptor gene. Eur J Biochem 197:23–28

Kornfeld S (1992) Structure and function of the mannose 6-phosphate/insulinlike growth factor II receptors. Annu Rev Biochem 61:307–330

Kornfeld S, Sly WS (1995) I-cell disease and pseudo-hurler polydystrophy: disorders of lysosomal enzyme phosphorylation and localization. McGraw-Hill, New York

Li M, Distler JJ, Jourdian GW (1990) The aggregation and dissociation properties of a low molecular weight mannose 6-phosphate receptor from bovine testis. Arch Biochem Biophys 283:150–157

Li MM, Jourdian GW (1991) Isolation and characterization of the two glycosylation isoforms of low molecular weight mannose 6-phosphate receptor from bovine testis. Effect of carbohydrate components on ligand binding. J Biol Chem 266:17621–17630

Lobel P, Dahms NM, and Kornfeld S (1988) Cloning and sequence analysis of the cation-independent mannose 6-phosphate receptor. J Biol Chem 263:2563–2570

Ma ZM, Grubb JH, Sly WS (1991) Cloning, sequencing, and functional characterization of the murine 46-kDa mannose 6-phosphate receptor. J Biol Chem 266:10589–10595

Marron-Terada PG, Bollinger KE, Dahms NM (1998a) Characterization of truncated and glycosylation-deficient forms of the cation-dependent mannose 6-phosphate receptor expressed in baculovirus-infected insect cells. Biochemistry 37:17223–17229

Marron-Terada PG, Brzycki-Wessell MA, Dahms NM (1998b) The two mannose 6-phosphate binding sites of the insulin-like growth factor-II/mannose 6-phosphate receptor display different ligand binding properties. J Biol Chem 273:22358–22366

Munier-Lehmann H, Mauxion F, Bauer U, Lobel P, Hoflack B (1996) Re-expression of the mannose 6-phosphate receptors in receptor-deficient fibroblasts. Complementary function of the two mannose 6-phosphate receptors in lysosomal enzyme targeting. J Biol Chem 271:15166–15174

Neufeld EF (1991) Lysosomal storage diseases. Annu Rev Biochem 60:257–280

Pohlmann R, Nagel G, Schmidt B, Stein M, Lorkowski G, Krentler C, Cully J, Meyer HE, Grzeschik KH, Mersmann G, Hasilik A, von Figura K (1987) Cloning of a cDNA encoding the human cation-dependent mannose 6-phosphate-specific receptor. Proc Natl Acad Sci USA 84:5575–5579

Rao PH, Murty VV, Gaidano G, Hauptschein R, Dalla-Favera R, Chaganti RS (1994) Subregional mapping of 8 single copy loci to chromosome 6 by fluorescence in situ hybridization. Cytogenet Cell Genet 66:272–273

Riederer MA, Soldati T, Shapiro AD, Lin J, Pfeffer SR (1994) Lysosome biogenesis requires Rab9 function and receptor recycling from endosomes to the trans-Golgi network. J Cell Biol 125:573–582

Roberts DL, Weix DJ, Dahms NM, Kim JJ (1998) Molecular basis of lysosomal enzyme recognition: three-dimensional structure of the cation-dependent mannose 6-phosphate receptor. Cell 93:639–648

Sahagian GG (1984) The mannose 6-phosphate receptor: function, biosynthesis and translocation. Biol Cell 51:207–214

Schweizer A, Kornfeld S, Rohrer J (1996) Cysteine 34 of the cytoplasmic tail of the cation-dependent mannose 6-phosphate receptor is reversibly palmitoylated and required for normal trafficking and lysosomal enzyme sorting. J Cell Biol 132:577–584

Schweizer A, Kornfeld S, Rohrer J (1997) Proper sorting of the cation-dependent mannose 6-phosphate receptor in endosomes depends on a pair of aromatic amino acids in its cytoplasmic tail. Proc Natl Acad Sci USA 94:14471–14476

Szebenyi G, Rotwein P (1994) The mouse insulin-like growth factor II/cation-independent mannose 6-phosphate (IGF-II/MPR) receptor gene: molecular cloning and genomic organization. Genomics 19:120–129

Tong PY, Kornfeld S (1989) Ligand interactions of the cation-dependent mannose 6-phosphate receptor. Comparison with the cation-independent mannose 6-phosphate receptor. J Biol Chem 264:7970–7975

Tong PY, Gregory W, Kornfeld S (1989) Ligand interactions of the cation-independent mannose 6-phosphate receptor. The stoichiometry of mannose 6-phosphate binding. J Biol Chem 264:7962–7969

Varki A, Kornfeld S (1980) Structural studies of phosphorylated high mannose-type oligosaccharides. J Biol Chem 255:10847–10858 von Figura K, Hasilik A (1986) Lysosomal enzymes and their receptors. Annu Rev Biochem 55:167–193

Waheed A, Hille A, Junghans U, von Figura K (1990) Quaternary structure of the Mr 46,000 mannose 6-phosphate specific receptor: effect of ligand, pH, and receptor concentration on the equilibrium between dimeric and tetrameric receptor forms. Biochemistry 29:2449–2455

Weis WI, Drickamer K (1996) Structural basis of lectin-carbohydrate recognition. Annu Rev Biochem 65:441–473

Wendland M, Hille A, Nagel G, Waheed A, von Figura K, Pohlmann R (1989) Synthesis of a truncated Mr 46,000 mannose 6-phosphate receptor that is secreted and retains ligand binding. Biochem J 260:201–206

Wendland M, von Figura K, Pohlmann R (1991a) Mutational analysis of disulfide bridges in the Mr 46,000 mannose 6-phosphate receptor. Localization and role for ligand binding. J Biol Chem 266:7132–7136

Wendland M, Waheed A, Schmidt B, Hille A, Nagel G, von Figura K, Pohlmann R (1991b) Glycosylation of the Mr 46,000 mannose 6-phosphate receptor. Effect on ligand binding, stability, and conformation. J Biol Chem 266:4598–4604

Wendland M, Waheed A, von Figura K, Pohlmann R (1991c) Mr 46,000 mannose 6-phosphate receptor. The role of histidine and arginine residues for binding of ligand. J Biol Chem 266:2917–2923

Westlund B, Dahms NM, Kornfeld S (1991) The bovine mannose 6-phosphate/insulin-like growth factor II receptor. Localization of mannose 6-phosphate binding sites to domains 1–3 and 7–11 of the extracytoplasmic region. J Biol Chem 266:23233–23239

Galectins Structure and Function – A Synopsis

Hakon Leffler[1]

1 Introduction

Galectins are defined as proteins with at least one carbohydrate recognition domain (CRD) that has (1) certain conserved sequence elements, and (2) affinity for β-galactoside-containing glycoconjugates and saccharides (Barondes et al. 1994a). Almost all galectins discovered so far are small (14–36 kDa), soluble proteins with one or two CRDs. There are now ten known mammalian galectins (Fig. 1) and many in other species, including birds, amphibians, fish, worms, sponges, and a fungus (Barondes et al. 1994b; Kasai and Hirabayashi 1996; Cooper et al. 1997; Leffler 1997; Muller 1997). The sequences in the EST databanks suggest the existence of at least another eight in mammals as well as putative galectins in insects, plants, and viruses (Cooper and Barondes 1999).

Galectins are unusual among lectins in that they are synthesized as cytosolic proteins, without secretion signals, and frequently reside mainly in the cytosol, yet can be targeted for secretion by non-classical pathways (Barondes et al. 1994b; Hughes 1997), or translocation into the nucleus (Wang et al. 1995). Most attention has been given to the possible function of secreted galectins, since their carbohydrate specificity suggests that they would act by binding cell surface and extracellular glycoconjugates, but intracellular functions have also been studied.

A diverse array of functional effects of adding, deleting, or inhibiting galectins in various experimental systems have been observed. However, it is not, at present, possible to amalgamate these into a coherent biological role for galectins. Their basic function remains unknown! Instead, this review will be a collection of field notes and short stories on mainly mammalian galectins, which nevertheless are scientifically very intriguing, and also offer interesting possible new avenues into medical practice. Other recent reviews are: Barondes et al. (1994b); Colnot et al. (1996); Kasai and Hirabayashi (1996); Hughes (1997); Perillo et al. (1998); Chiariotti et al. (1999); Cooper and Barondes (1999); Rabinovich (1999); Vasta et al. (1999) and a theme issue on galectins (Hirabayashi 1997).

---

[1] Department of Microbiology, Immunology and Glycobiology (MIG), Institute of Laboratory Medicine, Lund University, Sölvegatan 23, SE 22362 Lund, Sweden

Results and Problems in Cell Differentiation, Vol. 33
Paul R. Crocker (Ed.): Mammalian Carbohydrate Recognition Systems
© Springer-Verlag Berlin Heidelberg 2001

Galectin-1   Galectin-2   GRIFIN

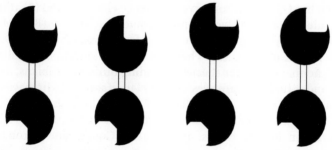

Galectin-4 Galectin-6 Galectin-8 Galectin-9

Galectin-3 Galectin-5 Galectin-7 Galectin-10

Fig. 1. Schematic of mammalian galectins and a galectin-related protein. Carbohydrate recognition domains (CRDs) are *filled* and other domains are *open* or *hatched*. Galectin-1 and -2 and GRIFIN are non-covalent dimers. GRIFIN is very similar to galectin-1 and -2, but so far no carbohydrate binding activity has been demonstrated. Galectin-4, -8, and -9 have two CRDs in one peptide chain (bi-CRD galectins)

1.1 Discovery of Galectins – Past and Present

The first galectins were discovered as part of the quest to find proteins recognizing cell surface carbohydrates thought to be involved in cell adhesion (Ginsburg and Neufeld 1969; Barondes 1970; Roseman 1970). Tissue or cell extracts were tested for agglutinating activities and subjected to affinity chromatography on immobilized β-galactosides (e.g. asialofetuin) and bound proteins eluted with lactose, a cheap β-galactoside that resembled known structures on the cell surface (Barondes 1984, 1997). In this way, a 14 kDa protein, now known as galectin-1, was isolated from electric eel (electrolectin) (Teichberg et al. 1975), chick tissues (Nowak et al. 1977), bovine lung, and other mammalian tissues (Briles et al. 1979; Harrison et al. 1984) and cancer cells (Raz and Lotan 1987). A few others, including galectin-3–5, were also first isolated by this approach (Roff and Wang 1983; Cerra et al. 1985; Raz and Lotan 1987; Leffler et al. 1989). After the cloning of galectins started, however, many proteins discovered based on other properties have been identified as galectins by sequence similarities. For example, galectin-3 was also discovered as an IgE-binding protein (Albrandt et al. 1987), a macrophage surface antigen (Ho and Springer 1982; Cherayil et al. 1989), and a laminin-binding protein (Woo et al. 1990). This and other research has broadened the spectrum of functions considered for galectins. Some galectins have been cloned by accident because antibodies used for screening bind to them via their saccharide chains (Gitt et al. 1995; Hadari et al. 1995). Now most new galectins are discovered by the finding of their signature sequences in expressed sequence tags (ESTs) or genomic sequence tagged sites (STSs) in the data banks (Cooper and Barondes 1999).

2 Structure, Specificity, and Endogenous Ligands

The CRD that defines a protein as a galectin is about 130 amino acids long and tightly folded into a globular domain. Within it is a shorter stretch of 40–60 amino acids that confers most of the carbohydrate binding. The CRD can be arranged in various ways together with other CRDs or peptide sequences (Fig. 1), and the galectins can occur as monomers, dimers or higher oligomers.

2.1 The Carbohydrate Recognition Domain (CRD) and Carbohydrate Binding Site

The X-ray crystal structures of many galectin CRDs have now been solved (Lobsanov et al. 1993; Bourne et al. 1994; Leonidas et al. 1995, 1998; Seetharaman et al. 1998; Rini and Lobsanov 1999). They have a highly conserved tight fold consisting of a sandwich of two anti-parallel β-sheets with six (S1–S6) and five (F1–F5) strands, respectively. Amino acid side chains on the six strand sheet forms the carbohydrate binding site. A similar fold has also been found in legume lectins (although in this case the carbohydrate binding site is different, Lobsanov et al. 1993), and in various other proteins, forming a structural super family (Rini 1995).

The details of the galectin carbohydrate binding site have been defined by binding assays for panels of saccharides or glycolipids (Leffler and Barondes 1986; Knibbs et al. 1993; Feizi et al. 1994), calorimetric analysis (Ramkumar et al. 1995; Gupta et al. 1996a; Surolia et al. 1997; Schwarz et al. 1998) and X-ray crystallography (see references above). Most galectins have a core binding site for lactose and related disaccharides formed by strands S4-S6. The Gal residue is most tightly bound and interacts with the protein along one side via hydrogen bonds involving OH on C4 and C6, the ring O, and van der Waals interaction between the hydrophobic patch formed by CH3-5 and a Trp or Tyr in the protein. The Glc residue also interacts, mainly via OH3, and for most galectins lactose binds with about 100-fold higher affinity (K_d 0.5 mM) than Gal alone. However, there is some room for variation of the Glc residue. If replaced by GlcNAc, the affinity goes up by a factor of about ten for some galectins, but not others, due to interactions involving the NAc group. Gal bound β1-3 to GlcNAc also binds well because here OH4 of GlcNAc takes the steric place of OH3 of GlcNAc in LacNAc (Galβ1-4GlcNAc). Galβ1-3GalNAc, where the corresponding OH4 is axial instead of equatorial, binds some galectins but not others (Leffler and Barondes 1986).

Some larger oligosaccharides bind e.g. galectin-3 with much higher affinity than lactose (K_d down to low μM; Leffler and Barondes 1986). This suggests that the carbohydrate binding site extends beyond the core Gal binding site. The X-ray crystal structures support this idea, since an extension of the groove formed by the core binding site is formed by strand S1-S3. The side chains lining this extended groove are much less conserved among galectins than the core binding site (Seetharaman et al. 1998). Hence, different galectins are expected to bind longer saccharides with different affinities.

If the extended groove is occupied by a saccharide, then the core Gal binding site should be occupied by an internal Gal-residue. This possibility was suggested for galectin-3 by binding assays (Leffler and Barondes 1986; Knibbs et al. 1993). Recently, occupancy of the extended binding site has been demonstrated by an X-ray crystal structure of the galectin-3 CRD in complex with lacto-N-neotetraose (Lobsanov and Rini 1997; Kanigsberg, Seetharaman, Leffler, Rini, unpubl.), and by NMR spectroscopy of galectin-3 CRD in complex with lacto-N-tetraose and Galα1-3LacNAc (Umemoto et al. 1996). Modeling also supports this notion (Henrick et al. 1998). It is interesting to note that the extended binding site can accommodate several different saccharide moieties, and the specificity is greater in the internal core binding site.

Binding to internal Gal-residues in the core site and occupancy of an extended binding site was also proposed to explain the preference of galectin-1 for poly-N-acetylglycosaminoglycans (Merkle and Cummings 1988; Zhou and Cummings 1993). However, a recent NMR-spectroscopy study suggested occupation of the core binding site only by the terminal disaccharide of lacto-N-neotetraose (Di Virgilio et al. 1999). This again supports the idea that the how and when of occupation of the extended binding site will vary among galectins.

Galectin-10 is an exception to the statements above. It binds lactose (Dyer and Rosenberg 1996) only weakly and lacks some of the conserved amino acids in the core binding site (Leonidas et al. 1995). Recently, mannose was found to be accommodated in the core site in an X-ray crystal structure, although the mannose was not bound the same way as Gal (Swaminathan et al. 1999). This surprising finding needs to be validated by demonstration of binding to a mannoside in solution or on a matrix. Nevertheless, it raises the important issue of galectin-like proteins with different carbohydrate binding specificity. Recently, a protein, GRIFIN, was identified in the eye lens and was shown to have high sequence similarity to galectin-1 but no lactose binding activity (Ogden et al. 1998). The legume lectins, which have the same fold as galectins, but no sequence similarity, bind mannose and other saccharide on a loop between β-strands corresponding to S5 and S6, and the binding involves divalent cations (Loris et al. 1998). Two proteins associated with intracellular vesicle transport, which bind mannose-containing saccharides (Carriere et al. 1999; Hara-Kuge et al. 1999), contain domains with low sequence similarity to galectin CRDs but enough to predict that they will have a similar fold (Fiedler and Simons 1994).

2.2 Domain Organization, Oligomerization, and Valency

Galectin-1 behaves as a stable dimer in most reports. However, hamster galectin-1 was found to dissociate into monomers with a K_d of about 7 μM (100 μg/ml; Cho and Cummings 1995). Other species of galectin-1 may also dissociate into monomers, but perhaps at lower concentrations. Galectin-1 may, hence, be dimeric and cross link ligands if present at a high enough concentration, which is well within the range found in nature (up to about 200 μg/ml). But it may also act as a monomer at lower concentrations as suggested (Giudicelli et al. 1997; Blaser et al. 1998). Galectin-2, and the galectin-like protein GRIFIN dimerize in a similar way as galectin-1. The X-ray crystal structures of galectin-1 and galectin-2 demonstrate hydrophobic dimer interfaces involving β-strands S1 and F1 from each subunit (Lobsanov et al. 1993; Liao et al. 1994).

Galectin-3 has one CRD at its C-terminal part linked to a domain consisting of about ten amino acid repeats, each 8–10 residues long and rich in Pro and Gly and with a conserved aromatic residue (the R-domain), and a short N-terminal sequence of unrelated structure (the N-domain; Herrmann et al. 1993). The number of repeats in the R-domain varies among species accounting for the size variations of galectin-3 (Mr ≈ 30 kDa for human and hamster, but 35 kDa for rat and mouse and higher for dog). Galectin-3 behaves as a

monomer in solution up to concentrations well above those found physiologically. However, when binding to ligands on a surface, it binds cooperatively and
appears to be able to cross-link the ligands, thus, acting as a di- or multimer.
This multimerization requires the N- and R-domains, since removal of these by
collagenase digestion of the R-domain produces the CRD alone which binds
carbohydrate but not cooperatively (Massa et al. 1993), and does not agglutinate cells (Hsu et al. 1992) or induce signals (Yamaoka et al. 1995). Subsequent
studies suggest various forms of self association of galectin-3, including primarily the N-terminal part (Mehul et al. 1994) but also the CRD (Yang et al.
1998). Moreover, the R-domain can be cross-linked by tissue transglutaminase
(Mehul et al. 1995; van den Brule et al. 1998), and degraded by matrix metalloproteinases (Herrmann et al. 1993; Mehul et al. 1994; Ochieng et al. 1998).

Galectin-4, -6, -8, and -9, behave as monomers in solution, but all have two
CRDs within one peptide, and, so are 'permanently' bivalent (Fig. 1). Since the
two CRDs differ, particularly in their extended binding sites, they may very
well prefer different ligands (Oda et al. 1993; Wasano and Hirakawa 1999).
Galectin-5 and -7 have one CRD and only a short additional sequence at the
N-terminus. They behave as monomers upon gel filtration at intermediate concentrations (Gitt et al. 1995; Madsen et al. 1995). However, galectin-5 can agglutinate cells, indicating that some type of cross-linking is possible. Moreover,
the X-ray crystal structure of galectin-7 revealed a possible dimer interface
(Leonidas et al. 1998), different from galectin-1 and -2, but similar to a hypo-
thetical dimer interface in the galectin-3 CRD (Seetharaman et al. 1998; Rini
and Lobsanov 1999).

Brewer et al. (Gupta et al. 1996b) have demonstrated that galectin-1, like
some plant lectins, can form ordered arrays with glycoproteins and, based on
this, sorting out of one glycoprotein from another can occur. This adds another
interesting level at which ligands selectivity might be determined, and an interesting possibility for galectin interaction with a cell surface (Pace et al. 1999).

Finally, galectin-10 adds a special case to galectin self aggregation. This
protein, which is expressed at extremely high levels in eosinophil and basophil
leukocytes (about 10% of soluble protein) can spontaneously form crystals in
tissue and secretion. These were discovered over a hundred years ago and
named Charcot-Leyden crystals (Leonidas et al. 1995).

In conclusion, it is fair to assume that all galectins are potentially capable
of cross-linking ligands and, hence, mediate cell adhesion and induce signals,
but there is also evidence of galectins acting as monomers.

2.3 Endogenous Galectin Glycoconjugate Ligands

Although galectins bind β-galactosides which are very common on glycoconjugates, they appear to bind only a few endogenous glycoproteins from a given
cell type. The first ligands identified by affinity chromatography of cell or tissue
extracts on immobilized galectin-1 were the poly-N-acetyllactosamine-

containing proteins; lysosomal membrane-associated proteins, Lamp I and II, and laminin (Do et al. 1990; Zhou and Cummings 1990). Since then, many other soluble or membrane glycoproteins with affinity for galectin-1 or galectin-3 have been isolated (a good table is given in Perillo et al. 1998). Thus, for example, galectin-1 was shown to bind CD43 and CD45, and galectin-3 was shown to bind laminin and Lamp I and II, like galectin-1, but also IgE, IgE-receptor, Mac-2-binding protein (a large, soluble multimeric glycoprotein with sequence similarities to scavenger receptors; Koths et al. 1993; Muller et al. 1999), and CD66 (Ig-domain-containing membrane glycoproteins of the carcinoembryonic antigen family).

The structural basis for the selective recognition of these ligands remains largely unknown. It is clear that poly-*N*-acetyllactosamine does not explain galectin binding to all these proteins, but it is not known what structural features do. It is also not clear to what extent galectin-1, -3 and perhaps other galectins prefer the same ligands or different ligands in a given cell type, since most reports describe only one galectin at a time. The differences in specificity for the small saccharides described above suggest that the different galectin CRDs should also prefer different sets of endogenous ligands; this was also suggested by a recent study of the two CRDs of galectin-4 (Wasano and Hirakawa 1999).

The functional consequence of galectin binding to a particular ligand is often difficult to delineate since galectin usually binds other ligands at the same time. A better understanding requires a careful integration of ligand binding with cell physiology, for example (Feuk-Lagerstedt et al. 1999; Pace et al. 1999).

3 Genes, Expression, and Targeting

3.1 Galectin Genes

The ten known mammalian galectins are encoded on different chromosomes but with some possible clusters: in human, galectin-1 and -2 are encoded at 22q13.1, galectin-3 at 14q21.3, galectin-4, -6, -7, -10 and five putative galectins at 19q13.1 and galectin-8 at 1q 41–44 (Cooper and Barondes 1999). The CRDs are encoded by three exons with relatively (but not perfectly) conserved boundaries. All the galectin signature residues, including all those forming the core ligand binding site, are encoded by the most conserved middle exon, which, thus, might be regarded as a cassette conferring carbohydrate binding activity when placed in the correct context (Gitt and Barondes 1991; Gitt et al. 1998b). As expected, the upstream portion of the genes have many regulatory elements. Their functions have been studied in some detail only for galectin-1 and galectin-3 (Chiariotti et al. 1999). For example, galectin-1 is induced by butyrate (Lu and Lotan 1999), and galectin-3 expression is regulated by serum response factors (Kadrofske et al. 1998), and p53 and other factors (Gaudin et al. 1997; Fogel et al. 1999).

3.2 Galectin Distribution in Cells and Tissues

The galectin composition varies between different cell types, but all cells appear to express at least one galectin. The average concentration of galectins in rat lung and intestine was about 30 and 70 µg/g of tissue, respectively (Cerra et al. 1985; Leffler et al. 1989). Each galectin tends to be expressed in a few, but different, cell types at relatively high concentration. For example, in kidney epithelial cells (MDCK) (Lindstedt et al. 1993) and colon epithelial T84 cells (Huflejt et al. 1997b) galectin-3 accounts for about 0.5% of soluble protein or around 100 µg/ml. Galectin-1 is abundant in adult muscle and other cells of mesodermal origin, galectin-3 is abundant in various epithelial cells and in macrophages, galectin-4 is confined to epithelial cells of the alimentary tract, and galectin-7 is found in epidermis (Colnot et al. 1996; Gitt et al. 1998a; Cooper and Barondes 1999; Timmons et al. 1999). During development, high expression is seen in a few selected cell types and stages for each galectin (Colnot et al. 1996, 1997; Gitt et al. 1998a).

On top of the above described constitutive expression, galectin expression can be induced by various stimuli (Chiariotti et al. 1999). For example, galectin-3 is modulated by inflammatory mediators (Cherayil et al. 1989; Liu 1993; Liu et al. 1995), and galectin-9 is induced by allergic stimulation in monocytes (Matsumoto et al. 1998) but galectin-7 is decreased by retinoic acid (Magnaldo et al. 1995). Cancer cells often have altered galectin expression.

3.3 Synthesis, Intracellular Targeting and Secretion

So far, no galectin has been identified with a secretion signal peptide. Thus, all known galectins have to be assumed to be synthesized as cytosolic proteins, a fact well established for galectin-1 and -3. Galectin-1 was shown to be synthesized on free ribosomes (Wilson et al. 1989), galectin-3 can be phosphorylated by casein kinase I (Huflejt et al. 1993), and both have an acetylated N-terminus (Clerch et al. 1988; Herrmann et al. 1993) – all features typical for cytosolic proteins but not secreted proteins. In contrast to classical secreted proteins, the galectins reside in the cytosol after their synthesis, frequently for most of the life of the cell. From there a fraction of the cytosolic galectin pool can be targeted to various subcompartments of the cytosol (Huflejt et al. 1997b), into the nucleus (Wang et al. 1991), into vesicles (Dvorak et al. 1996, 1997), or be secreted by non-classical (non-ER-Golgi) pathways (Cooper and Barondes 1990; Lindstedt et al. 1993; Hughes 1997).

A few other important proteins, besides galectins, are secreted by non-classical pathways – e.g. interleukin 1, bFGF, annexin, thioredoxin (Rubartelli and Sitia 1991). Non-classical secretion is often slow compared with classical secretion (e.g. 10% secreted in 8h; Lindstedt et al. 1993). Such slow secretion makes it difficult to rule out leakage of cytosolic protein from broken cells, a problem not encountered with classical secretion. This is probably one important reason why progress in elucidating the mechanism for non-classical secretion has been lagging.

One mechanism being considered is membrane blebbing (Cooper and Barondes 1990; Mehul and Hughes 1997). When galectin-1 secretion starts during muscle cell development, the galectin is first accumulated underneath the plasma membrane, then in small membrane evaginations, the blebs, which are about 2 μm in diameter. These are subsequently released from the cell and their membrane disintegrates, releasing the lectin. Nothing is known about what triggers this process, its mechanisms or how the protein to be secreted gets specifically targeted to the blebs.

ABC-transporter proteins are candidates for conducting non-classical protein secretion in mammals, since they do so in bacteria (Linton and Higgins 1998). However, there is no evidence for this in eukaryotes yet, although they transport short peptides both in yeast and mammals (Higgins 1995; Wandersman 1998). In yeast, another type of membrane protein, was identified as involved in secretion of transfected rat galectin-1, but no corresponding mammalian protein has been found (Cleves et al. 1996; Cleves 1997).

Mapping of the structural features of galectin-3 required for its secretion has begun; parts of the N- and R-domains were shown to trigger secretion when attached to a reporter protein (Menon and Hughes 1999). Secretion of galectin-3 occurs selectively at the apical surface in polarized epithelial cells like MDCK and Caco-2 (Lindstedt et al. 1993). This indicates selective intracellular targeting within the cytosol. Immunocytochemistry also supports this in other systems, e.g. galectin-3 in kidney (Sato et al. 1993), galectin-3 and -4 in T84 cells (Huflejt et al. 1997b), and galectin-4 in rat esophagus (Wasano and Hirakawa 1995).

4 Functional Effects

Most direct functional tests of galectins consist of adding purified galectin to some functional assay, e.g. readout for signaling or measurement of cell adhesion. Usually, the galectin concentration used is between 10 and 100 μg/ml, which is the low μM range. This represents low to moderate potency compared with many other biologically active proteins, but is within a reasonable range vis-a-vis the local galectin concentration expected in tissues (see above). In some cases 20–100-fold lower concentrations of galectins were effective, bringing them within the range of e.g. cytokines in potency (Allione et al. 1998; Matsumoto et al. 1998). In another case, different effects were observed at low or high concentrations of galectin-1 (Adams et al. 1996). How should this be interpreted? Do galectins have both high and low affinity receptors mediating different signals? Can one response overshadow another? Better dose-response curves are needed in most cases.

In another approach to galectin function, a galectin inhibitor, usually lactose or a related disaccharide such as thiodigalactoside, which is five to ten times more potent, is added to a functional assay. It is then argued that if inhibition occurs a galectin might be involved, whereas if inhibition does not occur a galectin is probably not involved. How much saccharide should be added? A low to moderate affinity interaction of galectin with a glycoconjugate receptor can usually be inhibited by over 90% by 10 mM lactose. However, model studies suggest that if the ligand has high affinity for the galectin and is presented in multivalent form, 10 mM lactose does not inhibit at all, and higher concentrations, e.g., 50–100 mM may be required (Qian and Leffler, unpubl.). Since one has to assume this to be the case also in a biological system, lack of inhibition by 10–20 mM lactose does not rule out a contribution of galectins to a particular function (Sasaki et al. 1998). More potent selective inhibitors of galectins are being developed (Sörme et al. 1999), since 50–100 mM saccharide can easily cause non-specific effects.

4.1 Cell Adhesion

As mentioned above, a possible role in cell adhesion was the early impetus leading to the discovery of galectins. Support for such a role come from localization of galectins at sites of adhesion (Cooper et al. 1991; Chiu et al. 1994; Wasano and Hirakawa 1995), and from in vitro cell adhesion assays. If coated on a surface, galectin-1, -3, and -4, for example (Mahanthappa et al. 1994; Huflejt et al. 1997b), have been shown to promote adhesion of various cells, presumably by binding to cell surface glycoproteins. If instead, the galectin is added to an assay system with known adhesion molecules coated on the surface, galectin-1 and galectin-3 were found to either inhibit (Cooper et al. 1991; Sato and Hughes 1992) or promote adhesion (Zhou and Cummings 1993; Kuwabara and Liu 1996) depending on conditions and cell type used. One mechanism proposed for the inhibitory effect was galectin binding to cell surface integrins and thereby down-regulating their function directly or via intracellular signaling (Gu et al. 1994).

The agglutination of red cells and other cells by galectins demonstrates a possible cell–cell adhesion mechanism, and cell–cell adhesion by endogenously produced galectin has also been demonstrated (Inohara and Raz 1995). Cell–cell adhesion mediated by binding of added soluble ligand glycoprotein to galectins at the cell surface has also been proposed (Inohara et al. 1996; Lotan and Raz 1988).

4.2 Galectin Induced Signaling

A variety of signals have been found to be induced when galectins are added to cells, mainly in the immune system. Induction of apoptosis in T-cells by galectin-1 has been the most extensively studied (Perillo et al. 1995, 1997, 1998; Vespa et al. 1999). Other signals include induction of oxidative burst in neutrophil leukocytes by galectin-3 (Yamaoka et al. 1995; Karlsson et al. 1998), regulation of cytokine secretion by galectin-3 (Jeng et al. 1994; Cortegano et al. 1998), chemotaxis in eosinophils by galectin-9 (Matsumoto et al. 1998), and mitogenesis by galectin-1 (Lipsick et al. 1980; Adams et al. 1996) and possibly galectin-3 (Inohara et al. 1998). Galectins probably act by cross-linking receptors at the cell surface, a common way for receptors to work, but there may be many other layers of complexity. As mentioned above, receptors proposed to mediate the signals have been isolated based on their affinity for galectin, but more than one candidate receptor is usually isolated, raising the question of which one mediates the signal. In the case of T-cell apoptosis induced by galectin-1, CD43, CD45, and CD7 were identified as the main receptors. Binding of galectn-1 to the T-cell surface induced interesting selective aggregation of CD45 and CD3, but not CD43 and CD7 (Pace et al. 1999), and it was proposed to be due to the formation of ordered lectin–glycoprotein arrays (Gupta et al. 1996b) mentioned above (Sect. 2.2). However, it is not known which receptor mediates the signal, and more than one signaling pathway may be involved.

Instead of looking for one signaling pathway, perhaps it is biologically more correct to regard intracellular signaling as the shift of a network of interacting signaling pathways from one state to another induced by multiplex inputs (Bhalla and Iyengar 1999). Such a view is especially attractive for signaling induced by lectins, including the many effects of plant lectins, which are specific for carbohydrates shared (as post-translational modifications) by subsets of glycoproteins, and hence expected to act through multiple receptors. This is in contrast to most other signaling substances (hormones, cytokines etc.) that bind specific receptors and are expected to act via one (or possibly an additional related receptor) in a given cell type.

The case of galectin-3 and neutrophil leukocytes illustrates another important aspect of signaling. Neutrophils from peripheral blood were found to be unresponsive to galectin-3, but neutrophils harvested from a model inflammatory site were highly responsive (Karlsson et al. 1998). Apparently the cells had undergone a change (priming) during their passage from blood to the inflammatory site, making them responsive to galectin-3. The priming of responsiveness to galectin-3 can be modeled in vitro using fMLP peptide, and was proposed to involve the mobilization of receptors initially stored in intracellular granules to the cell surface. Two types of galectin-3 binding proteins were found; isoforms of CD66 and the lamps. The CD66s were proposed as galectin receptors because they, and not the lamps, were stored in vesicles that moved to the cell surface upon priming (Feuk-Lagerstedt et al. 1999). The need to bring a cell from a non-responsive state into a responsive state, by one or more steps of pretreatments (priming), is common in biology and has also been observed for e.g. the apoptotic and growth inhibitory effects of galectin-1 (Allione et al. 1998; Vespa et al. 1999).

The above cases illustrate the difficulty in observing a functional effect of galectins. You have to guess what to measure, or discover it by chance. Then you have to find a convenient cell system to measure it in, and finally find out in what primed state to bring the cell into. Additional aspects to consider, not discussed in detail here, include time frame of cellular response (from seconds to days) and possible heterogeneity of the cell populations (is the response measured directly due to galectin or is it an indirect effect via a subpopulation of the cells?). With these considerations in mind, there are likely to be numerous galectin effects on cells, both immune and non-immune, that have not yet been discovered. Nevertheless, the observed effects suggest important roles of galectins in immunity and inflammation, and study of the signaling mechanisms may also provide tools for studies of other cell types.

4.3 Galectins in Apoptosis

As mentioned above, galectin-1 can induce apoptosis in T-cells when added as an extracellular protein to the cells (Perillo et al. 1995). The signaling is thought to involve CD45 and probably other membrane glycoproteins, and the T-cells have to be of the correct subset, or be primed into the correct stage to be responsive (Pace et al. 1999; Vespa et al. 1999). Galectin-9 can also induce apoptosis when added to T-cells (Wada et al. 1997).

Galectin-3 has been proposed to protect cells against apoptosis, but in this case when augmented intracellularly. When galectin-3 was overexpressed in cells, the cells grew faster and resisted apoptosis better than their mock-transfected counterparts; the mechanism was proposed to involve an interaction between galectin-3 and Bcl-2 (Yang et al. 1996; Akahani et al. 1997).

Galectin-7 also has some relationship to apoptosis. When the expression of over 7000 genes was assessed during early p53-induced apoptosis in a colon carcinoma, galectin-7 was the second most augmented, and it was one of only 14 genes strongly induced (Polyak et al. 1997). Recently, galectin-7 has also been found to be increased in apoptotic cells (sunburn cells) in skin after UV-radiation (Bernerd et al. 1999).

4.4 Galectins and Galectin Inhibitors In Vivo

Galectin-1 was found early on to have suppressive effects on the autoimmune disease myasthenia gravis when injected into rabbits; the idea for this experiment came from finding galectin-1 in the thymus, its interaction with thymocytes and the known relationship between the thymus and myasthenia gravis (Levi and Teichberg 1983; Levi et al. 1983). Subsequently, galectin-1 has been found to have a similar effect on two other autoimmune diseases – experimental allergic encephalomyelitis, a model of multiple sclerosis (Offner et al. 1990), and experimental rheumatoid arthritis (Rabinovich et al. 1999). The mechanism is now proposed to be induction of apoptosis in antigen-activated

T-cells by galectin-1 (Perillo et al. 1998). Perhaps one can regard galectin-1 as a booster of a normal turn-off mechanism in the immune response that fails in autoimmune disease (Rabinovich 1999).

In a recent, very elegant study (Rabinovich et al. 1999), mice were either given galectin-1 intraperitoneally, or given cells producing galectin-1. Although the latter protocol was most effective, the former also prevented the disease. The dose of $100\,\mu g/day$ is not very high, considering that the mouse (20–30 g body weight) probably already has many times that amount of galectin-1. Presumably, the effect is due to distribution of the administered galectin-1 to sites where it is not normally present in sufficient amounts. Perhaps most of the endogenous galectin-1 is sequestered inside cells, and its secretion is another important immunoregulatory event.

Some saccharides that inhibit galectins in vitro were found to have anti-tumor effects in mouse models (Inohara and Raz 1994; Glinsky et al. 1996). However, it is not known if the in vivo effect involved galectins.

4.5 Nuclear Functions

The finding of galectins in cell nuclei was a surprise (Wang et al. 1991) because the major hypothesis of early galectin research had to do with the function of cell surface carbohydrates. Nevertheless, the targeting in and out of the nucleus of galectin-3 and its possible functions there has been studied mainly by Wang et al. (1992). The nuclear localization of galectin-3 is induced in growing fibroblasts (Moutsatsos et al. 1987). However, the nuclear targeting can be decreased in some cancer cells (see Sect. 5.3.), and this may be why it is not observed in all cell lines. Galectin-3 interacts with ribonucleoproteins in the nucleus, and rigorous experiments suggest a role in RNA splicing (Dagher et al. 1995). Galectin-1 can also play a similar role in RNA splicing (Vyakarnam et al. 1997). Recently, a specific active export of galectin-3 from the nucleus has been demonstrated (Tsay et al. 1999), supporting the hypothesis that it can shuttle in and out of the nucleus and take part in transport of e.g. mRNA.

4.6 Other Galectin Effects

A clone encoding galectin-9 was isolated in a screen of a kidney cDNA library with a rabbit antiserum against uricase (Leal-Pinto et al. 1997). The goal was to find proteins involved in urate transmembrane transport, and to test this effect the model lipid membranes were treated with the recombinant galectin-9. Surprisingly, an effect indicating urate transport was observed, and a model of the protein was designed in which it is has four α-helical transmembrane segments (Leal-Pinto et al. 1999). The cloning of galectin-9 with anti-uricase antiserum can be due to binding via saccharides on the antibodies as found for galectin-5 and galectin-8 (see Sect. 1.2). However, the effect on lipid membranes is highly intriguing and, if true, would introduce a completely new viewpoint on galectins.

Galectins have been proposed to interact (directly or indirectly) with nonenzymatically glycated proteins, e.g., as found in diabetics (Vlassara et al. 1995; Thornalley 1998).

4.7 Galectin Null-Mutant Mice

Galectin-1 and galectin-3, as well as the double galectin-1/-3 null mutant mice have been generated (Poirier and Robertson 1993; Colnot et al. 1998a) All the mice were fully viable, reproduced and appeared generally as healthy as the normal mice of the same strain. No obvious phenotype was detected initially. However, careful examination guided by previous functional hypotheses for the galectins has revealed subtle effects. The galectin-1 null mutants have an altered guidance of neurons in the olfactory bulb (Puche et al. 1996). The galectin-3 null mutants have an altered recruitment/survival of neutrophils in inflammation (Colnot et al. 1998b), and an altered terminal differentiation of chondrocytes in fetal bone growth plates (Colnot, Sidhu, Baumnain, and Poirier, manscript in preparation). These, and other subtle effects of the galectin gene ablations will be very valuable in elucidating galectin function.

5 Biological Roles and Biomedical Use

5.1 Immunity and Inflammation

There are many possible roles of galectins in immunity. The apoptosis-inducing effect of galectin-1 has been proposed to contribute to T-cell selection in the thymus; galectin-1 is expressed by thymic epithelial cells, and immature T-cells, known to encounter such cells, are susceptible to its effect (Perillo et al. 1995). As mentioned above, the same effect may also be important in turning off an immune response (Gold et al. 1997; Perillo et al. 1998), since antigen-activated T-cells were found to be selectively susceptible to its effect (Vespa et al. 1999). The active galectin-1 in this case would be expected to come from the various cells in tissues that express it, such as fibroblasts or muscle cells. An important immunoregulatory step, then, may be the release of galectin-1 from such cells.

The research on galectin-3 in immunity has focused more on the inflammatory and allergic response: it is highly expressed in activated macrophages and it has various stimulating effects on neutrophils (Kuwabara and Liu 1996; Karlsson et al. 1998), monocytes (Jeng et al. 1994), basophils (Zuberi et al. 1994) etc. Perhaps it is an important endogenous modulator of these responses (Hughes 1997; Perillo et al. 1998).

Galectin-9 was found to be strongly induced in monocytes upon allergic antigen stimulation and acts as a potent chemotactic selectively for eosinophils (Matsumoto et al. 1998). Support for this occurring in vivo may be the finding of increased expression of galectin-9 in Hodgkin's lymphoma (Tureci et al. 1997) where eosinophils are a common feature.

Does, then, each galectin have a different immunological function? This is far from certain, because in most cases only one galectin has been examined. There are almost no comparisons of the effect of different galectins in the same system. Perhaps galectin-3 and -4 could also induce apoptosis in activated T-cells like galectin-1 and galectin-9.

5.2 Host–Pathogen Interaction

Many bacteria carry β-galactoside-containing saccharides on their surface, making it reasonable that they would interact with galectins (Mandrell et al. 1994). One group are the gram-negative non-enteric bacteria such as *Haemophilus* spp. and *Neisseria* spp. that, instead of LPS, have lipooligosaccharides (LOS) with shorter outer carbohydrate chains that often contain β-galactosides. Galectin-3 has been found to bind via its CRD to some such saccharides (Mandrell et al. 1994; Mey et al. 1996; Gupta et al. 1997). However, galectin-3 also binds the lipid-A moiety of LOS and LPS via its non-carbohydrate-binding N-terminal (Mey et al. 1996). These binding activities may be important for adhesion of the microbe to host cells, and various subsequent events, but no direct evidence for this has been reported yet.

5.3 Cancer

About 100 reports have suggested various connections between galectins and various neoplasias. Most compare galectin expression in normal tissue/cells with tumor tissue/cells of varying malignancy (see Andre et al. 1999 for a recent example). Here only a few examples will be given. An induction of galectin-1 in cultured cells from epithelial tumors has been easiest to observe, probably because the normal epithelial cells do not express galectin-1 (Ohannesian et al. 1994). However, in histopathological sections usually only a few of the cancer cells express galectin-1 (Lotan et al. 1991), even in the most malignant cancers, whereas an increased stromal expression has been reported (Sanjuan et al. 1997).

The correlation between galectin-3 expression and cancer malignancy is confounded by the fact that the cell of origin, usually epithelial, often expresses high levels of galectin-3. Hence, both increased (Lotz et al. 1993; Castronovo et al. 1996; Bresalier et al. 1997), and decreased (Lotz et al. 1993; Castronovo et al. 1996, 1999) expression has been observed. In tissues that do not normally express high levels of galectin-3, such as brain and thyroid, an increase of galectin-3 correlated with malignancy (Bresalier et al. 1997; Fernandez et al. 1997; Inohara et al. 1999).

Recently, a specific induction of a galectin-8-like protein in prostate cancer (Su et al. 1996), of galectin-9 in Hodgkin's lymphoma (Tureci et al. 1997) and galectin-4 in liver cancer (Kondoh et al. 1999) and breast cancer (Huflejt et al. 1997a) have been reported. Galectin-7 was found, by differential display, to be overexpressed in chemically induced rat mammary tumors (Lu et al. 1997).

There is also evidence for altered subcellular distribution of galectins in tumor cells. Most important for the functional hyopotheses described below is the evidence for increased cell surface exposure of galectin-1, -3, -8, and -9 (Lotan and Raz 1988; Su et al. 1996; Tureci et al. 1997). A loss of galectin-3 from the nucleus was found in colon cancer (Lotz et al. 1993).

The direct role of galectin expression on tumor development has been tested mainly for galectin-3. Cell lines transfected to increase galectin-3 expression gave much higher numbers of tumors in mice compared with controls (Raz et al. 1990; Bresalier et al. 1998).

The study of galectins in cancer was initiated by the finding by Raz and Lotan that some cancer cells are agglutinated by asialofetuin, in turn leading to their isolation of the β-galactoside binding lectins L14 and L31, now known to be galectin-1 and -3 (Raz and Lotan 1987). It was proposed that increased amounts of surface-exposed galectin would increase cell adhesion in turn, perhaps resulting in increased metastasis. This hypothesis is still reasonable, but there are many other possibilities based on the functional activities of galectins described above.

Secreted or cell surface galectin-1 could enhance tumors by suppressing anti-tumor immunity via its apoptotic effect on antigen activated T-cells (Perillo et al. 1998). Galectin-3 could be tumorigenic due its intracellular growth-promoting anti-apoptotic effect on the tumor cells themselves (Yang et al. 1996). Both could, conceivably, also act as paracrine mitogens (Adams et al. 1996; Yamaoka et al. 1996; Inohara et al. 1998).

5.4 Galectin Serology

There is evidence that galectin-3 and some of the bi-CRD galectins can be highly immunogenic. Anti-galectin-3 (anti-Mac-2) was one of the major monoclonal specificities found after immunization with macrophage membranes (Ho and Springer 1982). Similarly, galectin-8 (or a closely similar molecule) was a major specificity after immunization with prostate cancer membranes (Su et al. 1996). In vivo immunization to galectin-4 was indicated in colon carcinoma (Scanlan et al. 1998) and antibodies against galectin-9 were found to be frequent in patients with Hodgkin's lymphoma (Sahin et al. 1995; Tureci et al. 1997), and possibly also in some healthy individuals (Suk et al. 1999). Bi-CRD galectins of nematodes have also been found to be important pathogen antigens eliciting an immune response in their hosts (Klion and Donelson 1994; Greenhalgh et al. 1999). Autoantibodies to galectin-1 (Lutomski et al.

1997) and galectin-3 (Mathews et al. 1995) have also been reported. The possible clinical usefulness of detection of anti-galectin antibodies (galectin-serology) needs further exploration.

5.5 Tissue Organization and Repair

The function of galectins has been studied to a more limited extent outside the immune system. Here are some examples. Hughes et al. analyzed intriguing effects of galectin-3 on development and organization of kidney epithelium (Hughes 1997; Winyard et al. 1997; Bao and Hughes 1999). Cooper et al. suggested roles of galectin-1 in maturation and organization of muscle fibers (Cooper and Barondes 1990; Cooper et al. 1991; Gu et al. 1994). A role of galectin-3 in cartilage and bone formation/remodeling is suggested by high expression (Colnot et al. 1996; Aubin et al. 1995, 1996; Nurminskaya and Linsenmayer 1996) and alterations in galectin-3 null mutant mice (sect. 4.7). Galectin-1 and a lactose-binding 67 kDa protein was proposed to be involved in proper deposition of elastic fibers (Hinek et al. 1988). Even if the 67 kDa protein is not a galectin, the dramatic effect of lactose on elastic fiber deposition is intriguing.

5.6 Galectins in the Nervous System

Galectins have been isolated from brain (Bladier et al. 1989; Chadli et al. 1997; Zanetta 1998), but they seem to be found in only a few restricted cell types in the central nervous system. In the CNS, galectin-1 has been found mainly in olfactory neurons and has been proposed to act in neuron fasciculation and guidance based on experiments in tissue culture (Mahanthappa et al. 1994; Puche and Key 1995; Tenne-Brown et al. 1998) and the effects in galectin-1 null mutant mice (Puche et al. 1996). Galectin-3 is mainly associated with glia cells (Pesheva et al. 1998) and is induced in brain tumors (Bresalier et al. 1997) and microglial activation (Pesheva et al. 1998).

In peripheral nerve cells, galectin-1 and -3 were shown to have specific locations in dorsal root ganglia (Regan et al. 1986). Injury of peripheral nerve cells results in an induction of galectin-3 expression in the injured cell as well as the neighboring nerve cells (Cameron et al. 1993, 1997) and Schwann cells (Saada et al. 1996). It was proposed that galectin-3 helps provide macrophage-like properties to the Schwann cell and participates in scavenging of the debris from the injured cell (Saada et al. 1996; Be'eri et al. 1998).

6 Summary and Conclusions

The 20 or so galectins expected to be found in man, and their many possible functional effects promise a rich and fruitful research field in the future. At present, the biomedically most promising areas for use of galectins or their ligands are in inflammation, immunity, and cancer. Many good stories can be formulated, but the field lacks the cohesion of knowing basic galectin function. The only basic common denominators among galectins are β-galactoside binding, and the unusual combination of intra- and extracellular expression with non-classical secretion in between. Maybe that is all there is, and nature has used these properties for multiple, otherwise unrelated functions. Then again, maybe there is some deeper common function that has so far been overlooked. If it exists, this probably lies somewhere in the detailed integration of galectin activity in the complexities of cell physiology.

References

Adams L, Scott GK, Weinberg CS (1996) Biphasic modulation of cell growth by recombinant human galectin-1. Biochim Biophys Acta 1312:137–144

Akahani S, Nangia-Makker P, Inohara H, Kim HR, Raz A (1997) Galectin-3: a novel antiapoptotic molecule with a functional BH1 (NWGR) domain of Bcl-2 family. Cancer Res 57:5272–5276

Albrandt K, Orida NK, Liu FT (1987) An IgE-binding protein with a distinctive repetitive sequence and homology with an IgG receptor. Proc Natl Acad Sci USA 84:6859–6863

Allione A, Wells V, Forni G, Mallucci L, Novelli F (1998) Beta-galactoside-binding protein (beta GBP) alters the cell cycle, up-regulates expression of the alpha- and beta-chains of the IFN-gamma receptor, and triggers IFN-gamma-mediated apoptosis of activated human T lymphocytes. J Immunol 161:2114–2119

Andre S, Kojima S, Yamazaki N, Fink C, Kaltner H, Kayser K, Gabius HJ (1999) Galectins-1 and -3 and their ligands in tumor biology. Non-uniform properties in cell-surface presentation and modulation of adhesion to matrix glycoproteins for various tumor cell lines, in biodistribution of free and liposome-bound galectins and in their expression by breast and colorectal carcinomas with/without metastatic propensity. J Cancer Res Clin Oncol 125:461–474

Aubin JE, Liu F, Malaval L, Gupta AK (1995) Osteoblast and chondroblast differentiation. Bone 17:77S–83S

Aubin JE, Gupta AK, Bhargava U, Turksen K (1996) Expression and regulation of galectin 3 in rat osteoblastic cells. J Cell Physiol 169:468–480

Bao Q, Hughes RC (1999) Galectin-3 and polarized growth within collagen gels of wild-type and ricin-resistant MDCK renal epithelial cells. Glycobiology 9:489–495

Barondes SH (1970) Brain glycomacromolecules and interneuronal recognition. In: Schmitt FO (ed) The neurosciences: a second study program. Rockefeller University Press, New York, pp 747–760

Barondes SH (1984) Soluble lectins: a new class of extracellular proteins. Science 223:1259–1264

Barondes SH, Castronovo V, Cooper DN, Cummings RD, Drickamer K, Feizi T, Gitt MA, Hirabayashi J, Hughes C, Kasai K et al. (1994a) Galectins: a family of animal beta-galactoside-binding lectins (letter). Cell 76:597–598

Barondes SH, Cooper DN, Gitt MA, Leffler H (1994b) Galectins. Structure and function of a large family of animal lectins. J Biol Chem 269:20807–20810

Barondes SH (1997) Galectins: a personal overview. Trends Glycosci Glycotech 9:1–7

Be'eri H, Reichert F, Saada A, Rotshenker S (1998) The cytokine network of wallerian degeneration: IL-10 and GM-CSF. Eur J Neurosci 10:2707–2713

Bernerd F, Sarasin A, Magnaldo T (1999) Galectin-7 overexpression is associated with the apoptotic process in UVB-induced sunburn keratinocytes. Proc Natl Acad Sci USA 96:11329–11334

Bhalla US, Iyengar R (1999) Emergent properties of networks of biological signaling pathways (see comments). Science 283:381–387

Bladier D, Joubert R, Avellana-Adalid V, Kemeny JL, Doinel C, Amouroux J, Caron M (1989) Purification and characterization of a galactoside-binding lectin from human brain. Arch Biochem Biophys 269:433–439

Blaser C, Kaufmann M, Muller C, Zimmermann C, Wells V, Mallucci L, Pircher H (1998) Beta-galactoside-binding protein secreted by activated T cells inhibits antigen-induced proliferation of T cells. Eur J Immunol 28:2311–2319

Bourne Y, Bolgiano B, Liao DI, Strecker G, Cantau P, Herzberg O, Feizi T, Cambillau C (1994) Crosslinking of mammalian lectin (galectin-1) by complex biantennary saccharides. Nat Struct Biol 1:863–870

Bresalier RS, Yan PS, Byrd JC, Lotan R, Raz A (1997) Expression of the endogenous galactose-binding protein galectin-3 correlates with the malignant potential of tumors in the central nervous system. Cancer 80:776–787

Bresalier RS, Mazurek N, Sternberg LR, Byrd JC, Yunker CK, Nangia-Makker P, Raz A (1998) Metastasis of human colon cancer is altered by modifying expression of the beta-galactoside-binding protein galectin 3. Gastroenterology 115:287–296

Briles EB, Gregory W, Fletcher P, Kornfeld S (1979) Vertebrate lectins, comparison of properties of beta-galactoside-binding lectins from tissues of calf and chicken. J Cell Biol 81:528–537

Cameron AA, Dougherty PM, Garrison CJ, Willis WD, Carlton SM (1993) The endogenous lectin RL-29 is transynaptically induced in dorsal horn neurons following peripheral neuropathy in the rat. Brain Res 620:64–71

Cameron AA, Cliffer KD, Dougherty PM, Garrison CJ, Willis WD, Carlton SM (1997) Time course of degenerative and regenerative changes in the dorsal horn in a rat model of peripheral neuropathy. J Comp Neurol 379:428–442

Carriere V, Piller V, Legrand A, Monsigny M, Roche AC (1999) The sugar binding activity of MR60, a mannose-specific shuttling lectin, requires a dimeric state (in process citation). Glycobiology 9:995–1002

Castronovo V, Van Den Brule FA, Jackers P, Clausse N, Liu FT, Gillet C, Sobel ME (1996) Decreased expression of galectin-3 is associated with progression of human breast cancer. J Pathol 179:43–48

Castronovo V, Liu FT, van den Brule FA (1999) Decreased expression of galectin-3 in basal cell carcinoma of the skin. Int J Oncol 15:67–70

Cerra RF, Gitt MA, Barondes SH (1985) Three soluble rat beta-galactoside-binding lectins. J Biol Chem 260:10474–10477

Chadli A, LeCaer JP, Bladier D, Joubert-Caron R, Caron M (1997) Purification and characterization of a human brain galectin-1 ligand. J Neurochem 68:1640–1647

Cherayil BJ, Weiner SJ, Pillai S (1989) The Mac-2 antigen is a galactose-specific lectin that binds IgE. J Exp Med 170:1959–1972

Chiariotti L, Salvatore P, Benvenuto G, Bruni CB (1999) Control of galectin gene expression. Biochimie 81:381–388

Chiu ML, Parry DA, Feldman SR, Klapper DG, O'Keefe EJ (1994) An adherens junction protein is a member of the family of lactose-binding lectins. J Biol Chem 269:31770–31776

Cho M, Cummings RD (1995) Galectin-1, a beta-galactoside-binding lectin in Chinese hamster ovary cells. I. Physical and chemical characterization. J Biol Chem 270:5198–5206

Clerch LB, Whitney P, Hass M, Brew K, Miller T, Werner R, Massaro D (1988) Sequence of a full-length cDNA for rat lung beta-galactoside-binding protein: primary and secondary structure of the lectin. Biochemistry 27:692–699

Cleves AE (1997) Protein transports: the nonclassical ins and outs. Curr Biol 7:R318–320

Cleves AE, Cooper DN, Barondes SH, Kelly RB (1996) A new pathway for protein export in Saccharomyces cerevisiae. J Cell Biol 133:1017–1026

Colnot C, Ripoche MA, Scaerou F, Foulis D, Poirier F (1996) Galectins in mouse embryogenesis. Biochem Soc Trans 24:141–146

Colnot C, Ripoche MA, Fowlis D, Cannon V, Scaerou F, Cooper DNW, Poirier F (1997) The role of galectins in mouse development. Trends Glycosci Glycotech 45:31–40

Colnot C, Fowlis D, Ripoche MA, Bouchaert I, Poirier F (1998a) Embryonic implantation in galectin 1/galectin 3 double mutant mice. Dev Dyn 211:306–313

Colnot C, Ripoche MA, Milon G, Montagutelli X, Crocker PR, Poirier F (1998b) Maintenance of granulocyte numbers during acute peritonitis is defective in galectin-3-null mutant mice. Immunology 94:290–296

Cooper DN, Barondes SH (1990) Evidence for export of a muscle lectin from cytosol to extracellular matrix and for a novel secretory mechanism. J Cell Biol 110:1681–1691

Cooper DN, Barondes SH (1999) God must love galectins; he made so many of them (in process citation). Glycobiology 9:979–984

Cooper DN, Massa SM, Barondes SH (1991) Endogenous muscle lectin inhibits myoblast adhesion to laminin. J Cell Biol 115:1437–1448

Cooper DN, Boulianne RP, Charlton S, Farrell EM, Sucher A, Lu BC (1997) Fungal galectins, sequence and specificity of two isolectins from *Coprinus cinereus*. J Biol Chem 272:1514–1521

Cortegano I, del Pozo V, Cardaba B, de Andres B, Gallardo S, del Amo A, Arrieta I, Jurado A, Palomino P, Liu FT, Lahoz C (1998) Galectin-3 down-regulates IL-5 gene expression on different cell types. J Immunol 161:385–389

Dagher SF, Wang JL, Patterson RJ (1995) Identification of galectin-3 as a factor in pre-mRNA splicing. Proc Natl Acad Sci USA 92:1213–1217

Di Virgilio S, Glushka J, Moremen K, Pierce M (1999) Enzymatic synthesis of natural and ^{13}C enriched linear poly-*N*-acetyllactosamines as ligands for galectin-1. Glycobiology 9: 353–364

Do KY, Smith DF, Cummings RD (1990) LAMP-1 in CHO cells is a primary carrier of poly-*N*-acetyllactosamine chains and is bound preferentially by a mammalian S-type lectin. Biochem Biophys Res Commun 173:1123–1128

Dvorak AM, Ackerman SJ, Letourneau L, Morgan ES, Lichtenstein LM, MacGlashan DW Jr (1996) Vesicular transport of Charcot-Leyden crystal protein in tumor- promoting phorbol diester-stimulated human basophils. Lab Invest 74:967–974

Dvorak AM, MacGlashan DW Jr, Warner JA, Letourneau L, Morgan ES, Lichtenstein LM, Ackerman SJ (1997) Vesicular transport of Charcot-Leyden crystal protein in f-Met peptide-stimulated human basophils. Int Arch Allergy Immunol 113:465–477

Dyer KD, Rosenberg HF (1996) Eosinophil Charcot-Leyden crystal protein binds to beta-galactoside sugars. Life Sci 58:2073–2082

Feizi T, Solomon JC, Yuen CT, Jeng KC, Frigeri LG, Hsu DK, Liu FT (1994) The adhesive specificity of the soluble human lectin, IgE-binding protein, toward lipid-linked oligosaccharides. Presence of the blood group A, B, B-like, and H monosaccharides confers a binding activity to tetrasaccharide (lacto-*N*-tetraose and lacto-*N*-neotetraose) backbones. Biochemistry 33:6342–6349

Fernandez PL, Merino MJ, Gomez M, Campo E, Medina T, Castronovo V, Sanjuan X, Cardesa A, Liu FT, Sobel ME (1997) Galectin-3 and laminin expression in neoplastic and non-neoplastic thyroid tissue. J Pathol 181:80–86

Feuk-Lagerstedt E, Jordan ET, Leffler H, Dahlgren C, Karlsson A (1999) Identification of CD66a and CD66b as the major galectin-3 receptor candidates in human neutrophils. J Immunol 163:5592–5598

Fiedler K, Simons K (1994) A putative novel class of animal lectins in the secretory pathway homologous to leguminous lectins [letter]. Cell 77:625–626

Fogel S, Guittaut M, Legrand A, Monsigny M, Hebert E (1999) The tat protein of HIV-1 induces galectin-3 expression. Glycobiology 9:383–387

Gaudin JC, Arar C, Monsigny M, Legrand A (1997) Modulation of the expression of the rabbit galectin-3 gene by p53 and c-Ha-ras proteins and PMA. Glycobiology 7:1089–1098

Ginsburg V, Neufeld EF (1969) Complex heterosaccharides of animals. Annu Rev Biochem 38:371–388

Gitt MA, Barondes SH (1991) Genomic sequence and organization of two members of a human lectin gene family. Biochemistry 30:82–89

Gitt MA, Wiser MF, Leffler H, Herrmann J, Xia YR, Massa SM, Cooper DN, Lusis AJ, Barondes SH (1995) Sequence and mapping of galectin-5, a beta-galactoside-binding lectin, found in rat erythrocytes. J Biol Chem 270:5032–5038

Gitt MA, Colnot C, Poirier F, Nani KJ, Barondes SH, Leffler H (1998a) Galectin-4 and galectin-6 are two closely related lectins expressed in mouse gastrointestinal tract. J Biol Chem 273:2954–2960

Gitt MA, Xia YR, Atchison RE, Lusis AJ, Barondes SH, Leffler H (1998b) Sequence, structure, and chromosomal mapping of the mouse Lgals6 gene, encoding galectin-6. J Biol Chem 273:2961–2970

Giudicelli V, Lutomski D, Levi-Strauss M, Bladier D, Joubert-Caron R, Caron M (1997) Is human galectin-1 activity modulated by monomer/dimer equilibrium? (Letter.) Glycobiology 7:viii–x

Glinsky GV, Price JE, Glinsky VV, Mossine VV, Kiriakova G, Metcalf JB (1996) Inhibition of human breast cancer metastasis in nude mice by synthetic glycoamines. Cancer Res 56:5319–5324

Gold R, Hartung HP, Lassmann H (1997) T-cell apoptosis in autoimmune diseases: termination of inflammation in the nervous system and other sites with specialized immune-defense mechanisms. Trends Neurosci 20:399–404

Greenhalgh CJ, Beckham SA, Newton SE (1999) Galectins from sheep gastrointestinal nematode parasites are highly conserved. Mol Biochem Parasitol 98:285–289

Gu M, Wang W, Song WK, Cooper DN, Kaufman SJ (1994) Selective modulation of the interaction of alpha 7 beta 1 integrin with fibronectin and laminin by L-14 lectin during skeletal muscle differentiation. J Cell Sci 107:175–181

Gupta D, Cho M, Cummings RD, Brewer CF (1996a) Thermodynamics of carbohydrate binding to galectin-1 from Chinese hamster ovary cells and two mutants. A comparison with four galactose-specific plant lectins. Biochemistry 35:15236–15243

Gupta D, Kaltner H, Dong X, Gabius HJ, Brewer CF (1996b) Comparative cross-linking activities of lactose-specific plant and animal lectins and a natural lactose-binding immunoglobulin G fraction from human serum with asialofetuin. Glycobiology 6:843–849

Gupta SK, Masinick S, Garrett M, Hazlett LD (1997) *Pseudomonas aeruginosa* lipopolysaccharide binds galectin-3 and other human corneal epithelial proteins. Infect Immun 65:2747–2753

Hadari YR, Paz K, Dekel R, Mestrovic T, Accili D, Zick Y (1995) Galectin-8. A new rat lectin, related to galectin-4. J Biol Chem 270:3447–3453

Hara-Kuge S, Ohkura T, Seko A, Yamashita K (1999) Vesicular-integral membrane protein, VIP36, recognizes high-mannose type glycans containing α1–2 mannosyl residues in MDCK cells. Glycobiology 9:833–839

Harrison FL, FitzGerald JE, Catt JW (1984) Endogenous beta-galactoside-specific lectins in rabbit tissues. J Cell Sci 72:147–162

Henrick K, Bawumia S, Barboni EA, Mehul B, Hughes RC (1998) Evidence for subsites in the galectins involved in sugar binding at the nonreducing end of the central galactose of oligosaccharide ligands: sequence analysis, homology modeling and mutagenesis studies of hamster galectin-3. Glycobiology 8:45–57

Herrmann J, Turck CW, Atchison RE, Huflejt ME, Poulter L, Gitt MA, Burlingame AL, Barondes SH, Leffler H (1993) Primary structure of the soluble lactose binding lectin L-29 from rat and dog and interaction of its non-collagenous proline-, glycine-, tyrosine-rich sequence with bacterial and tissue collagenase. J Biol Chem 268:26704–26711

Higgins CF (1995) The ABC of channel regulation. Cell 82:693–696

Hinek A, Wrenn DS, Mecham RP, Barondes SH (1988) The elastin receptor: a galactoside-binding protein. Science 239:1539–1541

Hirabayashi J (ed) (1997) Recent topic on galectins. Trends Glycosci Glycotech 9: [multiple articles]

Ho MK, Springer TA (1982) Mac-2, a novel 32, 000 Mr mouse macrophage subpopulation-specific antigen defined by monoclonal antibodies. J Immunol 128:1221–1228

Hsu DK, Zuberi RI, Liu FT (1992) Biochemical and biophysical characterization of human recombinant IgE-binding protein, an S-type animal lectin. J Biol Chem 267:14167–14174

Huflejt ME, Turck CW, Lindstedt R, Barondes SH, Leffler H (1993) L-29, a soluble lactose-binding lectin, is phosphorylated on serine 6 and serine 12 in vivo and by casein kinase I. J Biol Chem 268:26712–26718

Huflejt ME, Gerardts J, Elliot ML, Leffler H, Liu FT (1997a) Galectin-4 is induced in human breast tumors and is localized to sites of cell adhesion in cultured cells. Am Assoc Cancer Res, San Diego, CA

Huflejt ME, Jordan ET, Gitt MA, Barondes SH, Leffler H (1997b) Strikingly different localization of galectin-3 and galectin-4 in human colon adenocarcinoma T84 cells. Galectin-4 is localized at sites of cell adhesion. J Biol Chem 272:14294–14303

Hughes RC (1997) The galectin family of mammalian carbohydrate-binding molecules. Biochem Soc Trans 25:1194–1198

Inohara H, Raz A (1994) Effects of natural complex carbohydrate (citrus pectin) on murine melanoma cell properties related to galectin-3 functions. Glycoconj J 11:527–532

Inohara H, Raz A (1995) Functional evidence that cell surface galectin-3 mediates homotypic cell adhesion. Cancer Res 55:3267–3271

Inohara H, Akahani S, Koths K, Raz A (1996) Interactions between galectin-3 and Mac-2-binding protein mediate cell–cell adhesion. Cancer Res 56:4530–4534

Inohara H, Akahani S, Raz A (1998) Galectin-3 stimulates cell proliferation. Exp Cell Res 245:294–302

Inohara H, Honjo Y, Yoshii T, Akahani S, Yoshida J, Hattori K, Okamoto S, Sawada T, Raz A, Kubo T (1999) Expression of galectin-3 in fine-needle aspirates as a diagnostic marker differentiating benign from malignant thyroid neoplasms. Cancer 85:2475–2484

Jeng KC, Frigeri LG, Liu FT (1994) An endogenous lectin, galectin-3 (epsilon BP/Mac-2), potentiates IL-1 production by human monocytes. Immunol Lett 42:113–116

Kadrofske MM, Openo KP, Wang JL (1998) The human LGALS3 (galectin-3) gene: determination of the gene structure and functional characterization of the promoter. Arch Biochem Biophys 349:7–20

Karlsson A, Follin P, Leffler H, Dahlgren C (1998) Galectin-3 activates the NADPH-oxidase in exudated but not peripheral blood neutrophils. Blood 91:3430–3438

Kasai K, Hirabayashi J (1996) Galectins: a family of animal lectins that decipher glycocodes. J Biochem (Tokyo) 119:1–8

Klion AD, Donelson JE (1994) OvGalBP, a filarial antigen with homology to vertebrate galactoside-binding proteins. Mol Biochem Parasitol 65:305–315

Knibbs RN, Agrwal N, Wang JL, Goldstein IJ (1993) Carbohydrate-binding protein 35. II. Analysis of the interaction of the recombinant polypeptide with saccharides. J Biol Chem 268:14940–14947

Kondoh N, Wakatsuki T, Ryo A, Hada A, Aihara T, Horiuchi S, Goseki N, Matsubara O, Takenaka K, Shichita M, Tanaka K, Shuda M, Yamamoto M (1999) Identification and characterization of genes associated with human hepatocellular carcinogenesis (in process citation). Cancer Res 59:4990–4996

Koths K, Taylor E, Halenbeck R, Casipit C, Wang A (1993) Cloning and characterization of a human Mac-2-binding protein, a new member of the superfamily defined by the macrophage scavenger receptor cysteine-rich domain. J Biol Chem 268:14245–14249

Kuwabara I, Liu FT (1996) Galectin-3 promotes adhesion of human neutrophils to laminin. J Immunol 156:3939–3944

Leal-Pinto E, Tao W, Rappaport J, Richardson M, Knorr BA, Abramson RG (1997) Molecular cloning and functional reconstitution of a urate transporter/channel. J Biol Chem 272:617–625

Leal-Pinto E, Cohen BE, Abramson RG (1999) Functional analysis and molecular modeling of a cloned urate transporter/channel. J Membr Biol 169:13–27

Leffler H (1997) Introduction to galectins. Trends Glycosci Glycotech 45:9–19

Leffler H, Barondes SH (1986) Specificity of binding of three soluble rat lung lectins to substituted and unsubstituted mammalian beta-galactosides. J Biol Chem 261:10119–10126

Leffler H, Masiarz FR, Barondes SH (1989) Soluble lactose-binding vertebrate lectins: a growing family. Biochemistry 28:9222–9229

Leonidas DD, Elbert BL, Zhou Z, Leffler H, Ackerman SJ, Acharya KR (1995) Crystal structure of human Charcot-Leyden crystal protein, an eosinophil lysophospholipase, identifies it as a new member of the carbohydrate-binding family of galectins. Structure 3:1379–1393

Leonidas DD, Vatzaki EH, Vorum H, Celis JE, Madsen P, Acharya KR (1998) Structural basis for the recognition of carbohydrates by human galectin-7. Biochemistry 37:13930–13940

Levi G, Teichberg VI (1983) Selective interactions of electrolectins from eel electric organ and mouse thymus with mouse immature thymocytes. Immunol Lett 7:35–39

Levi G, Tarrab-Hazdai R, Teichberg VI (1983) Prevention and therapy with electrolectin of experimental autoimmune myasthenia gravis in rabbits. Eur J Immunol 13:500–507

Liao DI, Kapadia G, Ahmed H, Vasta GR, Herzberg O (1994) Structure of S-lectin, a developmentally regulated vertebrate beta- galactoside-binding protein. Proc Natl Acad Sci USA 91:1428–1432

Lindstedt R, Apodaca G, Barondes SH, Mostov KE, Leffler H (1993) Apical secretion of a cytosolic protein by Madin-Darby canine kidney cells. Evidence for polarized release of an endogenous lectin by a nonclassical secretory pathway. J Biol Chem 268:11750–11757

Linton KJ, Higgins CF (1998) The *Escherichia coli* ATP-binding cassette (ABC) proteins (see comments). Mol Microbiol 28:5–13

Lipsick JS, Beyer EC, Barondes SH, Kaplan NO (1980) Lectins from chicken tissues are mitogenic for Thy-1 negative murine spleen cells. Biochem Biophys Res Commun 97:56–61

Liu FT (1993) S-type mammalian lectins in allergic inflammation. Immunol Today 14:486–490

Liu FT, Hsu DK, Zuberi RI, Kuwabara I, Chi EY, Henderson WR, Jr. (1995) Expression and function of galectin-3, a beta-galactoside-binding lectin, in human monocytes and macrophages. Am J Pathol 147:1016–1028

Lobsanov YD, Rini JM (1997) Galectin structure. Trends Glycosci Glycotech 45:145–154

Lobsanov YD, Gitt MA, Leffler H, Barondes SH, Rini JM (1993) X-ray crystal structure of the human dimeric S-Lac lectin, L-14-II, in complex with lactose at 2.9-Å resolution. J Biol Chem 268:27034–27038

Loris R, Hamelryck T, Bouckaert J, Wyns L (1998) Legume lectin structure. Biochim Biophys Acta 1383:9–36

Lotan R, Raz A (1988) Endogenous lectins as mediators of tumor cell adhesion. J Cell Biochem 37:107–117

Lotan R, Matsushita Y, Ohannesian D, Carralero D, Ota DM, Cleary KR, Nicolson GL, Irimura T (1991) Lactose-binding lectin expression in human colorectal carcinomas. Relation to tumor progression. Carbohydr Res 213:47–57

Lotz MM, Andrews CW Jr, Korzelius CA, Lee EC, Steele GD Jr, Clarke A, Mercurio AM (1993) Decreased expression of Mac-2 (carbohydrate binding protein 35) and loss of its nuclear localization are associated with the neoplastic progression of colon carcinoma. Proc Natl Acad Sci USA 90:3466–3470

Lu J, Pei H, Kaeck M, Thompson HJ (1997) Gene expression changes associated with chemically induced rat mammary carcinogenesis. Mol Carcinog 20:204–215

Lu Y, Lotan R (1999) Transcriptional regulation by butyrate of mouse galectin-1 gene in embryonal carcinoma cells. Biochim Biophys Acta 1444:85–91

Lutomski D, Joubert-Caron R, Lefebure C, Salama J, Belin C, Bladier D, Caron M (1997) Anti-galectin-1 autoantibodies in serum of patients with neurological diseases. Clin Chim Acta 262:131–138

Madsen P, Rasmussen HH, Flint T, Gromov P, Kruse TA, Honore B, Vorum H, Celis JE (1995) Cloning, expression, and chromosome mapping of human galectin-7. J Biol Chem 270:5823–5829

Magnaldo T, Bernerd F, Darmon M (1995) Galectin-7, a human 14 kDa S-lectin, specifically expressed in keratinocytes and sensitive to retinoic acid. Dev Biol 168:259–271

Mahanthappa NK, Cooper DN, Barondes SH, Schwarting GA (1994) Rat olfactory neurons can utilize the endogenous lectin, L-14, in a novel adhesion mechanism. Development 120:1373–1384

Mandrell RE, Apicella MA, Lindstedt R, Leffler H (1994) Possible interaction between animal lectins and bacterial carbohydrates. Methods Enzymol 236:231–254

Massa SM, Cooper DN, Leffler H, Barondes SH (1993) L-29, an endogenous lectin, binds to glycoconjugate ligands with positive cooperativity. Biochemistry 32:260–267

Mathews KP, Konstantinov KN, Kuwabara I, Hill PN, Hsu DK, Zuraw BL, Liu FT (1995) Evidence for IgG autoantibodies to galectin-3, a beta-galactoside-binding lectin (Mac-2, epsilon binding protein, or carbohydrate binding protein 35) in human serum. J Clin Immunol 15:329–337

Matsumoto R, Matsumoto H, Seki M, Hata M, Asano Y, Kanegasaki S, Stevens RL, Hirashima M (1998) Human ecalectin, a variant of human galectin-9, is a novel eosinophil chemoattractant produced by T lymphocytes. J Biol Chem 273:16976–16984

Mehul B, Hughes RC (1997) Plasma membrane targetting, vesicular budding and release of galectin 3 from the cytoplasm of mammalian cells during secretion. J Cell Sci 110:1169–1178

Mehul B, Bawumia S, Martin SR, Hughes RC (1994) Structure of baby hamster kidney carbohydrate-binding protein CBP30, an S-type animal lectin. J Biol Chem 269:18250–18258

Mehul B, Bawumia S, Hughes RC (1995) Cross-linking of galectin 3, a galactose-binding protein of mammalian cells, by tissue-type transglutaminase. FEBS Lett 360:160–164

Menon RP, Hughes RC (1999) Determinants in the N-terminal domains of galectin-3 for secretion by a novel pathway circumventing the endoplasmic reticulum-Golgi complex. Eur J Biochem 264:569–576

Merkle RK, Cummings RD (1988) Asparagine-linked oligosaccharides containing poly-N-acetyl-lactosamine chains are preferentially bound by immobilized calf heart agglutinin. J Biol Chem 263:16143–16149

Mey A, Leffler H, Hmama Z, Normier G, Revillard JP (1996) The animal lectin galectin-3 interacts with bacterial lipopolysaccharides via two independent sites. J Immunol 156:1572–1577

Moutsatsos IK, Wade M, Schindler M, Wang JL (1987) Endogenous lectins from cultured cells: nuclear localization of carbohydrate-binding protein 35 in proliferating 3T3 fibroblasts. Proc Natl Acad Sci USA 84:6452–6456

Muller SA, Sasaki T, Bork P, Wolpensinger B, Schulthess T, Timpl R, Engel A, Engel J (1999) Domain organization of Mac-2 binding protein and its oligomerization to linear and ring-like structures. J Mol Biol 291:801–813

Muller WE (1997) Origin of metazoan adhesion molecules and adhesion receptors as deduced from cDNA analyses in the marine sponge *Geodia cydonium*: a review. Cell Tissue Res 289:383–395

Nowak TP, Kobiler D, Roel LE, Barondes SH (1977) Developmentally regulated lectin from embryonic chick pectoral muscle. Purification by affinity chromatography. J Biol Chem 252:6026–6030

Nurminskaya M, Linsenmayer TF (1996) Identification and characterization of up-regulated genes during chondrocyte hypertrophy. Dev Dyn 206:260–271

Ochieng J, Green B, Evans S, James O, Warfield P (1998) Modulation of the biological functions of galectin-3 by matrix metalloproteinases. Biochim Biophys Acta 1379:97–106

Oda Y, Herrmann J, Gitt MA, Turck CW, Burlingame AL, Barondes SH, Leffler H (1993) Soluble lactose-binding lectin from rat intestine with two different carbohydrate-binding domains in the same peptide chain. J Biol Chem 268:5929–5939

Offner H, Celnik B, Bringman TS, Casentini-Borocz D, Nedwin GE, Vandenbark AA (1990) Recombinant human beta-galactoside binding lectin suppresses clinical and histological signs of experimental autoimmune encephalomyelitis. J Neuroimmunol 28:177–184

Ogden AT, Nunes I, Ko K, Wu S, Hines CS, Wang AF, Hegde RS, Lang RA (1998) GRIFIN, a novel lens-specific protein related to the galectin family. J Biol Chem 273:28889–28896

Ohannesian DW, Lotan D, Lotan R (1994) Concomitant increases in galectin-1 and its glycoconjugate ligands (carcinoembryonic antigen, lamp-1, and lamp-2) in cultured human colon carcinoma cells by sodium butyrate. Cancer Res 54:5992–6000

Pace KE, Lee C, Stewart PL, Baum LG (1999) Restricted receptor segregation into membrane microdomains occurs on human T cells during apoptosis induced by galectin-1. J Immunol 163:3801–3811

Perillo NL, Pace KE, Seilhamer JJ, Baum LG (1995) Apoptosis of T cells mediated by galectin-1. Nature 378:736–739

Perillo NL, Uittenbogaart CH, Nguyen JT, Baum LG (1997) Galectin-1, an endogenous lectin produced by thymic epithelial cells, induces apoptosis of human thymocytes. J Exp Med 185:1851–1858

Perillo NL, Marcus ME, Baum LG (1998) Galectins: versatile modulators of cell adhesion, cell proliferation, and cell death. J Mol Med 76:402–412

Pesheva P, Urschel S, Frei K, Probstmeier R (1998) Murine microglial cells express functionally active galectin-3 in vitro. J Neurosci Res 51:49–57

Poirier F, Robertson EJ (1993) Normal development of mice carrying a null mutation in the gene encoding the L14 S-type lectin. Development 119:1229–1236

Polyak K, Xia Y, Zweier JL, Kinzler KW, Vogelstein B (1997) A model for p53-induced apoptosis (see comments). Nature 389:300–305

Puche AC, Key B (1995) Identification of cells expressing galectin-1, a galactose-binding receptor, in the rat olfactory system. J Comp Neurol 357:513–523

Puche AC, Poirier F, Hair M, Bartlett PF, Key B (1996) Role of galectin-1 in the developing mouse olfactory system. Dev Biol 179:274–287

Rabinovich GA (1999) Galectins: an evolutionarily conserved family of animal lectins with multifunctional properties; a trip from the gene to clinical therapy (in process citation). Cell Death Differ 6:711–721

Rabinovich GA, Daly G, Dreja H, Tailor H, Riera CM, Hirabayashi J, Chernajovsky Y (1999) Recombinant galectin-1 and its genetic delivery suppress collagen-induced arthritis via T cell apoptosis. J Exp Med 190:385–398

Ramkumar R, Surolia A, Podder SK (1995) Energetics of carbohydrate binding by a 14 kDa S-type mammalian lectin. Biochem J 308:237–241

Raz A, Lotan R (1987) Endogenous galactoside-binding lectins: a new class of functional tumor cell surface molecules related to metastasis. Cancer Metastasis Rev 6:433–452

Raz A, Zhu DG, Hogan V, Shah N, Raz T, Karkash R, Pazerini G, Carmi P (1990) Evidence for the role of 34-kDa galactoside-binding lectin in transformation and metastasis. Int J Cancer 46:871–877

Regan LJ, Dodd J, Barondes SH, Jessell TM (1986) Selective expression of endogenous lactose-binding lectins and lactoseries glycoconjugates in subsets of rat sensory neurons. Proc Natl Acad Sci USA 83:2248–2252

Rini JM (1995) X-ray crystal structures of animal lectins. Curr Opin Struct Biol 5:617–621

Rini JM, Lobsanov YD (1999) New animal lectin structures. Curr Opin Struct Biol 9:578–584

Roff CF, Wang JL (1983) Endogenous lectins from cultured cells. Isolation and characterization of carbohydrate-binding proteins from 3T3 fibroblasts. J Biol Chem 258:10657–10663

Roseman S (1970) The synthesis of complex carbohydrates by multiglycosyltransferase systems and their potential function in intercellular adhesion. Chem Phys Lipids 5:270–297

Rubartelli A, Sitia R (1991) Interleukin 1 beta and thioredoxin are secreted through a novel pathway of secretion. Biochem Soc Trans 19:255–259

Saada A, Reichert F, Rotshenker S (1996) Granulocyte macrophage colony stimulating factor produced in lesioned peripheral nerves induces the up-regulation of cell surface expression of MAC-2 by macrophages and Schwann cells. J Cell Biol 133:159–167

Sahin U, Tureci O, Schmitt H, Cochlovius B, Johannes T, Schmits R, Stenner F, Luo G, Schobert I, Pfreundschuh M (1995) Human neoplasms elicit multiple specific immune responses in the autologous host. Proc Natl Acad Sci USA 92:11810–11813

Sanjuan X, Fernandez PL, Castells A, Castronovo V, van den Brule F, Liu FT, Cardesa A, Campo E (1997) Differential expression of galectin 3 and galectin 1 in colorectal cancer progression (see comments). Gastroenterology 113:1906–1915

Sasaki T, Brakebusch C, Engel J, Timpl R (1998) Mac-2 binding protein is a cell-adhesive protein of the extracellular matrix which self-assembles into ring-like structures and binds beta1 integrins, collagens and fibronectin. EMBO J 17:1606–1613

Sato S, Hughes RC (1992) Binding specificity of a baby hamster kidney lectin for H type I and II chains, polylactosamine glycans, and appropriately glycosylated forms of laminin and fibronectin. J Biol Chem 267:6983–6990

Sato S, Burdett I, Hughes RC (1993) Secretion of the baby hamster kidney 30-kDa galactose-binding lectin from polarized and nonpolarized cells: a pathway independent of the endoplasmic reticulum-Golgi complex. Exp Cell Res 207:8–18

Scanlan MJ, Chen YT, Williamson B, Gure AO, Stockert E, Gordan JD, Tureci O, Sahin U, Pfreundschuh M, Old LJ (1998) Characterization of human colon cancer antigens recognized by autologous antibodies. Int J Cancer 76:652–658

Schwarz FP, Ahmed H, Bianchet MA, Amzel LM, Vasta GR (1998) Thermodynamics of bovine spleen galectin-1 binding to disaccharides: correlation with structure and its effect on oligomerization at the denaturation temperature. Biochemistry 37:5867–5877

Seetharaman J, Kanigsberg A, Slaaby R, Leffler H, Barondes SH, Rini JM (1998) X-ray crystal structure of the human galectin-3 carbohydrate recognition domain at 2.1-Å resolution. J Biol Chem 273:13047–13052

Su ZZ, Lin J, Shen R, Fisher PE, Goldstein NI, Fisher PB (1996) Surface-epitope masking and expression cloning identifies the human prostate carcinoma tumor antigen gene PCTA-1, a member of the galectin gene family. Proc Natl Acad Sci USA 93:7252–7257

Suk K, Hwang DY, Lee MS (1999) Natural autoantibody to galectin-9 in normal human sera. J Clin Immunol 19:158–165

Surolia A, Swaminathan CP, Ramkumar R, Podder SK (1997) Unusual structural stability and ligand induced alterations in oligomerization of a galectin. FEBS Lett 409:417–420

Swaminathan GJ, Leonidas DD, Savage MP, Ackerman SJ, Acharya KR (1999) Selective recognition of mannose by the human eosinophil Charcot-Leyden crystal protein (galectin-10): a crystallographic study at 1.8 Å resolution. Biochemistry 38:13837–13843

Sörme P, Qian Y, Leffler H, Nilsson U (1999) Approaches towards low-molecular weight galectin inhibitors XVth international symposium on glycoconjugates (glyco XV), Tokyo

Teichberg VI, Silman I, Beitsch DD, Resheff G (1975) A beta-D-galactoside binding protein from electric organ tissue of *Electrophorus electricus*. Proc Natl Acad Sci USA 72:1383–1387

Tenne-Brown J, Puche AC, Key B (1998) Expression of galectin-1 in the mouse olfactory system. Int J Dev Biol 42:791–799

Timmons PM, Colnot C, Cail I, Poirier F, Magnaldo T (1999) Expression of galectin-7 during epithelial development coincides with the onset of stratification. Int J Dev Biol 43:229–235

Thornalley PJ (1998) Cell activation by glycated proteins. AGE receptors, receptor recognition factors and functional classification of AGEs. Cell Mol Biol (Noisy-le-grand) 44:1013–1023

Tsay YG, Lin NY, Voss PG, Patterson RJ, Wang JL (1999) Export of galectin-3 from nuclei of digitonin-permeabilized mouse 3T3 fibroblasts. Exp Cell Res 252:250–261

Tureci O, Schmitt H, Fadle N, Pfreundschuh M, Sahin U (1997) Molecular definition of a novel human galectin which is immunogenic in patients with Hodgkin's disease. J Biol Chem 272:6416–6422

Umemoto K, Carver J, Leffler H (1996) The binding of galectin-3 as studied by NMR. 11th Rinshoken international conference, Tokyo

van den Brule FA, Liu FT, Castronovo V (1998) Transglutaminase-mediated oligomerization of galectin-3 modulates human melanoma cell interactions with laminin. Cell Adhes Commun 5:425–435

Vasta GR, Quesenberry M, Ahmed H, O'Leary N (1999) C-type lectins and galectins mediate innate and adaptive immune functions: their roles in the complement activation pathway. Dev Comp Immunol 23:401–420

Vespa GN, Lewis LA, Kozak KR, Moran M, Nguyen JT, Baum LG, Miceli MC (1999) Galectin-1 specifically modulates TCR signals to enhance TCR apoptosis but inhibits IL-2 production and proliferation. J Immunol 162:799–806

Vlassara H, Li YM, Imani F, Wojciechowicz D, Yang Z, Liu FT, Cerami A (1995) Identification of galectin-3 as a high-affinity binding protein for advanced glycation end products (AGE): a new member of the AGE-receptor complex. Mol Med 1:634–646

Vyakarnam A, Dagher SF, Wang JL, Patterson RJ (1997) Evidence for a role for galectin-1 in pre-mRNA splicing. Mol Cell Biol 17:4730–4737

Wada J, Ota K, Kumar A, Wallner EI, Kanwar YS (1997) Developmental regulation, expression, and apoptotic potential of galectin-9, a beta-galactoside binding lectin. J Clin Invest 99:2452–2461

Wandersman C (1998) Protein and peptide secretion by ABC exporters. Res Microbiol 149:163–170

Wang JL, Laing JG, Anderson RL (1991) Lectins in the cell nucleus. Glycobiology 1:243–252

Wang JL, Werner EA, Laing JG, Patterson RJ (1992) Nuclear and cytoplasmic localization of a lectin–ribonucleoprotein complex. Biochem Soc Trans 20:269–274

Wang L, Inohara H, Pienta KJ, Raz A (1995) Galectin-3 is a nuclear matrix protein which binds RNA. Biochem Biophys Res Commun 217:292–303

Wasano K, Hirakawa Y (1995) Rat intestinal galactoside-binding lectin L-36 functions as a structural protein in the superficial squamous cells of the esophageal epithelium. Cell Tissue Res 281:77–83

Wasano K, Hirakawa Y (1999) Two domains of rat galectin-4 bind to distinct structures of the intercellular borders of colorectal epithelia. J Histochem Cytochem 47:75–82

Wilson TJ, Firth MN, Powell JT, Harrison FL (1989) The sequence of the mouse 14 kDa beta-galactoside-binding lectin and evidence for its synthesis on free cytoplasmic ribosomes. Biochem J 261:847–852

Winyard PJ, Bao Q, Hughes RC, Woolf AS (1997) Epithelial galectin-3 during human nephrogenesis and childhood cystic diseases. J Am Soc Nephrol 8:1647–1657

Woo HJ, Shaw LM, Messier JM, Mercurio AM (1990) The major non-integrin laminin binding protein of macrophages is identical to carbohydrate binding protein 35 (Mac-2). J Biol Chem 265:7097–7099

Yamaoka A, Kuwabara I, Frigeri LG, Liu FT (1995) A human lectin, galectin-3 (epsilon bp/Mac-2), stimulates superoxide production by neutrophils. J Immunol 154:3479–3487

Yamaoka K, Ingendoh A, Tsubuki S, Nagai Y, Sanai Y (1996) Structural and functional characterization of a novel tumor-derived rat galectin-1 having transforming growth factor (TGF) activity: the relationship between intramolecular disulfide bridges and TGF activity. J Biochem (Tokyo) 119:878–886

Yang RY, Hsu DK, Liu FT (1996) Expression of galectin-3 modulates T-cell growth and apoptosis. Proc Natl Acad Sci USA 93:6737–6742

Yang RY, Hill PN, Hsu DK, Liu FT (1998) Role of the carboxyl-terminal lectin domain in self-association of galectin-3. Biochemistry 37:4086–4092

Zanetta JP (1998) Structure and functions of lectins in the central and peripheral nervous system. Acta Anat (Basel) 161:180–195

Zhou Q, Cummings RD (1990) The S-type lectin from calf heart tissue binds selectively to the carbohydrate chains of laminin. Arch Biochem Biophys 281:27–35

Zhou Q, Cummings RD (1993) L-14 lectin recognition of laminin and its promotion of in vitro cell adhesion. Arch Biochem Biophys 300:6–17

Zuberi RI, Frigeri LG, Liu FT (1994) Activation of rat basophilic leukemia cells by epsilon BP, an IgE-binding endogenous lectin. Cell Immunol 156:1–12

Structure and Function of CD44: Characteristic Molecular Features and Analysis of the Hyaluronan Binding Site

Jürgen Bajorath[1]

1 Synopsis

CD44 is a cell surface glycoprotein belonging to the cartilage link protein family. It is a major receptor for hyaluronan (HA), a component of the extracellular matrix, but also binds other ligands. The N-terminal extracellular link domain of CD44 is critical for HA recognition. In the immune system, CD44 plays a role both as a cell adhesion molecule and as a signaling molecule. Many isoforms of CD44 can be expressed and post-translationally modified in a cell-specific manner. This structural diversity correlates with a diverse array of functional properties, some of which are not well understood.

In this chapter, molecular features and functions of CD44 are reviewed and studies designed to identify CD44 residues important for HA binding are discussed. These investigations combine site-specific mutagenesis, comparative protein model building, and a variety of ligand binding experiments. In the absence of an experimentally determined structure, mapping of residues important for HA binding on a three-dimensional model of CD44 has provided a reasonable picture of the binding site. Opportunities and limitations of the approach are discussed. The HA binding site in CD44 is analyzed and compared with carbohydrate binding sites in calcium-dependent lectins.

2 Cell Adhesion Proteins

2.1 Representative Families and Characteristic Features

Fig. 1 shows a schematic representation of major families of cell surface proteins that are implicated in cell adhesion and, at least in part, signal transduction and cell activation. Many of these proteins have important functions in the immune system. Ligand binding domains often belong to protein families or superfamilies, for which prototypic three-dimensional (3-D) structures have been determined. Prominent among these proteins are the immunoglobulin superfamily (IgSF; Williams and Barclay 1988) and the calcium-dependent (C-type) lectin superfamily (Drickamer 1988). CD44 is a member of the cartilage

[1] Department of Biological Structure, University of Washington, Seattle, Washington 98195, USA

Results and Problems in Cell Differentiation, Vol. 33
Paul R. Crocker (Ed.): Mammalian Carbohydrate Recognition Systems
© Springer-Verlag Berlin Heidelberg 2001

link protein (CLP) family of extracellular matrix- and HA-binding proteins. These proteins are also called hyaloadherins (Toole 1990).

As indicated in Fig. 1, many immune cell surface molecules are single-path transmembrane proteins. Their extracellular regions consist of multiple and, in part, independent domains that are only loosely tethered to the transmembrane segment. Thus, in contrast to integral membrane proteins, extracellular domains of these cell surface proteins can often be expressed in soluble recombinant form and studied in vitro. Consequently, significant progress has been made in understanding the molecular basis of cell adhesion and signal transduction over the past few years. Many 3-D structures of extracellular binding domains have been determined by X-ray crystallography or nuclear magnetic resonance (NMR) spectroscopy (Bajorath 1998).

3 Molecular Structure of CD44

3.1 Cloning of CD44 and Domain Organization

In independent studies, cDNA clones encoding a 90 kDa form of CD44 were isolated (Goldstein et al. 1989; Stamenkovic et al. 1989). This form of CD44 is expressed on hematopoietic cells and is therefore called CD44H. Analysis of the deduced amino acid sequence of CD44H revealed that CD44 is a type I

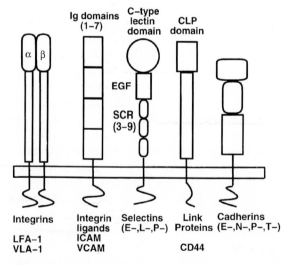

Fig. 1. Schematic representation of cell adhesion protein families. The membrane is represented as a *horizontal bar*. Extracellular domains are compared, and domain types are labeled: *Ig* immunoglobulin-like, *C-type lectin* calcium-dependent lectin, *EGF* epidermal growth factor, *SCR* complement receptor-like, *CLP* cartilage link protein-like domain

transmembrane protein (i.e., the N-terminus is located outside the cell) with an extracellular domain of 248 amino acids, a 21-residue transmembrane segment, and a cytoplasmic domain consisting of 72 residues. The N-terminal ~130 residues display sequence similarity to cartilage link proteins. These residues form the CLP (or link homology) domain. Fig. 2 shows the domain organization of CD44. The 341 residues of CD44H account for a molecular mass of only ~37 kDa. The apparent molecular mass of 90 kDa is due to extensive N-linked and O-linked glycosylation of the extracellular region, as further discussed below.

Another isoform of CD44 was cloned from epithelial cells and is termed CD44E (Stamenkovic et al. 1991). Compared with CD44H, this 150 kDa form displays an insert of 135 amino acids in the extracellular region. Like CD44H, this isoform is heavily glycosylated, including additions of glycosaminoglycans (heparan, chondroitin, and keratan sulfate; Stamenkovic et al. 1991). CD44 is therefore designated a proteoglycan.

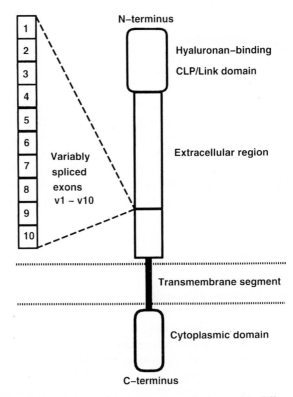

Fig. 2. Molecular organization of CD44. Variably spliced exons v1-v10 can be inserted in different combinations into the membrane-proximal extracellular region, giving rise to many CD44 isoforms. The relative dimensions of depicted domains do not strictly correlate with the number of amino acids

3.2 Genomic Structure and Isoforms

Genomic cloning of CD44 has revealed that CD44 consists of at least 21 exons. Nineteen exons (e1-e19) were initially identified (Screaton et al. 1992) and two additional exons, e6a (located between e5 and e6; Screaton et al. 1993) and e13a (located between e13 and e14; Yu and Toole 1996) were identified later. At least 11 of these exons can be variably spliced, e6a-e14 (corresponding to v1-v10) and e13a (corresponding to v9a), and give rise to a variety of CD44 isoforms. However, not all possible isoforms are expressed. The position for the insertion of variably spliced exons v1–v10 is located in the membrane-proximal part of the extracellular region (Fig. 2). This is approximately where N-terminal sequence homology between CD44 molecules from different species ends. CD44 isoforms are expressed in a cell-specific manner, and at least 30 different isoforms have been characterized to date. All isoforms contain exons e1–e5. Constitutively expressed exons encode the N-terminal signal sequence and the cartilage link protein-like domain. CD44H (which does not contain variably spliced exons) and CD44E are predominant isoforms of human CD44.

3.3 Glycosylation

The structural diversity of CD44 is further increased by extensive and, in part, isoform-specific post-translational modifications including N- and O-linked glycosylation and glycosaminoglycan attachment. The glycosylation pattern of CD44 isoforms, including CD44H, can vary substantially, depending on the CD44-expressing cell type. Furthermore, variably spliced exons can introduce many new glycosylation sites. For example, the constitutively expressed exon e5 contains two Ser-Gly motifs that support the synthesis of chondroitin but not heparan sulfate, whereas exon v3 contains a Ser-Gly-Ser-Gly motif that supports the synthesis of both chondroitin and heparan sulfate (Zhang et al. 1995).

4 Biological Functions and Ligands of CD44

4.1 Functional Diversity

CD44 is functionally as well as structurally highly diverse. It binds to components of the extracellular matrix (Lesley et al. 1993) and mediates homotypic cell adhesion and aggregation (St. John et al. 1990). In addition, it can act as a signaling molecule (Taher et al. 1996) and is capable of inducing the expression of chemokines (McKee et al. 1996). CD44 isoforms containing the heparan sulfate-decorated exon v3 are capable of recruiting and presenting heparin-binding growth factors (Bennett et al. 1995a). Compared with epithelial cells, tumor cell lines often display differential expression of CD44 (Sy et al. 1997), and isoforms containing exon v6 were found to play a significant role in tumor metastasis (Günthert et al. 1991). CD44 has also been shown to trigger the adhesion of leukocytes to sites of inflammation (Clark et al. 1996; DeGrendele et al. 1996). In murine models of inflammation, anti-CD44 monoclonal antibodies (mAbs) can block leukocyte extravasation (Mikecz et al. 1995).

4.2 Hyaluronan and Other Ligands

Several ligands for CD44 isoforms have been identified, including HA and chondroitin sulfate (Aruffo et al. 1990; Miyake et al. 1990), collagen (Carter and Wayner 1988), the heparin-binding domain of fibronectin (Jalkanen and Jalkanen 1992), and osteopontin, a cytokine and inducer of cellular chemotaxis (Weber et al. 1996). CD44 is the major cell surface receptor for HA (Aruffo et al. 1990), a glycosaminoglycan and component of the extracellular matrix (Lesley et al. 1993). Many functions of CD44 in the context of cell adhesion and activation can be attributed to HA binding (Miyake et al. 1990; Aruffo 1996) but the physiological relevance of other CD44-ligand interactions is less clear.

HA is a polymer consisting of D-glucuronic acid and N-acetyl-D-glucosamine disaccharide units (Toole 1990). Fig. 3 shows the structure of an HA trisaccharide (Holmbeck et al. 1995). For effective binding, CD44 requires at least an HA hexasaccharide (Underhill et al. 1983). Interestingly, the size of

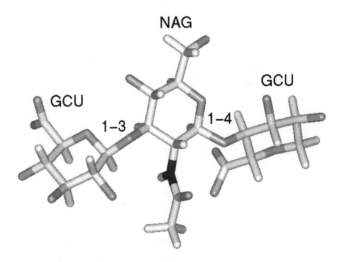

Fig. 3. Three-dimensional structure of a hyaluronan (HA) fragment (PDB id code '1hua'). The trisaccharide consists of two D-glucuronic, acid (*GCU*) units and N-acetyl-D-glucosamine (*NAG*). Carbon and hydrogen atoms are shown in *light gray*, oxygen atoms in *darker gray*, and nitrogen in *black* (hydrogen atoms attached to oxygens are omitted). HA is a polymer consisting of repeating [GCU-NAG] disaccharide units. The GCU-NAG-GCU trisaccharide contains two distinct glycosidic bonds, labeled *1–3* and *1–4*, respectively

recognized HA fragments appears to determine whether CD44 acts as an adhesion or signaling molecule. The picture that emerges is that recognition of polymeric HA correlates with adhesive functions, whereas binding of small HA fragments triggers CD44-mediated signaling and cell activation (Aruffo 1996). The presence of small HA fragments in the body may result from tissue damage. Recognition of such fragments by CD44 and cell activation may therefore help to elicit an appropriate immune response.

4.3 Regulation of Hyaluronan Binding

The regulation of HA binding is a particularly complex and not well understood aspect of CD44 biology. The level of CD44 expression on the cell surface and its distribution can certainly influence ligand binding. However, not all cells that express CD44H or other isoforms can bind HA, regardless of expression levels. Binding to some of these inactive cell lines can be induced, by an unknown mechanism, with anti-CD44 mAbs (Lesley et al. 1995). HA binding ability strongly depends on the cellular context, and many of the observed differences in binding can be attributed to isoform- and cell-specific differences in CD44 glycosylation. Reduced levels of both N- and O-linked glycosylation, including chondroitin sulfate attachment, lead to improved HA binding (Bennett et al. 1995b; Katoh et al. 1995; Lesley et al. 1995) but genetic disruption of N-linked glycosylation sites abolishes binding (Bartolazzi et al. 1996). Post-translational sulfation of CD44 was found to induce HA binding ability in the context of leukocyte adhesion (Maiti et al. 1998), similar to effects observed with anti-CD44 mAbs. Moreover, mutations or deletions in all regions of CD44, including the extracellular, transmembrane, and cytoplasmic domains, can affect HA binding (Kincade et al. 1997). A major difficulty in rationalizing these findings is that they may be due to a variety of direct or indirect effects. These include changes of the HA binding site or other structural perturbations, aggregation of CD44, or steric effects (making it easier or more difficult to access the binding site). Thus, HA binding to CD44 can be modulated in many ways.

4.4 Hyaluronan Binding Domain

The presence of a CLP (link) domain is a common feature of hyaloadherins (Toole 1990). Site-specific mutagenesis of positively charged residues in CD44 that are part of characteristic sequence motifs shared by hyaloadherins (Yang et al. 1994) identified one residue in the CLP domain, R41, as essential for the interaction with HA (Peach et al. 1993). Mutation of R41 to alanine was sufficient to abolish HA binding in enzyme-linked immunosorbent assays (ELISA). This finding is consistent with the idea that the link domain contains the HA binding site. However, mutation of basic residues in sequence motifs outside the link domain also reduced binding (Peach et al. 1993). Thus, it is possible that these (and other) residues stabilize the interaction with polymeric HA and contribute to high-avidity binding.

5 The Link Protein Module

5.1 Three-Dimensional Structure of TSG-6

A 3-D structure of CD44 has not yet been reported, but the structure of the link domain of TSG-6 (Lee et al. 1992), another hyaloadherin, has been determined by NMR spectroscopy (Kohda et al. 1996). The link domain of TSG-6, including residues 36–133, was expressed in *Escherichia coli* and therefore not glycosylated. However, the expressed domain specifically bound HA (Kohda et al. 1996). The TSG-6 structure forms a compact module with a core consisting of two three-stranded anti-parallel beta-sheets and two alpha-helices. Fig. 4 shows a schematic representation of the link module.

Unexpectedly, the structure was found to be similar to the X-ray structure of the C-type lectin domain of E-selectin (Graves et al. 1994), despite the lack of sequence similarity. The selectins (see also Fig. 1) are a family of cell adhesion molecules (Lasky 1995) and members of the C-type lectin superfamily (Drickamer 1988). Selectins play a critical role in the adhesion of leukocytes to activated vascular endothelium at sites of inflammation (Lasky 1995). As discussed above, a similar function has also been established for CD44 (DeGrendele et al. 1996). Fig. 4 shows a comparison of the link module of TSG-6 and the C-type lectin domain of E-selectin. The link module consists of ~90 residues and the C-type lectin domain of ~120 residues. Several extended regions of non-classical secondary structure in E-selectin and loops involved in the formation of a calcium binding site are absent in TSG-6. The selectins recognize tetrasaccharide structures of the sialyl-Lewis[x] type that are specifically presented on cell surface glycoproteins (Lasky 1995). Carbohydrate recognition by C-type lectins, including the selectins, is strictly calcium-dependent, whereas link proteins do not require calcium for binding.

5.2 Molecular Model of the Link Module of CD44

TSG-6 and CD44 are closely related. Fig. 5 shows a sequence comparison of the link domains that takes the 3-D structure of TSG-6 into account. TSG-6 has two disulfide bonds and the cysteine positions are conserved in CD44. Also conserved are residues that form the hydrophobic core of TSG-6. The sequence similarity between TSG-6 and CD44 in the aligned region is ~50%, and this level of similarity corresponds to the distinct structural similarity of

Fig. 4. Comparison of link and C-type lectin domains. Stereo view of the structures in ribbon representation (*thick line* TSG-6, *thin line* E-selectin). The functionally important calcium position in E-selectin is depicted as a cross. Alpha-carbon atoms of the link homology domain of TSG-6 (PDB id code '1tsg') and the C-type lectin domain of E-selectin ('1esl') were optimally superposed. The cumulative root mean square (rms) deviation for comparison of 74 matching residue positions in E-selectin and TSG-6 is ~2.9 Å. The orientations at the top and bottom are related by ~90° rotation around the vertical axis

protein core regions (Chothia and Lesk 1986). Therefore, it was possible to generate a 3-D model of the CD44 link domain on the basis of the TSG-6 structure by comparative model building (Bajorath et al. 1998).

Comparative computer modeling extrapolates from known structures to generate molecular models of homologous proteins (Greer 1990; Bajorath et al. 1993). Critical initial steps are the generation of topologically correct sequence alignments of template and target proteins and the identification of structurally conserved regions. The CD44 model was built on the basis of the sequence alignment shown in Fig. 5. Well-defined secondary structure

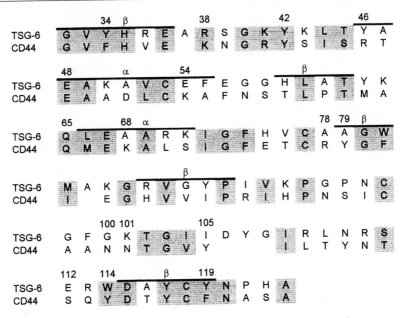

Fig. 5. Sequence comparison of the link modules in TSG-6 and CD44. Identical and conservatively replaced residue positions are *shaded*. Alpha-helices and beta-strands thought to be conserved in TSG-6 and CD44 are *overlined* and labeled. Residue numbers are given for CD44. Taking conservative residue replacements into account, the sequence similarity between the link modules in TSG-6 and CD44 is ~50%

elements of TSG-6 were assembled to form the core of the CD44 model. Side chain replacements were modeled using rotamer libraries (Ponder and Richards 1987). Loops and other non-conserved backbone regions were constructed by extracting suitable backbone fragments (Jones and Thirup 1986) from the Brookhaven Protein Data Bank (Bernstein et al. 1977) or, alternatively, by conformational search calculations (Bruccoleri and Karplus 1987). The model was refined by energy minimization and, finally, the compatibility of the CD44 sequence with the modeled structure was confirmed by energy profile analysis (Sippl 1993). Fig. 6 shows the CD44 molecular model and a comparison with TSG-6.

6 Analysis of the Hyaluronan Binding Site in CD44

6.1 Mutagenesis Strategy and Experimental Approach

On the basis of the CD44 molecular model, a detailed mutagenesis analysis of the hyaluronan binding site was carried out (Bajorath et al. 1998). Residues, including R41, were selected to screen the surface of the CD44 model and

Fig. 6. Comparison of the TSG-6 structure and the CD44 molecular model. A stereo view of ribbon representations of TSG-6 (*thin line*) and CD44 (*thick line*) is shown after optimal super-position of the alpha-carbon traces. The rms deviation for 90 corresponding residue pairs in TSG-6 and CD44 is ~1.5 Å. The orientation is similar to the top view of Fig. 4

subjected to site-specific mutagenesis. Both drastic and conservative mutations were carried out. For example, mutation of R46 to serine (R46S) introduces a significant change as it removes the large and positively charged site chain at this position. By contrast, mutations Y42F and F119Y test the removal or addition of a single hydroxyl group and are thus conservative.

Mutants were constructed using polymerase chain reaction (PCR) methods, transiently expressed in COS cells (a transformed African green monkey kidney cell line) as CD44-immunoglobulin (Ig) fusion proteins (Edwards and Aruffo 1993), and purified from cell supernatants by protein A affinity chromatography. In CD44-Ig fusion proteins, the extracellular region of CD44H is genetically linked to the constant domains (hinge region, CH2, CH3) of a human IgG antibody. This allows for efficient purification from cell supernatants and easy detection of the proteins using peroxidase-labeled mAbs against the IgG constant domains. Expressed CD44-Ig mutant proteins were then tested in ELISA assays, relative to wild type CD44, for binding to conformationally sensitive mAbs against the CD44 link domain and for binding to immobilized HA. Binding experiments with conformationally sensitive anti-CD44 mAbs (i.e., mAbs with no reactivity against wild type CD44 under denaturing conditions in Western blots) were carried out to identify proteins with significant structural perturbations as a consequence of the mutation. Table 1 summarizes the mutagenesis and binding experiments. In several rounds of mutagenesis, a total of 17 residues were selected and 24 mutants were generated. Twenty one mutant proteins were characterized; only three mutants were not expressed in sufficient quantities. Four of 21 mutant proteins tested displayed reduced or non-detectable mAb binding.

Table 1. Summary of CD44 mutagenesis and binding experiments. 'mAb binding' reports binding of mutant proteins to the conformationally sensitive monoclonal anti-CD44 antibody BU75 (Bajorath et al. 1998) and 'HA binding' reports binding to hyaluronan. A score of + represents binding comparable to wild type CD44, +/- represents intermediate, and - non-detectable binding. Residues are classified, on the basis of mutagenesis, as important for '3-D structure' (i.e., mutant proteins show reduced or abolished mAb and HA binding) or 'HA binding' (i.e., mutant proteins show mAb binding comparable to wild type but reduced or abolished HA binding). Bold face indicates residues classified as critical for HA binding (i.e., mutant proteins display no detectable HA binding)

Residue	Mutant	mAb binding	HA binding	Importance
wild type		+	+	
F34	F34A	−	−	3-D structure
	F34Y	+	+	
K38	K38R	+	+/−	HA binding
	K38S	Not	expressed	
R41	R41A	+	−	HA binding
Y42	Y42F	+	−	HA binding
	Y42S	+/−	−	
R46	R46S	+	+	
E48	E48S	Not	expressed	
K54	K54S	+	+	
Q65	Q65S	+/−	−	3-D structure
K68	K68S	+	+/−	HA binding
R78	R78K	Not	expressed	
	R78S	+	−	HA binding
Y79	Y79F	+	−	HA binding
N100	N100 A	+	+/−	HA binding
	N100R	+	+/−	
N101	N101S	+	+/−	HA binding
Y105	Y105F	+	+	HA binding
	Y105S	+	+/−	
S112	S112R	+	+	
Y114	Y114F	+	+	
F119	F119A	+/−	−	3-D structure
	F119Y	+	+	

6.2 Classification of Targeted Residues

On the basis of the mAb and HA binding profiles of mutant proteins, residues were classified according to their importance for structure or ligand binding. The classification is shown in Table 1. For example, the conservative mutation

F34Y did not affect mAb or HA binding, whereas the non-conservative mutation F34 A abolished both mAb and ligand binding. Therefore, F34 is classified
as primarily important for the structural integrity of CD44. The conservative
mutation K38R did not affect mAb binding but significantly reduced HA
binding. Residue K38 is therefore classified as important but not essential for
HA recognition. Mutant protein Y79F showed mAb binding comparable to wild
type but did not show detectable ligand binding. Thus, in this case, removal of
a single hydroxyl group was sufficient to completely abolish HA binding and,
accordingly, Y79 is classified as critical for the CD44-HA interaction.

6.3 Mapping and Characterization of the Binding Site

In Fig. 7, mutated residues are shown on the CD44 model, and residues critical or important for HA binding are highlighted. These residues cluster and
form a patch on the relatively flat protein surface. Three residues, classified as
important for structural integrity (F34, Q65, F119), map to the C-terminus of
the link module where the interface with the rest of the extracellular region is
formed. Residue N100 is a potential N-linked glycosylation site in CD44, and
its proximity to other important residues emphasizes the potential role of glycosylation in the regulation of HA binding (as discussed above). As shown in
Fig. 8, the cluster of critical arginine and tyrosine residues (which is thought
to form the center of the binding site), and the adjacent group of residues that

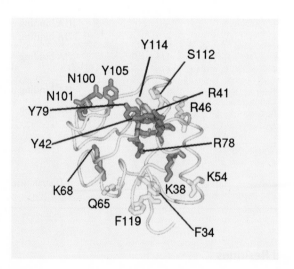

Fig. 7. CD44 residues targeted by site-specific mutagenesis. The residues are displayed on a solid
ribbon representation of the CD44 model. The orientation of the model is similar to Fig. 6.
Mutated residues are also numbered in Fig. 5. CD44 residues identified as critical or important
for hyaluronan binding, as explained in the text, are shown in *dark gray*

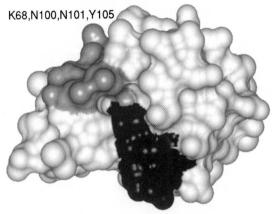

K68,N100,N101,Y105

R41,Y42,R78,Y79

Fig. 8. Hyaluronan binding surface in CD44. A solid molecular surface of the model is shown. The view is from the top, i.e., obtained from Fig. 7 by approximately 90° rotation around the horizontal axis. Residues that contribute to hyaluronan binding are color-coded (*black* critical for binding, *dark gray* important but not essential). These residues form a coherent binding surface. The binding site, as defined on the basis of mutagenesis, consists of a region critical for binding (formed by residues R41, Y42, R78, and Y79) and adjacent regions that support binding

support binding, form a coherent carbohydrate binding surface. The dimensions of this site are consistent with binding of a hexasaccharide or a somewhat larger structure.

The nature of the identified binding surface allows some conclusions to be drawn regarding CD44-HA interactions. The binding surface is extensive, consistent with the size of the ligand, but does not display a prominent cleft or cavity that would shield bound HA from the solvent environment. Together with the entropy penalty required to (at least partially) immobilize HA upon binding, this suggests that CD44-HA interactions have low intrinsic affinity. Thus, subtle changes in the molecular environment may suffice to regulate binding. Productive CD44-mediated cellular interactions may require avidity-increasing effects such as, for example, CD44 clustering and/or binding of multiple receptors to polymeric HA.

6.4 Opportunities and Limitations

The analysis described above illustrates the opportunities of structure-based design and interpretation of experiments. Three-dimensional mapping of important residues and analysis of their spatial arrangement are indispensable tools in the identification and characterization of binding sites. Since no experimentally determined structure of CD44 was available, a comparative model of the link module was used to guide the analysis. Molecular models are less accurate than experimental structures and it is therefore important to critically evaluate the consistency of obtained results (e.g., can surface residues be selected with confidence, are mutant proteins correctly folded, do important residues form a coherent binding site, are effects of mutations predictable?). Modeling studies cannot accurately predict atomic details of the CD44-HA interaction, and prediction of a CD44-ligand complex was therefore not attempted. Furthermore, it is important to note that results of mutagenesis experiments, be they based on an experimental or modeled structure, may be ambiguous in some cases. On one hand, mutagenesis analysis can only identify residues that make a significant energetic contribution to binding, and these residues are usually only a subset of residues forming a protein-ligand interface (Jin et al. 1992). On the other hand, even small structural perturbations due to mutation, the presence of which may not be detectable with conformationally sensitive mAbs, can implicate residues in binding that are not involved in protein-ligand contacts.

What level of accuracy can be expected in a successful model-based binding site analysis? Studies on the selectins, which were conceptually similar to those on CD44, may provide a reasonable reference. Fig. 9 shows a comparison of a molecular model of the C-type lectin domain of E-selectin and the X-ray structure (Bajorath et al. 1995). The model was built based on the crystal structure of the mannose binding protein (MBP; Weis et al. 1992) in the presence of only ~25% sequence identity. At this low level of similarity, meaningful sequence alignments are difficult to generate (more so than in the case of TSG-6 and CD44). Nevertheless, the E-selectin/MBP alignment, obtained by focusing on residues important for the integrity of the C-type lectin fold, was

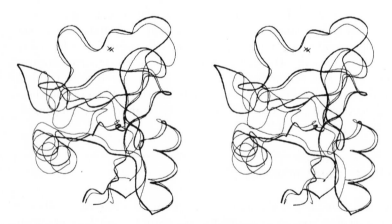

Fig. 9. Comparison of a molecular model and X-ray structure of E-selectin. The stereo view of E-selectin is similar to Fig. 4. Predicted and experimental calcium positions are shown as crosses. Alpha-carbon positions of the model (PDB id code '1kja', *thick line*) and X-ray structure ('1esl', *thin line*) were optimally superposed. The rms deviation for 109 matching residue pairs is ~1.6 Å.

overall ~93% correct (Bajorath et al. 1993). Fig. 10 shows the comparison of the predicted and experimentally determined carbohydrate binding site in E-selectin and illustrates that a reasonable picture was obtained on the basis of the model. The binding site was correctly mapped and only a few residues in the binding site region displayed significant deviations from the experimental structure.

7 Comparison of Carbohydrate Binding Sites

7.1 Link Proteins and C-Type Lectins

Of the nine residues in CD44 identified as critical or important for HA binding, only Y42 is conserved in TSG-6. In addition, R41 in CD44 is conservatively replaced by lysine. In contrast, K68, R78, and Y79 in CD44 correspond to alanine residues in TSG-6. Thus, even if corresponding regions in link modules are utilized to mediate HA binding, atomic details of hyaloadherin-HA interactions may differ substantially. Accordingly, the size of recognized HA fragments may vary, binding affinities may differ, and HA binding to hyaloadherins may be regulated in different ways.

The topological similarity of the link module and the C-type lectin fold suggests also to compare carbohydrate binding sites in these proteins. Fig. 11 shows a comparison of the binding sites in MBP, E-selectin, and CD44. In contrast to E-selectin and CD44, MBP is not a cell surface but a serum protein.

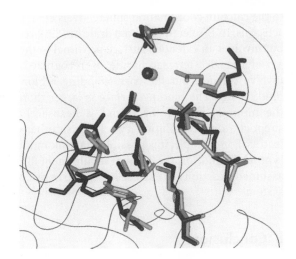

Fig. 10. Carbohydrate binding site in E-selectin. The close-up view compares the positions and conformations of predicted (*gray*) and experimentally determined (*black*) residues involved in ligand binding on a ribbon outline of the X-ray structure. These residues were identified by mutagenesis. Calcium positions are shown as *spheres*

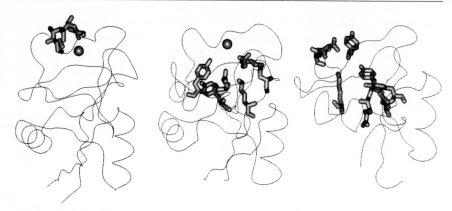

Fig. 11. Comparison of carbohydrate binding sites. From the left to the right, ribbon representations are shown of X-ray structures of two C-type lectins, the rat mannose binding protein (PDB id code '2msb') and human E-selectin ('1esl'), and of the molecular model of human CD44. The orientations of the domains correspond. Calcium positions are shown as *spheres*. Mannose binding protein is shown in complex with di-mannose that directly binds to the calcium. Residues identified by mutagenesis as important for ligand binding to E-selectin and CD44 are displayed

MBP is involved in innate immune responses in mammals by recognizing mannose on the surface of circulating pathogens (Drickamer and Taylor 1993). Thus, MBP may be an ancient molecule. The comparison shows that spatially corresponding regions in these proteins are utilized to recognize diverse carbohydrate ligands. In MBP, interactions with mannose are essentially limited to the calcium coordination sphere (Weis et al. 1992) which is conserved in the selectins. In E-selectin, carbohydrate binding requires the calcium binding site but involves an extended surface proximal to the calcium, since a larger ligand, a sialyl-Lewis^X tetrasaccharide, is recognized. In CD44, binding does not involve calcium but the corresponding region forms the center of the HA binding site. The binding surface is larger than in E-selectin, consistent with the minimal size of the ligand, a hexasaccharide. These findings further support a direct evolutionary relationship between link proteins and C-type lectins. For example, it is possible that link modules have diverged from the C-type lectin fold, lost the calcium requirement for carbohydrate recognition, and assumed functions as extracellular matrix receptors, also outside the immune system.

8 Conclusions

The molecular organization and possible functions of CD44 are complex. Many investigations have focused on the role of CD44 isoforms in cell adhesion, inflammation, or tumor metastasis. We are now beginning to understand the structural basis of ligand binding. Molecular model-based mutagenesis analysis of CD44 has provided a reasonable, albeit approximate, picture of the HA binding site. However, concerning the structure and function of CD44, there are still more questions than answers. For example, what does the structure of the CD44-HA complex look like? Do regions outside the link module participate in HA binding and, if so, how? What are the molecular mechanisms that regulate HA binding? Are variably spliced exons independent folding units, and what are their structures? Finally, how does binding of shorter HA fragments trigger signaling, while binding of HA polymers confers adhesive functions? It is anticipated that current research activities on CD44 and other hyaloadherins will soon help to answer some of these questions.

Acknowledgements and dedication. I would like to thank Alejandro Aruffo, Bristol-Myers Squibb, for our collaborative efforts to better understand the structure and function of cell surface proteins including CD44.

I wish to dedicate this contribution to the memory of my grandfather, Otto Maschlanka, who has inspired me in many ways.

References

Aruffo A (1996) CD44: one ligand, two functions. J Clin Invest 98:2191–2192

Aruffo A, Stamenkovic IA, Melnick M, Underhill CB, Seed B (1990) CD44 is the principal cell surface receptor for hyaluronate. Cell 61:1303–1313

Bajorath J (1998) Cell surface receptors and adhesion molecules, three-dimensional structures. In: Delves PJ, Roitt EM (eds) Encyclopedia of immunology, 2nd edn. Academic Press, London, pp 515–520

Bajorath J, Stenkamp RE, Aruffo A (1993) Knowledge-based model building of proteins: concepts and examples. Protein Sci 2:1798–1810

Bajorath J, Stenkamp RE, Aruffo A (1995) Comparison of a protein model with its X-ray structure: the ligand binding domain of E-selectin. Bioconjug Chem 6:3–6

Bajorath J, Greenfield B, Munro SB, Day AJ, Aruffo A (1998) Identification of CD44 residues important for hyaluronan binding and delineation of the binding site. J Biol Chem 273:338–343

Bartolazzi A, Nocks A, Aruffo A, Spring F, Stamenkovic I (1996) Glycosylation of CD44 is implicated in CD44-mediated cell adhesion to hyaluronan. J Cell Biol 132:1199–1208

Bennett KL, Jackson DG, Simon JC, Tanczos E, Peach RJ, Modrell B, Stamenkovic I, Plowman G, Aruffo A (1995a) CD44 isoforms containing exon v3 are responsible for the presentation of heparin-binding growth factors. J Cell Biol 128:687–698

Bennett KL, Modrell B, Greenfield B, Bartolazzi A, Stamenkovic I, Peach R, Jackson DG, Spring F, Aruffo A (1995b) Regulation of CD44 binding to hyaluronan by glycosylation of variably spliced exons. J Cell Biol 131:1623–1633

Bernstein FC, Koetzle TF, Williams GJB, Meyer EF Jr, Brice MD, Rodgers JR, Kennard O, Shimanouchi T, Tasumi M (1977) The protein data bank: a computer-based archival file for macromolecular structures. J Mol Biol 112:535–542

Bruccoleri RE, Karplus M (1987) Prediction of folding of short polypeptide segments by uniform conformational sampling. Bioploymers 26:137–168

Carter WG, Wayner EA (1988) Characterization of the class III collagen receptor, a phosphorylated, transmembrane glycoprotein expressed in nucleated human cells. J Biol Chem 263:4193–4201

Clark RA, Alon R, Springer TA (1996) CD44 and hyaluronan-dependent rolling interactions of lymphocytes on tonsillar stroma. J Cell Biol 134:1075–1087

Chothia C, Lesk AM (1986) The relation between the divergence of sequence and structure in proteins. EMBO J 5:823–826

DeGrendele HC, Estess P, Picker LJ, Siegelman MH (1996) CD44 and its ligand hyaluronate mediate rolling under physiological flow: a novel lymphocyte-endothelial cell primary adhesion pathway. J Exp Med 183:1119–1130

Drickamer K (1988) Two distinct classes of carbohydrate recognition domains in animal lectins. J Biol Chem 263:9557–9560

Drickamer K, Taylor ME (1993) Biology of animal lectins. Annu Rev Cell Biol 9:237–264

Edwards CP, Aruffo A (1993) Current applications of COS cell based transient expression systems. Curr Opin Cell Biol 4:558–563

Goldstein LA, Zhou DF, Picker LJ, Minty CN, Bargatze RF, Ding JF, Butcher EC (1989) A human lymphocyte homing receptor, the hermes antigen, is related to cartilage proteoglycan core and link proteins. Cell 56:1063–1072

Graves BJ, Crowther RL, Chandran C, Rumberger JM, Li S, Huang KS, Presky DH, Familletti PC, Wolitzky BA, Burns DK (1994) Insight into E-selectin/ligand interaction from the crystal structure and mutagenesis of the lec/EGF domains. Nature 367:532–538

Greer J (1990) Comparative modeling methods: applications to the family of serine proteases. Proteins Struct Funct Genet 7:317–334

Günthert U, Hofmann M, Rudy W, Reber S, Zoller M, Haussmann I, Matzku S, Wenzel A, Ponta H, Herrlich P (1991) A new variant of glycoprotein CD44 confers metastatic potential to rat carcinoma cells. Cell 65:13–24

Holmbeck SM, Petillo PA, Lerner LE (1995) The solution conformation of hyaluronan: a combined NMR and molecular dynamics study. Biochemistry 33:14246–14255

Jalkanen S, Jalkanen M (1992) CD44 binds to the COOH-terminal heparin-binding domain of fibronectin. J Cell Biol 116:817–825

Jin L, Fendly BM, Wells JA (1992) High resolution functional analysis of antibody-antigen interactions. J Mol Biol 226:851–865

Jones TA, Thirup S (1986) Using known substructures in protein model building and crystallography. EMBO J 5:819–822

Katoh S, Zheng Z, Ortani K, Shimozato T, Kincade PW (1995) Glycosylation of CD44 negatively regulates its recognition of hyaluronan. J Exp Med 182:419–429

Kohda D, Morton CJ, Parkar AA, Hatanaka H, Inagaki FM, Campbell ID, Day AJ (1996) Solution structure of the link module: a hyaluronan-binding domain involved in extracellular matrix stability and cell migration. Cell 86:767–775

Kincade PW, Zheng Z, Katoh S, Hanson L (1997) The importance of cellular environment to function of the CD44 matrix receptor. Curr Opin Cell Biol 9:635–642

Lasky LA (1995) Selectin-carbohydrate interactions and the initiation of the inflammatory response. Annu Rev Biochem 64:113–139

Lee T, Wisniewski H-G, Vilcek J (1992) A novel secretory tumour necrosis factor-inducible protein (TSG-6) is a member of the family of hyaluronate binding proteins, closely related to the adhesion receptor CD44. J Cell Biol 116:545–557

Lesley J, Hyman R, Kincade PW (1993) CD44 and its interaction with extracellular matrix. Adv Immunol 54:271–335

Lesley J, English N, Perschl A, Gregoroff J, Hyman R (1995) Variant cell lines selected for alterations in the function of the hyaluronan receptor CD44 show differences in glycosylation. J Exp Med 182:431–437

Maiti A, Maki G, Johnson P (1998) TNF-alpha induction of CD44-mediated leukocyte adhesion by sulfation. Science 282:941–943

McKee CM, Penno MB, Cowman M, Bao C, Noble PW (1996) Hyaluronan (HA) fragments induce chemokine gene expression in alveolar macrophages. The role of HA size and CD44. J Clin Invest 98:2403–2413

Mikecz K, Brennan FR, Kim JH, Glant TT (1995) Anti-CD44 treatment abrogates tissue oedema and leukocyte infiltration in murine arthritis. Nature Med 1:558–563

Miyake K, Underhill CB, Lesley J, Kincade PW (1990) Hyaluronate can function as a cell adhesion molecule and CD44 participates in hyaluronate recognition. J Exp Med 172:69–75

Peach RJ, Hollenbaugh D, Stamenkovic I, Aruffo A (1993) Identification of hyaluronic acid binding sites in the extracellular domain of CD44. J Cell Biol 122:257–264

Ponder JW, Richards FM (1987) Tertiary templates for proteins. Use of packing criteria in the enumeration of allowed sequences for different structural classes. J Mol Biol 193:775–791

Screaton GR, Bell MV, Jackson DG, Cornelis FB, Gerth U, Bell JI (1992) Genomic structure of DNA encoding the lymphocyte homing receptor CD44 reveals at least 12 alternatively spliced exons. Proc Natl Acad Sci USA 89:12160–12164

Screaton GR, Bell MV, Bell JI, Jackson DG (1993) The identification of a new alternative exon with highly restricted tissue expression in transcripts encoding the mouse Pgp-1 (CD44) homing receptor. Comparison of all 10 variable exons between mouse, human, and rat. J Biol Chem 268:12235–12238

Sippl MJ (1993) Recognition of errors in three-dimensional structures of proteins. Proteins Struct Funct Genet 17:355–362

St John T, Meyer J, Idzerda R, Gallatin WM (1990) Expression of CD44 confers a new adhesive prototype on transfected cells. Cell 62:45–52

Stamenkovic I, Aruffo A, Amiot M, Seed B (1989) A lymphocyte molecule implicated in lymph node homing is a member of cartilage link protein family. Cell 56:1057–1062

Stamenkovic I, Aruffo A, Amiot M, Seed B (1991) The hematopoietic and epithelial forms of CD44 are distinct polypeptides with different adhesion potentials for hyaluronate-bearing cells. EMBO J 10:343–348

Sy M-S, Mori H, Liu D (1997) CD44 as a marker in human cancers. Curr Opin Oncol 9:108–112

Taher TEI, Smit L, Griffioen AW, Schilder-Tol EJM, Borst J, Pals ST (1996) Signaling through CD44 is mediated by tyrosine kinases – association with p56^{lck} in T lymphocytes. J Biol Chem 271:2863–2867

Toole BP (1990) Hyaluronan and its binding proteins, the hyaloadherins. Curr Opin Cell Biol 2:839–844

Underhill CB, Chi-Rosso G, Toole BP (1983) Effects of detergent solubilization on the hyaluronate-binding protein from membranes of simian virus 40-transformed 3T3 cells. J Biol Chem 258:8086–8091

Weber GF, Ashkar S, Glimcher MJ, Cantor H (1996) Receptor-ligand interaction between CD44 and osteopontin (Eta-1). Science 271:509–512

Weis WI, Drickamer K, Hendrickson WA (1992) Structure of a C-type mannose-binding protein complexed with an oligosaccharide. Nature 360:127–134

Williams AF, Barclay AN (1988) The immunoglobulin superfamily – domains for cell surface recognition. Annu Rev Immunol 6:381–406

Yang B, Yang BL, Savani RC, Turley EA (1994) Identification of a common hyaluronan binding motif in the hyaluronan binding proteins RHAMM, CD44 and link protein. EMBO J 13:286–296

Yu Q, Toole BP (1996) A new alternatively spliced exon between v9 and v10 provides a molecular basis for synthesis of soluble CD44. J Biol Chem 271:20603–20607

Zhang L, David G, Esko JD (1995) Repetitive Ser-Gly sequences enhance heparan sulfate assembly in proteoglycans. J Biol Chem 270:27127–27135

Structure and Function of the Macrophage Mannose Receptor

Maureen E. Taylor[1]

1 Functions and Biological Ligands of the Mannose Receptor

1.1 Identification and Localization of the Mannose Receptor

The mannose receptor acts as a molecular scavenger by mediating Ca^{2+}-dependent recognition and internalization of glycoconjugates terminating in mannose, N-acetylglucosamine or fucose. The receptor was identified when it was found that glycoproteins terminating in GlcNAc or mannose, including lysosomal enzymes, are rapidly cleared from the bloodstream by the liver (Schlesinger et al. 1976). The mannose receptor was found to be located on hepatic endothelial cells and Kupffer cells but not on hepatocytes (Schlesinger et al. 1978). The receptor has since been found on most types of tissue macrophages, including those of the placenta, but not on circulating monocytes (Shepherd et al. 1982). The retinal pigmented epithelium, a phagocytic cell layer, also expresses the mannose receptor (Shepherd et al. 1991). More recently, the mannose receptor has been identified on CD1-positive dendritic cells and Langerhan's cells (Sallusto et al. 1995; Condaminet et al. 1998).

1.2 Roles of the Mannose Receptor in the Immune Response

Recognition of sugars such as mannose and GlcNAc by the mannose receptor allows discrimination between self and non-self, since these monosaccharides are not common at the termini of mammalian cell-surface or serum glycoproteins, but are frequently found on the surfaces of microorganisms. Thus, the receptor is able to recognize many pathogens and play a role in the innate immune response by mediating opsonin-independent phagocytosis. The mannose receptor has been implicated in uptake of a wide variety of pathogenic micro-organisms, including *Pneumocystis carinii* (Ezekowitz et al. 1991), *Mycobacterium tuberculosis* (Schlesinger 1993), *Candida albicans* (Ezekowitz et al. 1990) and *Leishmania donovani* (Chakraborty and Das 1988).

[1] Glycobiology Institute, Department of Biochemistry, University of Oxford, South Parks Road, Oxford OX1 3QU, UK

Results and Problems in Cell Differentiation, Vol. 33
Paul R. Crocker (Ed.): Mammalian Carbohydrate Recognition Systems
© Springer-Verlag Berlin Heidelberg 2001

The ability of the mannose receptor to mediate internalisation of soluble glycoconjugates and viruses as well as entire pathogens such as bacteria and fungi suggests that the receptor could play a role in facilitating antigen uptake and processing in the adaptive immune response, as well as mediating direct uptake of pathogens in the innate immune response. The function of the receptor would be to mediate enhanced uptake and processing of soluble glycoconjugates released from pathogens and thereby enhance association of degradation fragments with MHC class II molecules for presentation to T cells. The peptide antigens appearing on the surface of antigen-presenting cells would not necessarily themselves be conjugated to carbohydrates; the presence of carbohydrates characteristic of potentially harmful structures would allow selective uptake mediated by the mannose receptor, so that peptide components could then be released to instruct the adaptive immune response. In this way, the mannose receptor could serve as a link between the innate and the adaptive immune systems.

Identification of the mannose receptor on dendritic cells and Langerhan's cells, which are the main antigen-presenting cells, as well as macrophages, is consistent with a role in antigen processing. Some indirect evidence for involvement of the mannose receptor in processing of lipoglycan and peptide antigens has been obtained. The mycobacterial lipoglycan lipoarabinomannan (LAM) is presented to T cells in complex with CD1b, a specialized MHC-like molecule (Prigozy et al. 1997). LAM can bind to the mannose receptor, and LAM-dependent proliferation of T cells is inhibited by mannan, another ligand for the receptor. These results suggest that the mannose receptor is responsible for transporting LAM to intracellular vesicles where it is loaded onto CD1b. Mannosylation of peptides has been shown to enhance their presentation by dendritic cells, but involvement of the mannose receptor in this process has not yet been directly demonstrated (Engering et al. 1997; Tan et al. 1997).

1.3 Clearance of Soluble Endogenous Ligands by the Mannose Receptor

Some endogenous proteins bearing high-mannose oligosaccharides are cleared from the circulation by the mannose receptor. These are typically proteins released in response to pathological events that must be removed from the circulation once they have acted, to prevent tissue damage occurring. Endogenous proteins cleared by the mannose receptor include lysosomal hydrolases and myeloperoxidase secreted at sites of inflammation (Stahl and Schlesinger 1980; Shepherd and Hoidal 1990), tissue plasminogen activator (Otter et al. 1991) and the C-terminal propeptide of type I procollagen (Smedsrod et al. 1990). Although other receptors are involved in tissue plasminogen activator clearance, approximately 50% is removed from the circulation by the mannose receptor (Biessen et al. 1997). Soluble ligands bound to the receptor are internalized via clathrin-coated pits into endosomes, where the ligand-receptor complex is uncoupled. The ligand is targeted to the lysosomes for destruction, while the receptor returns to the cell surface (Wileman et al. 1984).

In addition to its role in clearing endogenous proteins with high mannose-type oligosaccharides, the mannose receptor has recently been shown to mediate clearance from the circulation of pituitary hormones such as lutropin. Clearance of lutropin appears to be mediated by liver mannose receptor through recognition of sulphated N-acetylgalactosamine present on the oligosaccharides of the hormone (Fiete et al. 1997). Recognition of sulphated N-acetylgalactosamine by the mannose receptor is discussed elsewhere in this volume.

2 Structure of the Mannose Receptor

2.1 Primary Structure

Mannose receptors have been isolated from macrophages (Haltiwanger and Hill 1986), placenta (Lennartz et al. 1987), retinal pigment epithelium (Shepherd et al. 1991) and liver (Otter et al. 1992). The receptor from each source consists of a single subunit with a molecular weight of about 175,000. The receptor is a glycoprotein, with both N- and O-linked oligosaccharides (Lennartz et al. 1989). Mannose receptors from human placenta (Taylor et al. 1990) and human monocyte-derived macrophages (Ezekowitz et al. 1990) have been cloned and sequenced and found to be identical. It seems likely that the mannose receptors expressed by other cells will also have identical sequences.

The mannose receptor cDNA encodes a protein of 1438 amino acids (after removal of an N-terminal signal sequence). The receptor is oriented as a type I transmembrane protein (COOH terminus inside the cell) with a 45 amino acid cytoplasmic tail (Fig. 1). The extracellular portion of the receptor consists of three types of cysteine-rich domains. The N-terminal cysteine-rich domain shows no similarity to other known sequences. The second domain resembles the type II repeats of fibronectin (Kornblihtt et al. 1985). The rest of the extracellular region of the receptor consists of eight domains related in sequence to the C-type carbohydrate-recognition domains (CRDs) of animal lectins (Drickamer 1993).

2.2 Features of Individual Domains

2.2.1 The Cytoplasmic Tail

The 45-amino acid cytoplasmic tail of the mannose receptor is not homologous to the tails of any other known receptors, although it does contain a single tyrosine residue in a context similar to that found in other endocytic recep-

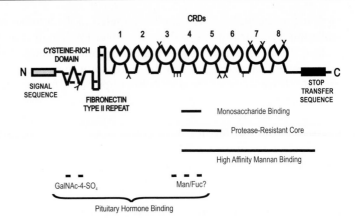

Fig. 1. Domain organization of the mannose receptor. The major domains of the receptor are shown diagrammatically. Potential or actual sugar attachment sites to asparagine (*Y*) and threonine (*I*) are marked. CRDs involved in ligand binding are indicated *below* the structure

tors. This tyrosine residue is critical for high efficiency localization of some endocytic receptors into clathrin-coated pits (Chen et al. 1990). The mannose receptor cytoplasmic tail contains the sequence Phe-Glu-Asn-Thr-Leu-Tyr which is similar to the internalization signal Phe-X-Asn-Pro-X-Tyr (where X indicates that this residue can be changed without affecting internalization) in the cytoplasmic tail of the low density lipoprotein receptor. The cytoplasmic tail of the mannose receptor also contains several serine and threonine residues that are potential sites of phosphorylation, although none is in a context typical of substrates for well-studied kinases. Deletion of the cytoplasmic tail abolishes the ability of the mannose receptor to phagocytose yeast (Ezekowitz et al. 1990). Mutation of the single tyrosine residue in the cytoplasmic tail of the receptor reduces the efficiency of both phagocytosis and endocytosis, indicating that this residue may form part of the internalization signal in both processes (Kruskal et al. 1992).

2.2.2 The N-Terminal Cysteine-Rich Domain and the Fibronectin Type II Repeat

The N-terminal 139 amino acid residues of the mannose receptor form a domain unique to the mannose receptor and the three other members of the mannose receptor family (discussed below). Although the domain is termed cysteine-rich owing to the presence of six cysteine residues, it is not related in sequence to the cysteine-rich domains of growth factor receptors or the low density lipoprotein receptor. This domain has been shown to contain an *N*-acetylgalactosamine 4-sulphate binding site involved in recognition of pituitary hormones (Fiete et al. 1998; discussed elsewhere in this Vol.).

The fibronectin type II repeat of the mannose receptor is closest in sequence to the second of these domains found in fibronectin (Kornblihtt et al. 1985). Fibronectin type II repeats are found in a few other proteins, including coagulation factor XII (Cool et al. 1985) the cation-independent mannose 6-phosphate receptor (Lobel et al. 1988), and gelatinases A and B. In the gelatinases, and possibly in fibronectin, these domains are involved in gelatin binding (Banyai et al. 1994). No function has yet been associated with this domain in the mannose receptor.

2.2.3 The C-Type Carbohydrate Recognition Domains

The mannose receptor belongs to a large family of animal lectins, known as C-type lectins, that mediate Ca^{2+}-dependent sugar recognition through homologous C-type CRDs. Well-characterized members of the C-type lectin superfamily include the mannose binding proteins, the asialoglycoprotein receptor and the selectin cell adhesion molecules (Drickamer and Taylor 1993; Weis et al. 1998). C-type CRDs are found in association with a wide variety of effector domains. However, the mannose receptor and the three other members of the mannose receptor family are the only proteins known to contain more than one C-type CRD in a single polypeptide.

C-type CRDs are domains of about 120 amino acids with a characteristic pattern of conserved residues, including two cysteine residues that form disulphide bonds and several aliphatic and aromatic amino acids that make up the hydrophobic core of the domain (Drickamer 1993). Crystal structures of two C-type CRDs, those of rat serum mannose binding protein (MBP-A) and rat liver mannose-binding protein (MBP-C) have been solved in complex with sugar ligands (Weis et al. 1992; Ng et al. 1996). The C-type CRD of E-selectin has also been crystallized, but without bound sugar (Graves et al. 1994). Examination of these crystal structures, combined with other physical techniques and mutagenesis has established some of the molecular mechanisms involved in sugar recognition by C-type CRDs (Weis and Drickamer 1996). These studies have identified key residues associated with ligation of Ca^{2+} and sugar by the domains. The absence of these key residues from many of the more divergent C-type CRDs suggests that not all of these domains will bind sugars. Domains such as those found in several natural killer cell receptors which are predicted to fold in the same way as the C-type CRDs, but do not contain the residues for Ca^{2+} and sugar binding have been termed C-type-lectin-like domains (Weis et al. 1998).

The C-type CRDs of the mannose receptor are quite divergent from the prototypes of the family, but each of them contains the conserved residues responsible for maintaining the C-type CRD fold. Thus, it is likely that the CRDs of the mannose receptor will fold in a similar way. Each of the CRDs of the mannose receptor contains at least some of the residues shown to be important for Ca^{2+} and sugar binding by the CRDs of MBP and E-selectin. These residues are best conserved in CRDs 4 and 5.

3 Mechanisms of Carbohydrate Binding by the Mannose Receptor

3.1 Roles of Individual Domains

Expression of portions of the receptor in several different systems has been used to determine which domains are involved in binding and endocytosis of ligands (Taylor et al. 1992; Taylor and Drickamer 1993). Results of these studies are summarized in Fig. 1. Initial experiments indicated that the N-terminal cysteine-rich domain and the fibronectin type II repeat are not essential for binding and endocytosis of glycoproteins. Fibroblasts expressing a truncated receptor with an extracellular portion consisting of CRDs 1–8 are able to bind and endocytose mannose-terminated ligands as efficiently as fibroblasts expressing the intact receptor. These results indicate that the carbohydrate-binding activity of the receptor resides within the CRDs.

Results from in vitro translation studies suggest that CRDs 1–3 do not bind carbohydrates and that the binding activity of the receptor resides in CRDs 4–8. Only CRD-4 shows sugar binding activity when expressed alone, but a fragment consisting of CRDs 5–8 retains sugar binding activity, indicating that other CRDs in this fragment must have weak affinity for sugars. Localization of the carbohydrate-binding activity of the receptor to CRDs 4–8 was confirmed by experiments showing that cells expressing a truncated receptor consisting of domains 4–8 are able to endocytose a neoglycoprotein, mannose-BSA, as efficiently as cells expressing the intact receptor. Cells expressing a truncated receptor consisting of CRDs 5–8 can also endocytose mannose-BSA, but at a much slower rate than cells expressing CRDs 4–8 or the intact receptor, highlighting the importance of CRD 4. Cells expressing CRDs 6–8 are not able to internalize mannose-BSA, indicating that CRD 5 is also important.

CRD-4 binds mannose, GlcNAc, fucose and glucose with affinities in the millimolar range (Taylor et al. 1992). The intact receptor also has specificities for these monosaccharides with dissociation constants in the millimolar range. Thus, a single CRD can mimic the monosaccharide-binding properties of the whole receptor. However, CRD 4 binds only poorly to glycoproteins such as invertase and mannan and thus cannot account for the binding of the receptor to natural ligands. Other CRDs within the 4–8 fragment are required for binding to oligosaccharides. CRDs 4–5 show much higher affinity for ligands such as yeast mannan and invertase than does CRD-4 alone. These two domains form a protease-resistant ligand-binding core, essential for high affinity binding of multivalent ligands (Taylor and Drickamer 1993). However, the affinity for yeast mannan increases as more CRDs are added with the full affinity of the receptor for this ligand only being achieved when CRDs 4–8 are present. Thus, as well as CRDs 4 and 5, accessory domains 6, 7 and 8 are necessary for binding to some ligands. High affinity binding to natural ligands is achieved through multiple weak interactions with several CRDs within the mannose receptor polypeptide.

3.2 Molecular Mechanism of Monosaccharide Binding to the Fourth Carbohydrate-Recognition Domain

Since CRD-4 is the smallest piece of the receptor that retains the ability to interact with sugars, an understanding of the molecular mechanism of sugar binding by this CRD is a first step towards understanding how the whole receptor recognizes its natural ligands. A detailed picture of how CRD-4 binds to sugar and Ca^{2+} has been obtained by a combination of ligand binding assays, site-directed mutagenesis and NMR (Mullin et al. 1994, 1997; Hitchen et al. 1998). CRD-4 of the mannose receptor has specificity for mannose, GlcNAc and fucose, like the CRDs of MBP-A and MBP-C, which have been well characterized by crystallography, NMR and mutagenesis (Weis and Drickamer 1996). Ligand-binding studies on CRD-4 show that some aspects of the mode of binding of sugar and Ca^{2+} by CRD-4 are similar to those of the MBP CRDs, but others are different, indicating that mannose-binding by the mannose receptor probably evolved separately from mannose-binding by other C-type lectins.

3.2.1 Interaction of Ca^{2+} with CRD-4

Results from a solid phase-binding assay and a protease resistance assay show that CRD-4 requires two Ca^{2+} for sugar binding, and that a ternary complex is formed between protein, sugar and Ca^{2+} (Mullin et al. 1994). The stability of Ca^{2+} binding is pH-dependent, and a conformational change in CRD-4 due to loss of Ca^{2+} binding at low pH probably contributes to release of glycoconjugates by the mannose receptor in endosomes.

The fact that CRD-4 requires two Ca^{2+} for sugar binding is surprising, since sequence comparisons with other C-type CRDs suggest that it might only bind one Ca^{2+}. In the crystal structure of the CRD of MBP-A in complex with an oligosaccharide ligand, one Ca^{2+} (designated Ca^{2+} 2) ligates directly to the sugar, while the other (designated Ca^{2+} 1) is thought to be necessary for the correct positioning of the loops forming the sugar-binding site (Weis et al. 1992). Alignment of the sequences of CRD-4 and MBP-A shows that all of the residues ligating Ca^{2+} 2 in MBP-A are present in CRD-4, suggesting that one of the two Ca^{2+} bound to CRD-4 is ligated at a conserved site (Fig. 2).

At this conserved Ca^{2+} binding site in MBP-A, each side-chain forms a hydrogen bond to hydroxyl groups 3 or 4 of mannose, as well as ligating Ca^{2+},

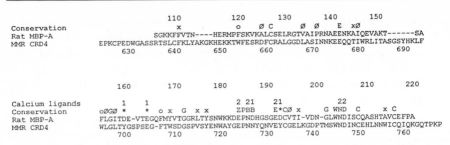

Fig. 2. Alignment of sequences of CRD-4 and the MBP-A CRD. The sequence of the CRD of MBP-A in the crystal structure (Weis et al. 1992) is aligned with the sequence of CRD-4 of the mannose receptor. The numbering of residues of the intact MBP-A is shown at the *top*. The numbering of residues of the intact mannose receptor (MMR) is shown *underneath* the CRD-4 sequence. The sequences are 28% identical. Conservation: residues that are conserved amongst C-type lectins are shown. Invariant residues are shown in single letter code; positions that are conserved in character are shown with the following code: *x* aliphatic or aromatic, *o* aromatic, Ø aliphatic, * side chain with carbonyl oxygen (D, N, E or Q). Calcium ligands: residues involved in ligating Ca^{2+} at sites 1 or 2 in the MBP-A crystal structure are indicated by *1* or *2* above the residue

which also ligates directly to hydroxyl groups 3 and 4 of mannose (Weis et al. 1992). Thus, a ternary complex between protein, sugar and Ca^{2+} is formed (Fig. 3). The same type of ternary complex is seen in the crystal structure of the CRD of MBP-C (Ng et al. 1996). The conservation of these Ca^{2+} ligands in CRD-4, combined with the fact that a ternary complex is also formed in CRD-4 suggests that the interaction of protein, Ca^{2+} and sugar at this site in CRD-4 is very similar to that seen in the mannose-binding proteins (Mullin et al. 1994).

Ca^{2+} 1 in MBP-A is ligated by two glutamate and two aspartate residues. Of these four residues, only one glutamate residue is conserved in CRD-4, with two tyrosine residues, and an asparagine residue found at the three other positions (Fig. 2). Mutation of each of these four residues to alanine indicated that only the asparagine (Asn^{728}) is likely to be involved in ligation of a second Ca^{2+} to CRD-4 (Mullin et al. 1997). CRD-4 with the change Asn^{728} to Ala binds sugar weakly with first order dependence on Ca^{2+}, indicating that one Ca^{2+} binding site is lost. The fact that this mutated domain binds sugar only weakly suggests that, as in MBP-A, binding of the second Ca^{2+} to CRD-4 enhances sugar binding at the other Ca^{2+}. Alanine scanning mutagenesis identified two other asparagine residues (Asn^{731} and Asn^{750}) and one glutamic acid residue (Glu^{737}) that are probably involved in ligating the second Ca^{2+} to CRD-4. Sequence comparisons with other C-type CRDs suggests that the proposed mode of binding of the second Ca^{2+} to CRD-4 is unique.

3.2.2 Involvement of a Stacking Interaction in Sugar Binding to CRD-4

Although it is probable that the major interaction between sugars and CRD-4 of the mannose receptor is via direct ligation to a conserved Ca^{2+}, there are likely to be other contacts between the protein and sugar. Ring current shifts

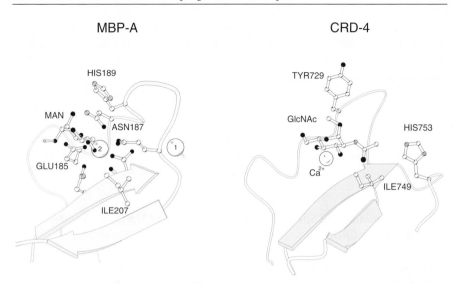

Fig. 3. Monosaccharide binding to CRD-4 and the MBP-A CRD. The *left* panel shows mannose bound to the CRD of MBP-A, as seen in the crystal structure (Weis et al. 1992). The *right* panel shows a model of α-methyl GlcNAc bound to CRD-4 based on the crystal structures of the CRD of rat MBP-C in complex with methyl glycosides of mannose, fucose and GlcNAc (Ng et al. 1996). Changes to CRD-4 residues Tyr[729], Ile[749] and His[753] were introduced at the equivalent positions in MBP-C. α-methyl GlcNAc is shown bound in the orientation seen in the MBP-C crystals, and predicted to be the orientation in CRD-4. Carbon atoms are *white*, nitrogen atoms are *gray* and oxygen atoms are *black*

seen in the ^{1}H NMR spectra of methyl glycosides of mannose, GlcNAc and fucose in the presence of CRD-4, and site-directed mutagenesis indicate that a stacking interaction with a tyrosine residue (Tyr[729]) is also involved in binding of sugars to CRD-4. This interaction contributes about 25% of the total free energy of binding to mannose. C5 and C6 of mannose interact with Tyr[729], whereas C2 of GlcNAc is closest to this residue, indicating that these two sugars bind to CRD-4 in opposite orientations (Mullin et al. 1997).

Stacking interactions between aromatic residues and sugars are a common feature of protein-carbohydrate complexes (Weis and Drickamer 1996). In all structures of proteins in complex with galactose, including plant lectins, bacterial toxins, animal S-type lectins, and the galactose-binding mutant of MBP-A, the non-polar B-face of galactose stacks against tryptophan or phenylalanine residues. However, although there is a precedent for stacking of mannose and GlcNAc in plant lectins, stacking interactions with these sugars are not seen in the CRDs of MBP-A or MBP-C (Weis and Drickamer 1996). Thus, CRD-4 of the mannose receptor provides the first example of this type of interaction with these sugars in an animal lectin.

Sequence comparisons with other mannose/GlcNAc-specific C-type CRDs suggest that use of a stacking interaction in binding of these sugars is probably unique to CRD-4 of the mannose receptor. Of the mannose/fucose/GlcNAc-specific C-type CRDs, none, apart from MBP-A, has an aromatic residue at the position equivalent to Tyr^{729} (Drickamer 1993). The histidine residue, His^{189}, at this position in MBP-A, however, is splayed away from bound mannose and makes only an edge-wise contact with the 2 hydroxyl group (Weis et al. 1992). Only the β-carbon of His^{189} in MBP-A contributes significantly to the binding energy (Iobst et al. 1994). It seems that a difference between CRD-4 and MBP-A in the arrangement of the loops in this area of the molecule must be responsible for positioning Tyr^{729} so that it can stack against bound sugars. This difference in the arrangement of loops around the sugar binding site could be brought about by the different modes of ligation of the second Ca^{2+} in the two proteins.

3.2.3 Determinants of Specificity and Orientation of Monosaccharides Bound to CRD-4

Mannose, GlcNAc and fucose are all predicted to bind at a common site on CRD-4. However, there are differences in the way each sugar interacts with CRD-4. NMR studies indicate that mannose bound to CRD-4 must be rotated 180° relative to bound GlcNAc (Mullin et al. 1997). Also, dissociation constants determined for binding of methyl glycosides indicate that CRD-4 binds α-methyl Man approximately twice as tightly as β-methyl fucose and α-methyl GlcNAc, but has a strong preference for α-methyl fucose, which is bound approximately ten times more tightly than β-methyl fucose (Mullin et al. 1997). Each sugar is predicted to ligate to the conserved Ca^{2+} ion via two equatorial hydroxyl groups, but other interactions of the sugar with the protein must determine the orientation of sugars in the binding-site as well as the relative affinities of CRD-4 for different sugars.

Evidence from NMR titrations and mutagenesis suggests that the acetamido group of GlcNAc bound to CRD-4 is close to His^{753} and Ile^{729}, and that C1 and C2 are close to Tyr^{729} (Hitchen et al. 1998). These data indicate that GlcNAc binds to CRD-4 in the orientation seen in crystal structures of the CRD of MBP-C (Fig. 3). Mannose binds to CRD-4 in the orientation seen in the CRD of MBP-A and is rotated 180° relative to GlcNAc bound to CRD-4. Interaction of the O-methyl group and C1 of α-methyl fucose with Tyr^{729} accounts for the strong preference of CRD-4 for this anomer of fucose (Hitchen et al. 1998). Both anomers of fucose bind to CRD-4 in the orientation seen in MBP-C.

3.3 Spatial Arrangement of Domains

As is the case in other C-type lectins, high affinity binding to the mannose receptor is achieved through clustering of multiple CRDs, each with weak affinity for single sugars. However, other C-type lectins have only a single CRD in each polypeptide and thus must form oligomers. Differences in domain organization between other C-type lectins, such as the mannose binding proteins and the mannose receptor, probably reflect their specificity for different ligands. The mannose receptor must recognize mammalian high mannose oligosaccharides as well as the sugars found on the surface of pathogens, and it is likely that the multiple different CRDs in a single polypeptide give the mannose receptor the ability to recognize the wide range of different structures required for its dual functions.

In order to understand how the mannose receptor is able to select carbohydrates that define pathogens or harmful glycoproteins, it is necessary to understand the spatial arrangement of the domains within the polypeptide, as well as how individual domains interact with sugars. Although progress has been made in understanding the interactions of individual CRDs of the receptor with sugars, nothing is known about how the CRDs are arranged spatially to match the geometric configurations of particular oligosaccharide ligands. The mannose receptor is unusual in that the domains that seem to be most important for ligand binding, CRDs 4 and 5, are located in the middle of the polypeptide chain, rather than at the end. One explanation for this phenomenon could be that the extracellular region of the receptor bends to adopt a U-shaped conformation (Fig. 4 right), rather than extending linearly from the cell

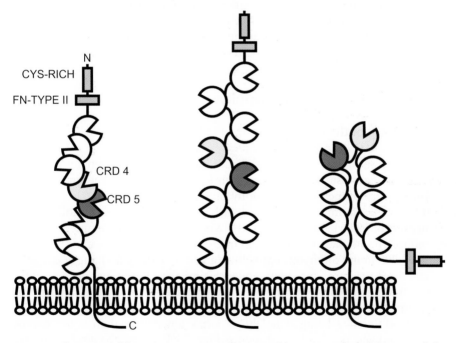

Fig. 4. Conformation of the mannose receptor. Three possible arrangements of the extracellular domains of the mannose receptor are shown. *Right* U-shaped model. *Centre* extended model. *Left* compact model, suggesting extensive contacts between CRDs. *CYS-RICH* cysteine-rich domain, *FN-TYPE II* fibronectin type II domain

surface. Such an arrangement would put CRDs 4 and 5 furthest from the membrane and closest to potential ligands in the extracellular medium.

It is not known to what extent the domains of the receptor interact with each other. The CRDs are separated by linker regions of about 10 to 20 amino acids (Taylor et al. 1990), each containing several proline residues. The linkers could confer an extended conformation, so that the CRDs might be spread out like beads on a string (Fig. 4 centre). Alternatively, the CRDs could be close together, with extensive contacts between them (Fig. 4 left). The fact that proteolysis of a fragment consisting of CRDs 4–5 does not result in the release of individual CRDs suggests that these two domains form extensive contacts and are not simply linked by a flexible tether (Taylor and Drickamer 1993). Such an arrangement may fix the orientation of the binding sites within these two CRDs. It is also hard to explain the high affinity of the receptor for some ligands without invoking co-operative interactions between domains. Only CRD-4 of the receptor has been shown to retain sugar-binding activity when expressed alone, with an affinity for mannose of about 3 mM (Taylor et al. 1992). Thus, each of the other CRDs must have affinities for monosaccharides that are even weaker. However, a synthetic cluster mannoside ligand containing only two α-mannose residues binds to the receptor with micromolar affinity, and the affinity increases to nanomolar with an increasing number of mannose residues, up to a maximum of six (Biessen et al. 1996). It is therefore plausible that binding of a single sugar residue to one CRD results in increased affinity for binding of sugars by neighboring CRDs

4 The Mannose Receptor Family

4.1 Members of the Family

Three other endocytic receptors showing the same overall domain organization as the mannose receptor have been identified, making the mannose receptor the prototype of a new family of receptors (Taylor 1997). The other members of the family are a widely distributed receptor for phospholipases A_2 (Ishizaki et al. 1994; Lambeau et al. 1994), a receptor of unknown function found on mouse endothelial cells and chondrocytes (Wu et al. 1996), and the receptor DEC-205 found on thymic epithelium and dendritic cells (Jiang et al. 1995).

The three new proteins show sequence similarity to the mannose receptor throughout their whole length. They are predicted to show the same membrane orientation and domain organization as the mannose receptor (Fig. 1), the only difference being that DEC-205 has ten, rather than eight, C-type CRDs. However, the level of sequence identity between the four members of the family is not high: about 30–35% when pairwise comparisons are made. The C-type CRDs of these proteins contain the residues responsible for maintaining the fold of the domain, but many of them are very divergent from the prototype C-type CRDs and probably do not mediate sugar binding.

4.2 Ligand Binding by Members of the Mannose Receptor Family

Of the four proteins, only the mannose receptor has been shown to bind gly-coconjugates in a Ca^{2+}-dependent manner. Examination of the sequence of the endothelial cell receptor suggests that CRDs 1 and 2 of this receptor might bind sugar and Ca^{2+} in a similar manner to other C-type lectins (Wu et al. 1996), but no ligand-binding studies have been reported for this receptor. None of the CRDs of the phospholipase A_2 receptor or of DEC-205 contain the residues involved in sugar- and Ca^{2+}-binding to prototype C-type CRDs. Thus, it is clear that if either of these two proteins bind Ca^{2+} or sugar, it must be by a different mechanism from that seen in other C-type lectins.

Binding and endocytosis of secretory phospholipase A_2 by the phospholi-pase A_2 receptor have been well characterized. Phospholipase A_2 binding is Ca^{2+}-independent, and is not sugar-mediated since these enzymes are not gly-cosylated (Ishizaki et al. 1994; Lambeau et al. 1994, 1995). Binding of some neo-glycoproteins has been demonstrated for the rabbit but not for the human phospholipase A_2 receptor, but the physiological relevance of such binding is unclear. As in the mannose receptor, the major binding site of the phospholi-pase A_2 receptor is in the middle of the polypeptide. CRDs 3–6 contain the phospholipase A_2 binding site, with CRD-5 being the most important (Nicolas et al. 1995).

Like the mannose receptor, DEC-205 is thought to have a role in enhancing uptake of antigens for processing and presentation to T-cells. Antibodies that bind to DEC-205 are presented to T cells more efficiently than non-specific antibodies (Jiang et al. 1995). The presence of multiple potential binding sites in DEC-205, as in the mannose receptor, could allow recognition of a wide variety of antigens. However, no potential physiological ligands have been identified for DEC-205, and no ligand-binding studies have been reported.

4.3 Evolution of the Mannose Receptor Family

It is likely that the four members of the mannose receptor family arose through duplication of a single gene, which itself would have been formed by exon shuf-fling and gene duplication to associate eight C-type carbohydrate-recognition domains with the N-terminal cysteine-rich domain and the fibronectin type II repeat. Sequence analysis and construction of dendrograms have allowed con-clusions to be drawn about the evolution of these receptors (Taylor 1997). It is clear from the dendrograms that duplication of the ancestral gene to form four proteins was a very early event in the evolution of the family. The most likely sequence of events in the evolution of the receptor family, implied by the sequence analysis, is summarized in Fig. 5.

Comparison of the sequences of the CRDs shows that, with the exception of CRDs 4 and 5 of the mannose receptor, each CRD of a given receptor is most similar to the equivalent CRD of the other three receptors. So, for example,

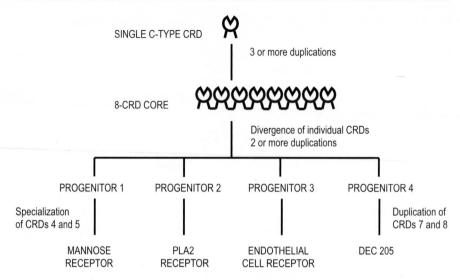

Fig. 5. Steps in the evolution of the mannose receptor family

CRD-1 of the mannose receptor is more closely related to the first CRD of the other three receptors than it is to the other seven CRDs of the mannose receptor. This finding implies that duplication of a primordial CRD occurred to form an eight-CRD core and that the eight CRD-encoding portions then diverged significantly from each other before the gene duplication events to form the four receptors. Three duplication events would be sufficient to form eight CRDs from the primordial CRD, with the first duplication yielding two CRDs, followed by duplication of two CRDs to give four and then duplication of four to give eight. However, it is also possible that the core of eight CRDs resulted from repeated duplication of individual CRDs.

CRDs 4 and 5 of the mannose receptor are more closely related to each other than to any other CRDs. Because these CRDs form the ligand-binding core of the mannose receptor, it is likely that the wide divergence of these domains is related to the evolution of their ligand-binding specificities. The probable origin of the two extra CRDs in DEC 205 can also be discerned from the sequence analysis. CRD-9 of DEC-205 is most similar to CRD-7 of DEC-205 and of the other receptors, while CRD-10 is most closely related to CRD-8 of each of the other receptors. Thus, it is likely that CRDs 9 and 10 of DEC-205 arose through a more recent duplication of the portion of the gene encoding CRDs 7 and 8.

5 Conclusions

The presence of multiple different carbohydrate-recognition domains in the mannose receptor gives it the flexibility needed to recognize the wide variety of oligosaccharides required for its diverse functions. It is clear that selection of ligands by the receptor depends both on the binding properties of individual domains within the protein and on the relative orientation of the multiple binding sites. Our increasing understanding of the structure of the receptor is revealing how specificity and affinity for oligosaccharide ligands is determined by the relative disposition of the binding sites and the orientation of sugars bound in each site.

References

Banyai L, Patthy L (1991) Evidence for the involvement of type II domains in collagen binding by 72-kDa type IV procollagenase. FEBS Lett 282:23–25

Biessen EAL, Noorman F, van Teijlingen ME, Kuiper J, Barret-Bergshoeff M, Rijken DC, van Berkel TJC (1996) Lysine-based cluster mannosides that inhibit ligand binding to the human mannose receptor at nanomolar concentration. J Biol Chem 271:28024–28030

Biessen EAL, van Teijlingen M, Vietsch H, Barret-Bergshoeff MM, Bijsterbosch MK, Rijken DC, van Berkel TJC, Kuiper J (1997) Antagonists of the mannose receptor and the LDL-receptor-related protein dramatically delay the clearance of tissue-type plasminogen activator. Circulation 95:46–52

Chakraborty P, Das PK (1988) Role of mannose/GlcNAc receptors in blood clearance and cellular attachment of *Leishmania donovani*. Mol Biochem Parasitol 28:55–62

Chen WJ, Goldstein JL, Brown MS (1990) NPXY, a sequence often found in cytoplasmic tails, is required for coated pit-mediated internalization of the low density lipoprotein receptor. J Biol Chem 265:3116–3123

Condaminet B, Peguet-Navarro J, Stahl PD, Dalbiez-Gauthier C, Schmitt D, Berthier-Vergnes O (1998) Human epidermal Langerhans cells express the mannose-fucose binding receptor. Eur J Immunol 28:3541–3551

Cool DE, Edgell CJ, Louie GV, Zoller MJ, Brayer GD, MacGillivray RT (1985) Characterization of human blood coagulation factor XII cDNA. Prediction of the primary structure of factor XII and the tertiary structure of beta-factor XIIa. J Biol Chem 260:13666–13676

Drickamer K (1993) Increasing diversity of animal lectin structures. Curr Opin Struct Biol 3:393–400

Drickamer K, Taylor ME (1993) Biology of animal lectins. Annu Rev Cell Biol 9:237–264

Engering AJ, Cella M, Fluisma D, Brockhaus M, Hoefsmit ECM, Lanzavecchia A, Pieters J (1997) The mannose receptor functions as a high capacity and broad specificity antigen receptor in human dendritic cells. Eur J Immunol 27:2417–2425

Ezekowitz RAB, Sastry K, Bailly P, Warner A (1990) Molecular characterization of the human macrophage mannose receptor: demonstration of multiple carbohydrate recognition-like domains and phagocytosis of yeasts in Cos-1 cells. J Exp Med 172:1785–1794

Ezekowitz RAB, Williams DJ, Koziel H, Armstrong MYK, Warner A, Richards FF, Rose RM (1991) Uptake of *Pneumocystis carinii* mediated by the macrophage mannose receptor. Nature 351:155–158

Fiete D, Beranek MC, Baenziger JU (1997) The macrophage/endothelial cell mannose receptor cDNA encodes a protein that binds oligosaccharides terminating with SO_4-4-GalNAcβ1, 4GlcNAcβ or Man at independent sites. Proc Natl Acad Sci USA 94:11256–11261

Fiete D, Beranek MC, Baenziger JU (1998) A cysteine-rich domain of the "mannose" receptor mediates GalNAc-4-SO$_4$ binding. Proc Natl Acad Sci USA 95:2089–2093

Graves BJ, Crowther RL, Chandran C, Rumberger JM, Li-S, Huang KS, Presky-DH, Familletti PC, Wolitzky BA, Burns DK (1994) Insight into E-selectin/ligand interaction from the crystal structure and mutagenesis of the lec/EGF domains. Nature 367:532–538

Haltiwanger RS, Hill RL (1986) Isolation of a rat alveolar macrophage lectin. J Biol Chem 261:7440–7444

Hitchen PG, Mullin NP, Taylor ME (1998) Orientation of sugars bound to the principal C-type carbohydrate-recognition domain of the macrophage mannose receptor. Biochem J 333:601–608

Iobst ST, Wormald MR, Weis WI, Dwek RA, Drickamer K (1994) Binding of sugar ligands to Ca (2+)-dependent animal lectins. I. Analysis of mannose binding by site-directed mutagenesis and NMR. J Biol Chem 269:15505–15511

Ishizaki J, Hanasaki K, Higashino K, Kishino J, Kikuchi N, Ohara O, Arita H (1994) Molecular cloning of pancreatic group I phospholipase A$_2$ receptor. J Biol Chem 269:5897–5904

Jiang W, Swiggard WJ, Heufler C, Peng M, Mirza A, Steinman RM, Nussenzweig MC (1995) The receptor DEC-205 expressed by dendritic cells and thymic epithelial cells is involved in antigen processing. Nature 375:151–155

Kornblihtt AR, Umezawa K, Vibe-Pedersen K, Baralle FE (1985) Primary structure of human fibronectin: differential splicing may generate at least 10 polypeptides from a single gene. EMBO J 4:1755–1759

Kruskal A, Sastry K, Warner A, Mathieu CE, Ezekowitz RAB (1992) Phagocytic chimeric receptors require both transmembrane and cytoplasmic domains from the mannose receptor. J Exp Med 176:1673–1680

Lambeau G, Ancian P, Barhanin J, Lazdunski M (1994) Cloning and expression of a membrane receptor for secretory phospholipases A$_2$. J Biol Chem 269:1575–1578

Lambeau G, Ancian P, Mattei M-G, Lazdunski M (1995) The human 180-kDa receptor for secretory phospholipases A$_2$. J Biol Chem 270:8963–8970

Lennartz MR, Cole FS, Shepherd VL, Wileman TE, Stahl PD (1987) Isolation and characterization of a mannose-specific endocytosis receptor from human placenta. J Biol Chem 262:9943–9944

Lennartz MR, Cole FS, Stahl PD (1989) Biosynthesis and processing of the mannose receptor in human macrophages. J Biol Chem 264:2385–2390

Lobel P, Dahms NM, Kornfeld S (1988) Cloning and sequence analysis of the cation-independent mannose 6-phosphate receptor. J Biol Chem 263:2563–2570

Mullin NP, Hall KT, Taylor ME (1994) Characterization of ligand binding to a carbohydrate-recognition domain of the macrophage mannose receptor. J Biol Chem 269:28405–28413

Mullin NP, Hitchen PG, Taylor ME (1997) Mechanism of Ca^{2+}- and monosaccharide-binding to a C-type carbohydrate-recognition domain of the macrophage mannose receptor. J Biol Chem 272:5668–5681

Ng K K-S, Drickamer K, Weis WI (1996) Structural analysis of monosaccharide recognition by rat liver mannose-binding protein. J Biol Chem 271:663–674

Nicolas J-P, Lambeau G, Lazdunski M (1995) Identification of the binding domain for secretory phospholipases A$_2$ on their M-type 180-kDa membrane receptor. J Biol Chem 270:28869–28873

Otter M, Barrett-Bergshoeff MM, Rijken DC (1991) Binding of tissue-type plasminogen activator by the mannose receptor. J Biol Chem 266:13931–13935

Otter M, Zockova P, Kuiper J, van-Berkel TJ, Barrett-Bergshoeff MM, Rijken DC (1992) Isolation and characterization of the mannose receptor from human liver potentially involved in the plasma clearance of tissue-type plasminogen activator. Hepatology 16:54–59

Prigozy TI, Sieling PA, Clemens D, Stewart PL, Behar SM, Porcelli SA, Brenner MB, Modlin RL, Kronnenberg M (1997) The mannose receptor delivers lipoglycan antigens to endosomes for presentation to T cells by CD1b molecules. Immunity 6:187–197

Sallusto F, Cella M, Danieli C, Lanzavecchia A (1995) Dendritic cells use macropinocytosis and the mannose receptor to concentrate macromolecules in the major histocompatibility complex class II compartment: downregulation by cytokines and bacterial products. J Exp Med 182:389–400

Schlesinger LS (1993) Macrophage phagocytosis of virulent but not attenuated strains of *Mycobacterium tuberculosis* is mediated by mannose receptors in addition to complement receptors. J Immunol 150:2920–2930

Schlesinger P, Rodman JS, Frey M, Lang S, Stahl P (1976) Clearance of lysosomal hydrolases following intravenous infusion. The role of the liver in the clearance of ß-glucuronidase and N-acetyl-β-D-glucosaminidase. Arch Biochem Biophys 177:606–614

Schlesinger P, Doebber TW, Mandell BF, White R, DeSchryver C, Rodman JS, Miller MJ, Stahl P (1978) Plasma clearance of glycoproteins with terminal mannose and GlcNAc by liver non-parenchymal cells. Biochem J 176:103–109

Shepherd VL, Hoidal JR (1990) Clearance of neutrophil-derived myeloperoxidase by the macrophage mannose receptor. Am J Respir Cell Mol Biol 2:335–340

Shepherd VL, Cambell TJ, Senior RM, Stahl PD (1982) Characterization of the mannose/fucose receptor on human mononuclear phagocytes. J Retic Endothel Soc 32:423–421

Shepherd VL, Tarnowski BI, McLaughlin BJ (1991) Isolation and characterization of a mannose receptor from human pigment epithelium. Invest Opthalmol Vis Sci 32:1779–17784

Smedsrod B, Melkko J, Risteli L, Risteli J (1990) Circulating C-terminal propeptide of type I collagen is cleared mainly via the mannose receptor in liver endothelial cells. Biochem J 271:345–350

Stahl PD, Schlesinger PH (1980) Receptor mediated-pinocytosis of mannose/GlcNAc-terminated glycoproteins and lysosomal enzymes by macrophages. Trends Biochem Sci 5:194–196

Tan MCAA, Momaas AM, Drihout JW, Jordens R, Onderwater JJM, Verwoerd D, Mulder AA, van der Heiden AN, Scheidegger D, Oomen LCJM, Ottenhoff THM, Tulp A, Neefjes JJ, Koning F (1997) Mannose receptor-mediated uptake of antigens strongly enhances HLA class II-restricted antigen presentation by cultured dendritic cells. Eur J Immunol 27:2426–2435

Taylor ME (1997) Evolution of a family of receptors containing multiple C-type carbohydrate-recognition domains. Glycobiology 7:R5–R8

Taylor ME, Drickamer K (1993) Structural requirements for high affinity binding of complex ligands by the macrophage mannose receptor. J Biol Chem 268:399–404

Taylor ME, Conary JT, Lennarz MR, Stahl PD, Drickamer K (1990) Primary structure of the mannose receptor contains multiple motifs resembling carbohydrate-recognition domains. J Biol Chem 265:12156–12162

Taylor ME, Bezouska K, Drickamer K (1992) Contribution to ligand binding by multiple carbohydrate-recognition domains in the macrophage mannose receptor. J Biol Chem 267:1719–1726

Weis WI, Drickamer K (1996) Structural basis of lectin-carbohydrate-recognition. Annu Rev Biochem 65:441–473

Weis WI, Taylor ME, Drickamer K (1998) The C-type lectin superfamily in the immune system. Immunol Rev 163:19–34

Weis WI, Drickamer K, Hendrickson WA (1992) Structure of a C-type mannose-binding protein complexed with an oligosaccharide. Nature 360:127–134

Wileman T, Boshans R, Stahl PD (1984) Uptake and transport of mannosylated ligands by alveolar macrophages. Studies on ATP-dependent receptor ligand dissociation. J Biol Chem 260:7387–7393

Wu K, Yuan J, Lasky LA (1996) Characterization of a novel member of the macrophage mannose receptor type C lectin family. J Biol Chem 271:21323–21330

The Man/GalNAc-4-SO$_4$-Receptor has Multiple Specificities and Functions

Alison Woodworth[1] and Jacques U. Baenziger[1]

1 Introduction

Carbohydrate-specific recognition by receptors contributes to a variety of different biological processes. These include: the folding and assembly of glycoproteins in the endoplasmic reticulum (Helenius 1994; Williams 1995; Trombetta and Helenius 1998), transport of glycoproteins to lysosomes (Kornfeld 1990, 1992), control of the circulatory half-life of individual glycoproteins (Drickamer 1991; Baenziger et al. 1992; Drickamer and Taylor 1993), endocytosis (Ashwell and Harford 1982; Drickamer and Taylor 1993), phagocytosis (Stahl and Ezekowitz 1998), cellular recognition during coagulation (McEver and Cummings 1997; Rosenberg et al. 1997), inflammation (Rosen and Bertozzi 1994; McEver et al. 1995; Lowe 1997), and development (Ioffe and Stanley 1994; Metzler et al. 1994; Chui et al. 1997). We recently described the first example of a carbohydrate-specific receptor, the Man/GalNAc-4-SO$_4$ receptor (Fiete et al. 1998), that utilizes two distinct regions with different structural motifs to bind completely different carbohydrate structures. Furthermore, different cell types, such as macrophages and hepatic endothelial cells, express forms of the Man/GalNAc-4-SO$_4$ receptor that differ with respect to their carbohydrate specificity. Macrophages express a mannose (Man) specific form of the receptor (Man-receptor), while hepatic endothelial cells express an N-acetylgalactosamine-4-SO$_4$ (GalNAc-4-SO$_4$) specific form of the receptor (GalNAc-4-SO$_4$-receptor; Fiete and Baenziger 1997; Fiete et al. 1997). The regulated expression of different specificities by the Man/GalNAc-4-SO$_4$-receptor indicates that the Man-receptor and the GalNAc-4-SO$_4$-receptor have different biological functions. Asparagine-linked carbohydrate structures terminating with the sequence SO$_4$–4GalNAcβ1,4GlcNAcβ1,2Manα (S4GGnM) are present on highly restricted populations of glycoproteins, including the pituitary glycoprotein hormones LH and TSH (Green and Baenziger 1988a,b; Baenziger and Green 1991; Stockell Hartree and Renwick 1992; Thotakura and Blithe 1995). The synthesis of these sulfated oligosaccharides is protein specific, and is both hormonally and developmentally regulated (S.M. Dharmesh and J.U. Baenziger, unpubl. obs.; Dharmesh and Baenziger 1993; Manzella et al. 1997). Further, recognition of terminal GalNAc-4-SO$_4$ on

[1] Department of Pathology, Washington University School of Medicine, 660 S. Euclid Ave., St. Louis, MO 63110

Results and Problems in Cell Differentiation, Vol. 33
Paul R. Crocker (Ed.): Mammalian Carbohydrate Recognition Systems
© Springer-Verlag Berlin Heidelberg 2001

the oligosaccharides of LH by the GalNAc-4-SO$_4$-receptor residing in hepatic endothelial cells (Fiete et al. 1991) plays a critical role in vivo by controlling the circulatory half-life of LH at key points during the ovulatory cycle (Baenziger et al. 1992; Baenziger 1996; Manzella et al. 1996). In contrast, recognition of carbohydrates terminating with Man, Fucose (Fuc), or N-acetylglucosamine (GlcNAc) by the Man-receptor, present on terminally differentiated macrophages, is thought to be essential for phagocytosis of pathogens such as yeasts, bacteria, and viruses (Fraser et al. 1998; Stahl and Ezekowitz 1998). Thus, the Man/GalNAc-4-SO$_4$-receptor, in the form of the Man-receptor or the GalNAc-4-SO$_4$-receptor, may fulfill distinct roles that reflect its ligand specificity as well as the setting in which it is expressed.

2 Oligosaccharides Terminating with the Sequence SO$_4$–4GalNAcβ1,4GlcNAcβ1,2Manα (S4GGnM) Are Found on the Pituitary Glycoprotein Hormones of All Vertebrates

Asn-linked oligosaccharides terminating with β1,4-linked GalNAc-4-SO$_4$, rather than the more commonly encountered sequence of sialic acidα2,3 or α2,6Galβ1,4, were first described in human, bovine, and ovine pituitary glycoprotein hormones (Green and Baenziger 1988a,b; Baenziger and Green 1991; Stockell Hartree and Renwick 1992; Manzella et al. 1996). We later demonstrated that β1,4-linked GalNAc-4-SO$_4$ is present on oligosaccharides of the pituitary glycoprotein hormones of all vertebrate species, including the α and β subunits of the salmon hormone GTH II (Manzella et al. 1995). More recently, it has been established that terminal β1,4-linked GalNAc-4-SO$_4$ can also be found on O-linked oligosaccharides (Siciliano et al. 1994). However, only a small number of glycoproteins bearing terminal β1,4-linked GalNAc-4-SO$_4$ on N or O-linked oligosaccharides have thus far been described (Van den Eijnden et al. 1995). The addition of terminal GalNAc-4-SO$_4$ to N-linked oligosaccharides requires the sequential action of two transferases: a protein-specific N-acetylgalactosaminyl-transferase (GalNAc-transferase) and a GalNAcβ1,4GlcNAcβ1,2Man-specific GalNAc-4-sulfotransferase. The protein-specific GalNAc-transferase recognizes a cluster of basic amino acid residues on the α subunit of the glycoprotein hormones. The basic residues critical for recognition as well as the glycosylation sites are conserved in the glycoprotein hormone α subunits of all vertebrates (Manzella et al. 1995). In addition, both transferases are expressed in the pituitaries of all vertebrate species tested from fish to man (Manzella et al. 1995). Thus, all three components required for modification of the oligosaccharides on glycoprotein hormones with terminal GalNAc-4-SO$_4$ are evolutionarily conserved.

Expression of the GalNAc-transferase and the GalNAc-4-sulfotransferase is highly regulated in the gonadotrophs that synthesize LH. Both activities are initially detected in the pituitaries of rats at post-natal day 21. At this time, both LH synthesis and release into the circulation increase dramatically (S.M.

Dharmesh and J.U. Baenziger, unpubl. obs.). Further, expression of the transferases responsible for modifying LH with GalNAc-4-SO$_4$ is also regulated by estrogen in the adult female. Thus, GalNAc-transferase and GalNAc-4-sulfotransferase levels rise and fall concomitantly with the levels of LH in the gonadotroph in response to estrogen (Dharmesh and Baenziger 1993). As a result, LH is selectively and efficiently modified with GalNAc-4-SO$_4$ throughout the ovulatory cycle.

Conservation of the structure of the sulfated oligosaccharides on vertebrate glycoprotein hormones from fish to man indicates that these structures confer a biological function that has been retained during the evolution of vertebrate species. This has required the co-ordinate expression in all vertebrates of: (1) the recognition determinant utilized by the protein-specific GalNAc-transferase, (2) the GalNAc-transferase, and (3) the GalNAc-4-sulfotransferase. The highly regulated expression of these transferases in gonadotrophs further supports the conclusion that oligosaccharides terminating with GalNAc-4-SO$_4$ play an integral role in reproduction. We have postulated that one important biological role of the terminal GalNAc-4-SO$_4$ structure, present on oligosaccharides of LH and TSH, is control of the hormone's circulatory half-life.

3 Terminal β1,4-Linked GalNAc-4-SO$_4$ Determines the Circulatory Half-life of LH

Forms of LH that differ in their pattern of terminal glycosylation can be obtained by expressing recombinant hormones in cells, like CHO (Chinese hamster ovary) cells, that have a complement of glycosyltransferases distinct from those expressed in gonadotrophs (see Table 1). Previous studies comparing native LH (terminal GalNAc-4-SO$_4$) and recombinant LH expressed in CHO cells (terminal sialic acidα2,3Gal) with desulfated LH (terminal GalNAc) and with desialylated LH (terminal Gal) revealed that the terminal sugar has only a modest impact on binding to and activation of the LH receptor (Smith et al. 1990). This implies that activation of the LH receptor, a G-protein-coupled receptor, is not dependent on the pattern of terminal glycosylation. In contrast, as shown in Table 1, terminal glycosylation has a dramatic impact on the circulatory half-life of LH (Baenziger et al. 1992). Following injection, recombinant LH is removed from the circulation at a rate of 1.7%/min. whereas native LH is removed at a rate of 7.3%/min. The rapid removal of native LH from the blood reflects the action of a GalNAc-4-SO$_4$-specific receptor that is expressed in hepatic endothelial cells (Fiete et al. 1991). Removal of either the terminal SO$_4$ or the terminal sialic acid to expose β1,4-linked GalNAc or Gal, respectively, results in clearance at a rate of 35%/min. This rapid clearance is mediated by the asialoglycoprotein receptor located in hepatocytes (Baenziger et al. 1992). Like LH, native and recombinant TSH expressed in CHO cells, which also bear oligosaccharides terminating with GalNAc-4-SO$_4$ and sialic acid respectively, display similar properties with respect to both the site and rate of clear-

Table 1. The site and rate of clearance of LH from the blood is determined by its pattern of terminal glycosylation. Differentially glycosylated forms of native and recombinant ^{125}I-labeled LH, isolated from pituitaries or CHO cells respectively, were injected into rats via the carotid artery and the rate of clearance monitored at specific time points. Terminal sulfate and sialic acid were removed enzymatically as described (Baenziger et al. 1992). The site of clearance was established by determining the amount of radiolabel present in each tissue at necropsy

Source	Terminal Sugars	Clearance Rate (%/min.)	Site of Clearance
CHO cells	Siaα2,6Galβ1,4GlcNAc-	1.7	Kidney
Gonadotrophs	SO_4-4-GalNAcβ1,4GlcNAc-	7.3	Liver (endothelial cells)
CHO cells	Galβ1,4GlcNAc-	35	Liver (hepatocytes)
Gonadotrophs	GalNAcβ1,4GlcNAc-	35	Liver (hepatocytes)

ance (Szkudlinski et al. 1995; Thotakura and Blithe 1995). This suggests that the presence of terminal GalNAc-4-SO_4 on TSH may also have important biological consequences.

We have proposed that modification of LH oligosaccharides with terminal GalNAc-4-SO_4 is essential for the optimal activation of the LH receptor, present in pre-ovulatory follicles, that induces ovulation and corpus luteum formation. LH induces ovulation by binding to the LH receptor, a seven-transmembrane G-protein-coupled receptor, and stimulating adenylyl cyclase activity (Segaloff and Ascoli 1993). β-arrestin-1 then binds the activated receptor, inducing desensitization of LH receptor induced adenylyl cyclase activity (Mukherjee et al. 1999). Upon desensitization, the LH receptor, with its bound ligand, is internalized and replaced with an unoccupied receptor on the cell surface (Dohlman et al. 1991; Lefkowitz 1993, 1998; Segaloff and Ascoli 1993). Thus, a key requirement for efficient activation of the LH receptor is the episodic rise and fall of circulating LH levels. One contributing factor to the dynamic circulatory LH levels, is the highly regulated release of LH from dense core granules in the pituitary. Yet regulated release of the hormone is not sufficient to produce the episodic rise and fall of LH seen in the blood. A relatively short circulatory half-life, characteristic of native LH modified with sulfated oligosaccharides, is also required to obtain the periodic rise and fall in LH levels needed to fully activate the LH receptor, which is G-protein coupled. Thus, removal of circulatory LH from the blood by the GalNAc-4-SO_4-specific receptor of hepatic endothelial cells at a highly specific rate results in optimal activation of the LH receptor in vivo (Fig. 1).

4 The GalNAc-4-SO_4-Receptor: A Receptor with Multiple Carbohydrate Specificities and Functions

Using isolated hepatic endothelial and parenchymal cells, we demonstrated that an abundant receptor (500,000 per cell) resides on the cell surface of the

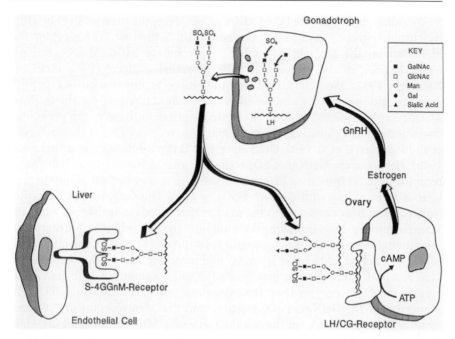

Fig. 1. The role of GalNAc-4-SO₄ and the GalNAc-4-SO₄-receptor in the ovulatory cycle. LH, when synthesized in gonadotrophs of the pituitary, is modified with GalNAc-4-SO₄. The LH is stored in dense core granules of gonadotrophs and is episodically secreted upon stimulation by the gonadotrophin releasing hormone (GnRH). Once in the blood-stream, the LH is transported to the ovary where it binds to and activates the G-protein-coupled LH receptor, resulting in the production of cAMP and ultimately increased levels of estrogen production. Binding to the LH receptor reflects predominantly protein-protein interactions. Following binding, cyclic-AMP production, induced by the occupied LH receptor, is down-regulated by β1-arrestin. The occupied LH receptor is then internalized and replaced by unoccupied receptor. The excess LH remaining in the circulation is removed by the GalNAc-4-SO₄-receptor (S-4GGnM-Receptor) in hepatic endothelial cells prior to the next episode of stimulated LH release from the pituitary

endothelial cells in liver. This receptor binds to the sulfated oligosaccharides on LH with an apparent k_d of 2.7×10^{-7}M (Fiete et al. 1991). Glycoproteins bound by the GalNAc-4-SO₄-receptor are rapidly internalized and then degraded in the lysosome. Like a number of other receptors that transport ligands to lysosomes, the GalNAc-4-SO₄-receptor requires a pH above 5.0 for binding. Thus, bound ligands dissociate in the acidic environment of the late endosome. However, despite the pH requirement, GalNAc-4-SO₄ binding is not dependent on divalent cations (Fiete et al. 1991). The linkage of the sulfate to the underlying GalNAc is, however, critical for recognition. Bovine albumin (BSA) chemically modified with multiple trisaccharides consisting of SO₄-4-GalNAcβ1,4GlcNAcβ1,2Manα (S4GGnM-BSA) is rapidly removed from the blood following injection, whereas BSA modified with multiple trisaccharides consisting of SO₄-3-GalNAcβ1,4GlcNAcβ1,2Manα (S3GGnM-BSA) is not cleared from the circulation (Fiete et al. 1991).

We subsequently purified the GalNAc-4-SO$_4$-receptor from rat liver by affinity chromatography using LH immobilized on Sepharose. The receptor was identified on the basis of its ability to precipitate S4GGnM-BSA, but not S3GGnM-BSA, in the presence of 10% polyethyleneglycol (PEG; Fiete and Baenziger 1997). We determined that the amino terminal sequence, peptide maps, peptide sequences, and the molecular weight obtained for the GalNAc-4-SO$_4$-receptor isolated from rat liver are identical to those of the previously characterized and cloned macrophage mannose receptor (Man receptor;Taylor et al. 1990; Harris et al. 1992; Drickamer and Taylor 1993; Stahl and Ezekowitz 1998). However, the GalNAc-4-SO$_4$-receptor isolated from liver and the Man-receptor isolated from lung display marked differences in their ability to bind ligands terminating with GalNAc-4-SO$_4$ or Man. The GalNAc-4-SO$_4$-receptor isolated from liver does not bind to Man immobilized on Sepharose, while the Man-receptor isolated from lung does not bind to LH-Sepharose. Furthermore, even though the GalNAc-4-SO$_4$-receptor precipitates Man-BSA, the interaction is much less efficient because 10- to 20-fold more GalNAc-4-SO$_4$-receptor than Man-receptor is required to precipitate the same amount of Man-BSA. Conversely, the Man-receptor from lung does not precipitate S4GGnM-BSA. The properties of the GalNAc-4-SO$_4$-receptor and the Man-receptor are summarized in Table 2. Thus, even though the GalNAc-4-SO$_4$-receptor and the Man-receptor appear to be antigenically identical and have the same amino acid sequence, they differ in their ability to bind oligosaccharides terminating with GalNAc-4-SO$_4$ and Man, GlcNAc, or Fuc. This raises the possibility that some form of cell-specific post-translational modification determines the specificity of this receptor.

Remarkably, the cDNA for the Man-receptor expressed in CHO cells produced a receptor able to mediate the binding and internalization of both S4GGnM-BSA and Man-BSA (Fiete et al. 1997). This confirmed that the GalNAc-4-SO$_4$-receptor is encoded by the same gene as the macrophage Man receptor and that their different specificities reflect some type of post-

Table 2. The GalNAc-4-SO$_4$-receptor and the Man-receptor differ in their ability to bind to ligands terminating in GalNAc-4-SO$_4$ or Mannose. The GalNAc-4-SO$_4$-receptor, isolated from liver, the Man-receptor isolated from lung, and the Fc chimera purified from CHO cells were examined for their ability to bind to LH-sepharose or Man-sepharose and to precipitate Man-BSA or S4GGnM-BSA. The Fc region of human IgG1 has been substituted for the transmembrane and cytosolic domains of the Man/GalNAc-4-SO$_4$-receptor in the Fc chimera. The efficiency of binding is indicated for the precipitation assay by +, +/−, and − for S4GGnM-BSA and Man-BSA, respectively

Ligand binding properties	GalNAc-4-SO$_4$-receptor	Man-receptor	Man/GalNAc-4-SO$_4$ Fc chimera
LH-sepharose	+	−	+
Man-sepharose	−	+	+
S4GGnM-BSA	+	−	+
Man-BSA	+/−	+	+

translational event. We have utilized the term Man/S4GGnM-receptor or Man/GalNAc-4-SO$_4$-receptor to indicate the dual specificity of this receptor (Fiete et al. 1997).

The Man/GalNAc-4-SO$_4$-receptor is a member of the 'multi-CRD receptor family' of C-type lectins (Drickamer and Taylor 1993; Taylor 1997). To date this family consists of four structurally related receptors: the Man/GalNAc-4-SO$_4$-receptor (Taylor et al. 1990; Fiete and Baenziger 1997; Fiete et al. 1997), DEC-205 (Jiang et al. 1995), the phospholipase A$_2$ receptor (PLA$_2$ receptor) (Lambeau et al. 1994), and a 'novel' lectin (Wu et al. 1996). Each receptor contains a terminal Cysteine-rich (Cys-rich) domain, a fibronectin type II repeat, eight to ten carbohydrate recognition domains (CRDs), a transmembrane domain (TM), and a short carboxy-terminal, cytosolic domain (see Fig. 2). The CRDs present in these receptors are homologous to those in other C-type lectins that display Ca^{2+}-dependent binding of carbohydrates (Drickamer 1996; Wu et al. 1996; Taylor 1997; Stahl and Ezekowitz 1998). Yet carbohydrate binding has only been demonstrated for Man/GalNAc-4-SO$_4$ and PLA$_2$ receptors (Di Jeso et al. 1992; Drickamer and Taylor 1993; Lambeau et al. 1994; Fiete et al. 1997). For example, binding of PLA$_2$ to its receptor is inhibited by BSA conjugates bearing Man, GlcNAc or Gal even though PLA$_2$ itself is not a glycoprotein (Lambeau et al. 1994; Nicolas et al. 1995). The biological significance of this inhibition is not yet known. Further, CRDs 4–8 of the Man/GalNAc-4-SO$_4$-receptor display a broad ligand-binding specificity. In particular, these CRDs can bind glycoconjugates with terminal Man, GlcNAc, or Fuc (Taylor et al. 1990; Taylor and Drickamer 1993; Fiete et al. 1998). Yet glycoconjugates that have multiple Man termini are the most effective ligands for CRDs 4–8 of the Man/GalNAc-4-SO$_4$-receptor, suggesting that multivalent interactions contribute to binding specificity and avidity (Taylor et al. 1992). C-type lectins containing a single CRD also bind multivalent ligands by forming oligomeric complexes that contain multiple CRDs (Drickamer 1993; Drickamer and Taylor 1993; Iida et al. 1999).

The regions of the Man/GalNAc-4-SO$_4$-receptor accounting for Man and GalNAc-4-SO$_4$ binding were identified using a chimeric protein in which the transmembrane and cytosolic domains of the Man/GalNAc-4-SO$_4$-receptor were replaced by the Fc region of human IgG1. The Man/GalNAc-4-SO$_4$-Fc chimera is synthesized and secreted as a soluble, dimeric protein when expressed in CHO cells. The chimera binds to both Man-Sepharose and LH-Sepharose, and precipitates Man-BSA and S4GGnM-BSA in the presence of PEG (Table 2). Utilizing deletion mutants of this Fc-chimera, we demonstrated that the N-terminal Cys-rich domain is both necessary and sufficient to mediate GalNAc-4-SO$_4$-specific binding (Fiete et al. 1998). Thus, the GalNAc-4-SO$_4$ binding site within the Cys-rich domain, and the Man binding domains located in CRDs 4–8, are physically distinct. The difference in binding sites explains the lack of competition between ligands containing terminal GalNAc-4-SO$_4$ and those containing terminal Man, GlcNAc, or Fuc (Fiete and Baenziger 1997; Fiete et al. 1998). It has not yet been determined whether a ligand bearing

Man/GalNAc-4-SO$_4$-Receptor

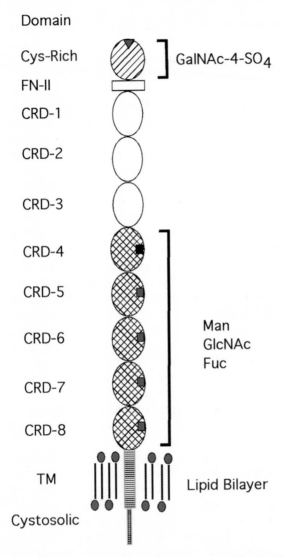

Fig. 2. Structure of Man/GalNAc-4-SO$_4$-receptor. The Man/GalNAc-4-SO$_4$-receptor consists of a Cys-rich domain, a fibronectin type II repeat domain (*Fn-II*), eight carbohydrate recognition domains (*CRDs*), a transmembrane domain (*TM*), and a cytosolic domain. The Cys-rich domain mediates binding of GalNAc-4-SO$_4$ while CRDs 4–8 mediate binding of Man, GlcNAc, and Fuc. The *filled triangle* and *filled squares* indicate the GalNAc-4-SO$_4$ and Man/GlcNAc/Fuc binding sites, respectively

terminal Man and a ligand with terminal GalNAc-4-SO$_4$ containing oligosaccharides are capable of binding simultaneously to their respective binding domains on the Man/GalNAc-4-SO$_4$-receptor.

The Cys-rich domain of the Man/GalNAc-4-SO$_4$-receptor represents a new carbohydrate binding motif that is distinct from previously described motifs present in the C-type lectins (Drickamer and Taylor 1993), I-type lectins (Powell and Varki 1995; Kelm et al. 1996), S-type lectins (Drickamer 1988), and P-type lectins (Drickamer and Taylor 1993; Wu et al. 1996; Taylor 1997). The other receptors in this family of proteins with multiple CRDs also have Cys-rich domains that are structurally related to that of the Man/GalNAc-4-SO$_4$-receptor. Although no binding activity has yet been characterized for the Cys-rich domains of other members of the 'multi CRD receptor family', they may also bind carbohydrates.

The Man/GalNAc-4-SO$_4$-receptor has the potential in vivo to bind two unrelated ligands through two separate binding sites. This differential binding is regulated so that, depending on the cell or tissue type, one binding specificity or the other is expressed. The molecular basis for the two forms of the Man/GalNAc-4-SO$_4$-receptor that are isolated from liver and lung, respectively, remains to be established. We have developed methods that allow us to examine frozen tissue sections for binding of fluorescently labeled ligands bearing terminal Man or GalNAc-4-SO$_4$. Under these conditions, alveolar macrophages bind those ligands bearing terminal Man but not those bearing terminal GalNAc-4-SO$_4$. Conversely, hepatic endothelial cells bind those ligands with terminal GalNAc-4-SO$_4$, but not those with terminal Man. We have found that the receptor also displays differential binding of Man- and GalNAc-4-SO$_4$-bearing ligands in other tissues. Thus, this pattern of ligand binding by two forms of the same receptor is regulated in a cell-specific manner (Y.L. Mi and J.U. Baenziger, unpublished observation). This regulated expression of different binding specificities indicates that the Man/GalNAc-4-SO$_4$-receptor has multiple functions that reflect the specificity of the receptor being expressed, as well as the structure of the carbohydrate present on the target glycoproteins.

5 Functional Significance of the Man/GalNAc-4-SO$_4$ Receptor

Both forms of the Man/GalNAc-4-SO$_4$ receptor mediate uptake and transport of circulating glycosylated ligands to lysosomes where they are degraded. When expressed by terminally differentiated macrophages, the Man receptor mediates phagocytosis of particulate ligands, including yeasts, bacteria, and viruses. The detailed structural features of the carbohydrates that account for binding and phagocytosis of pathogens by the Man receptor are not known. Yeast mannan, a large carbohydrate with multiple terminal Man residues, inhibits Man-receptor pathogen interaction. Thus, recognition of various potentially pathogenic organisms by this multifunctional receptor is most likely mediated through CRDs 4–8, the Man-specific binding site. Further, the repeating character of the carbohydrates found on the surface of pathogenic organisms may provide a feature that is selectively recognized by the Man receptor. The term 'pattern recognition' has been proposed to describe this form of recognition (Stahl and Ezekowitz 1998). In contrast, relatively few endogenous glycoproteins of mammals bear oligosaccharides with terminal Man, Fuc, or GlcNAc in a form that can be recognized by the Man receptor. As a result, the macrophage Man receptor may contribute to innate immunity by selectively recognizing carbohydrates on pathogens. Levels of the Man receptor are reduced in individuals infected with HIV and it has been proposed that this may account for increased susceptibility to infection with *Pnemocystis carinii* (Koziel et al. 1993, 1998). However, confirmation of the role of the macrophage Man receptor in innate immunity awaits an in vivo genetic demonstration.

The $GalNAc-4-SO_4$-specific form of the $Man/GalNAc-4-SO_4$-receptor found in hepatic endothelial cells has a distinct function from that of the Man-specific form in macrophages. The $GalNAc-4-SO_4$-receptor mediates the clearance of LH from the blood at a rate that precisely mirrors the requirement for episodic activation of the G-protein-coupled LH receptor in the ovary. Episodic activation is required because cyclic-AMP production ceases following $\beta 1$-arrestin binding to the activated LH receptor and internalization (Mukherjee et al. 1999). Since the same sulfated oligosaccharides are present on glycoprotein hormones of all vertebrates, it is likely that the same or a closely related form of the $Man/GalNAc-4-SO_4$-receptor is expressed by all vertebrate species and that it plays an integral role in reproduction.

The $GalNAc-4-SO_4$-receptor may bind glycoproteins in addition to the glycoprotein hormones LH and TSH. A number of glycoproteins that bear oligosaccharides with terminal $GalNAc-4-SO_4$ have been described (Skelton et al. 1992; Smith et al. 1992; Siciliano et al. 1994; Bergwerff et al. 1995; Green and Baenziger 1988a,b; Vanrooijen et al. 1998), and many of these glycoproteins contain potential GalNAc-transferase-specific peptide sequences (Skelton et al. 1992; Smith et al. 1992; Hooper et al. 1995; Manzella et al. 1997). Other glycoproteins that potentially contain carbohydrates with terminal $GalNAc-4-SO_4$ have been detected in tissue sections utilizing a chimeric protein consisting of the Cys-rich domain of the $Man/GalNAc-4-SO_4$-receptor and the constant region of human IgG1 (Cys-rich-Fc) and/or with a monoclonal antibody specific for terminal $GalNAc-4-SO_4$ (Smith et al. 1993; Martinez-Pomares et al. 1996; Martinez-Pomares and Gordon 1999; Y.L. Mi and J.U. Baenziger, unpubl. obs.). For example, metallophilic macrophages present in the marginal zone of the spleen and subcapsular macrophages in lymph nodes express glycoproteins reactive with the Cys-rich-Fc (Martinez-Pomares et al. 1996). The number and location of the Cys-rich-Fc reactive cells is altered upon antigenic stimulation, and it has been hypothesized that the Cys-rich region of the $Man/GalNAc-4-SO_4$-receptor may play a role in antigen transport and immunity (Martinez-Pomares et al. 1996; Martinez-Pomares and Gordon 1999). We have detected both Cys-rich-Fc reactive ligands and the Man/GalNAc-4-SO$_4$-receptor in a number of other settings (Y.L. Mi, A. Woodworth, J.U. Baenzinger, unpubl. obs.). This raises the possibility that different combinations of GalNAc-4-SO$_4$-containing proteins and the Man/GalNAc-4-SO$_4$-receptor have additional biological functions in these other settings.

A soluble form of the Man/GalNAc-4-SO$_4$-receptor is present in mouse serum (Martinez-Pomares et al. 1998). Cleavage by a metalloprotease between CRD 8 and the transmembrane domain accounts for production of soluble receptor in cultured macrophages. Neither the function nor the origin of the soluble form of the Man/GalNAc-4-SO$_4$-receptor in serum is known. Like the receptor isolated from lung, the soluble receptor found in serum binds to Man-Sepharose. However, its ability to bind to glycoproteins bearing terminal GalNAc-4-SO$_4$ has not been assessed. Thus, the origin of the soluble form, whether from hepatic endothelial cells or from macrophages, is not yet known.

The existence of multiple forms for the same receptor, that differ in specificity and are expressed by highly specialized cells, makes it essential to determine the molecular basis for these differences, the regulation of the expression of these different forms, and the in vivo functions of this receptor in different settings. There are several potential mechanisms that could account for the differences in binding activities of the Man/GalNAc-4-SO$_4$-receptor when expressed in different cells. For instance, individual domains of the receptor may be activated or inactivated by post-translational modifications, protease cleavage, or improper folding. Interaction with another membrane or cytosolic protein may be required to induce either Man or GalNAc-4-SO$_4$-specific binding by the receptor. Alternatively, there could be cell-specific expression of endogenous ligands for one of the binding sites of the Man/GalNAc-4-SO$_4$-receptor which block binding to other ligands containing either terminal Man or GalNAc-4-SO$_4$. Another attractive possibility is that the receptor exists in a monomeric form when binding Man-containing glycoconjugates and in a dimeric or multimeric form when binding GalNAc-4-SO$_4$-containing glycoproteins. Post-translational modifications and/or interaction with an accessory protein could determine if the receptor is able to form dimeric or multimeric species in a cell-specific manner.

The regulated ability to bind unrelated carbohydrate structures at two distinct sites to fulfill two or more completely different functions, hormone clearance and phagocytosis of pathogens, indicates the Man/GalNAc-4-SO$_4$-receptor has multiple functions. These remarkable properties of the Man/GalNAc-4-SO$_4$-receptor imply that the other members of this family of structurally related receptors may also be multi-functional and display properties that reflect their cell of origin. Additional ligands for these receptors must be identified before this issue can be effectively addressed.

The Man/GalNAc-4-SO$_4$-receptor is one component of a complex system mediating extracellular recognition. Four separate and highly regulated components are required for effective functioning of this receptor/ligand system.

First, glycoproteins bearing oligosaccharides terminating with GalNAc-4-SO$_4$ will be synthesized only if the protein to be modified contains the proper peptide recognition determinant for the protein-specific GalNAc-transferase. Second, the protein containing the recognition determinant must be expressed in cells that also express both the GalNAc-transferase and the GalNAc-4-sulfotransferase. Third, the glycoprotein containing terminal GalNAc-4-SO$_4$ must be secreted or expressed on the surface of cells that are in close proximity or have access to cells expressing the Man/GalNAc-4-SO$_4$-receptor. Fourth, the Man/GalNAc-4-SO$_4$-receptor must be in a form that recognizes terminal GalNAc-4-SO$_4$ rather than Man. Since we have obtained evidence for additional circumstances in which the GalNAc-4-SO$_4$-specific form of the receptor and its sulfated ligands are expressed in close proximity, it is possible that these receptor-ligand pairs may play a number of different roles in vivo.

6 Future Directions

It is becoming increasingly clear that unique terminal glycosylation patterns have important biological consequences that often can only be appreciated within the context of the intact organism. For example, the N-linked oligosaccharides of LH are selectively modified with terminal β1,4-linked GalNAc-4-SO$_4$ only when LH, the hormone-specific β1,4-GalNAc-transferase, and the GalNAc-4-sulfotransferase are expressed by the gonadotroph. LH bearing GalNAc-4-SO$_4$-terminal oligosaccharides is secreted and subsequently recognized by the Man/GalNAc-4-SO$_4$-receptor in hepatic endothelial cells when the receptor is in a form that recognizes GalNAc-4-SO$_4$, rather than the form, present in macrophages, that recognizes Man. Removal of LH from the circulation by the GalNAc-4-SO$_4$-receptor reduces its circulatory half-life and has an impact on activation of the LH receptor located in the ovary. There are numerous levels at which regulation could occur and ultimately have an impact on ovulation and other aspects of the reproductive cycle.

Recognition of GalNAc-4-SO$_4$-bearing structures on other glycoproteins may have different but equally important biological functions in other settings. Furthermore, the regulated expression of Man-specific binding activity by the Man/GalNAc-4-SO$_4$-receptor reflects a distinct function, perhaps related to innate immunity. Other members of this family of receptors with an N-terminal Cys-rich domain followed by multiple CRDs may also recognize more than one ligand, some of which may be carbohydrates, and may also have multiple functions. The temporal and spatial regulation of the synthesis of both sulfated oligosaccharides and the specificity of the Man/GalNAc-4-SO$_4$-receptor indicate that this complex system has several distinct biological roles. The challenge will be to identify and characterize these roles for the Man/GalNAc-4-SO$_4$-receptor as well as for other members of this multi-CRD family of receptors.

References

Ashwell G, Harford J (1982) Carbohydrate-specific receptors of the liver. Annu Rev Biochem 51:531–554

Baenziger JU (1996) Glycosylation: to what end for the glycoprotein hormones? (Editorial.) Endocrinology 137(5):1520–1522

Baenziger JU, Green ED (1991) Structure, synthesis, and function of the asparagine-linked oligosaccharides on pituitary glycoprotein hormones. In: Ginsberg V, Robbins PW (eds) Biology of carbohydrates, vol 3. JAI Press, London, pp 1–46

Baenziger JU, Kumar S, Brodbeck RM, Smith PL, Beranek MC (1992) Circulatory half-life but not interaction with the lutropin/chorionic gonadotropin receptor is modulated by sulfation of bovine lutropin oligosaccharides. Proc Natl Acad Sci USA 89:334–338

Bergwerff AA, Van Oostrum J, Kamerling JP, Vliegenthart JFG (1995) The major N-linked carbohydrate chains from human urokinase. The occurrence of 4-O-sulfated, (a2-6-sialylated or (a1-3)-fucosylated N-acetylgalactosamine(b1-4)-4-N-acetylgucosamine elements. Eur J Biochem 228:1009–1019

Chui D, Oh-Eda M, Liao YF, Panneerselvam K, Lal A, Marek KW, Freeze HH, Moremen KW, Fukuda MN, Marth JD (1997) Alpha-mannosidase-II deficiency results in dyserythropoiesis and unveils an alternate pathway in oligosaccharide biosynthesis. Cell 90(1):157–167

Dharmesh SM, Baenziger JU (1993) Estrogen modulates expression of the glycosyltransferases that synthesize sulfated oligosaccharides on lutropin. Proc Natl Acad Sci USA 90:11127–11131

Di Jeso B, Liguoro D, Ferranti P, Marinaccio M, Acquaviva R, Formisano S, Consiglio E (1992) Modulation of the carbohydrate moiety of thyroglobulin by thyrotropin and calcium in Fisher rat thyroid line-5 cells. J Biol Chem 267:1938–1944

Dohlman HG, Thorner J, Caron MG, Lefkowitz RJ (1991) Model systems for the study of seven-transmembrane-segment receptors. Annu Rev Biochem 60:653–688

Drickamer K (1988) Two distinct classes of carbohydrate-recognition domains in animal lectins. J Biol Chem 263:9557–9560

Drickamer K (1991) Clearing up glycoprotein hormones. Cell 67:1029–1032

Drickamer K (1993) Recognition of complex carbohydrates by Ca(2+)-dependent animal lectins (Review). Biochem Soc Trans 21(2):456–459

Drickamer K (1996) Ca(2+)-dependent sugar recognition by animal lectins (review, 20 refs). Biochem Soc Trans 24(1):146–150

Drickamer K, Taylor ME (1993) Biology of animal lectins (review). Annu Rev Cell Biol 9:237–264

Fiete D, Baenziger JU (1997) Isolation of the SO₄-4-GaINAcb1,4GlcNAcb1,2Mana-specific receptor from rat liver. J Biol Chem 272(23):14629–14637

Fiete D, Srivastava V, Hindsgaul O, Baenziger JU (1991) A hepatic reticuloendothelial cell receptor specific for SO₄-4GaINAcb1,4GlcNAcb1,2Mana that mediates rapid clearance of lutropin. Cell 67:1103–1110

Fiete D, Beranek MC, Baenziger JU (1997) The macrophage/endothelial cell mannose receptor cDNA encodes a protein that binds oligosaccharides terminating with SO₄-4-GaINAcb1,4GlcNAcb or Man at independent sites. Proc Natl Acad Sci USA 94:11256–11261

Fiete DJ, Beranek MC, Baenziger JU (1998) A cysteine-rich domain of the 'mannose' receptor mediates GaINAc-4-SO₄ binding. Proc Natl Acad Sci USA 95(5):2089–2093

Fraser IP, Koziel H, Ezekowitz RA (1998) The serum mannose-binding protein and the macrophage mannose receptor are pattern recognition molecules that link innate and adaptive immunity (review, 94 refs). Semin Immunol 10(5):363–372

Green ED, Baenziger JU (1988a) Asparagine-linked oligosaccharides on lutropin, follitropin, and thyrotropin: I. Structural elucidation of the sulfated and sialylated oligosaccharides on bovine, ovine, and human pituitary glycoprotein hormones. J Biol Chem 263:25–35

Green ED, Baenziger JU (1988b) Asparagine-linked oligosaccharides on lutropin, follitropin, and thyrotropin: II. Distributions of sulfated and sialylated oligosaccharides on bovine, ovine, and human glycoprotein hormones. J Biol Chem 263:36–44

Harris N, Super M, Rits M, Chang G, Ezekowitz RA (1992) Characterization of the murine macrophage mannose receptor: demonstration that the downregulation of receptor expression mediated by interferon-gamma occurs at the level of transcription. Blood 80:2363–2373

Helenius A (1994) How N-linked oligosaccharides affect glycoprotein folding in the endoplasmic reticulum. Mol Biol Cell 5:253–265

Hooper LV, Beranek MC, Manzella SM, Baenziger JU (1995) Differential expression of GalNAc-4-sulfotransferase and GalNAc-transferase results in distinct glycoforms of carbonic anhydrase VI in parotid and submaxillary glands. J Biol Chem 270:5985–5993

Iida S, Yamamoto K, Irimura T (1999) Interaction of human macrophage C-type lectin with O-linked N-acetylgalactosamine residues on mucin glycopeptides. J Biol Chem 274(16): 10697–10705

Ioffe E, Stanley P (1994) Mice lacking N-acetylglucosaminyltransferase I activity die at mid-gestation, revealing an essential role for complex or hybrid N-linked carbohydrates. Proc Natl Acad Sci USA 91:728–732

Jiang W, Swiggard WJ, Heufler C, Peng M, Mirza A, Steinman RM, Nussenzweig MC (1995) The receptor DEC-205 expressed by dendritic cells and thymic epithelial cells is involved in antigen processing. Nature 375(6527):151–155

Kelm S, Schauer R, Crocker PR (1996) The sialoadhesins–a family of sialic acid-dependent cellular recognition molecules within the immunoglobulin superfamily (review, 114 refs). Glycoconj J 13(6):913–926

Kornfeld S (1990) Lysosomal enzyme targeting. Biochem Soc Trans 18:367–374

Kornfeld S (1992) Structure and function of the mannose-6-phosphate/insulinlike growth factor II receptors. Annu Rev Biochem 61:307–330

Koziel H, Kruskal BA, Ezekowitz RAB, Rose RM (1993) HIV impairs alveolar macrophage mannose receptor function against Pneumocystis carinii. Chest 103 [Suppl]:111S–112S

Koziel H, Eichbaum Q, Kruskal BA, Pinkston P, Rogers RA, Armstrong MY, Richards FF, Rose RM, Ezekowitz RA (1998) Reduced binding and phagocytosis of Pneumocystis carinii by alveolar macrophages from persons infected with HIV-1 correlates with mannose receptor downregulation. J Clin Invest 102(7):1332–1344

Lambeau G, Ancian P, Barhanin J, Lazdunski M (1994) Cloning and expression of a membrane receptor for secretory phospholipase A_2. J Biol Chem 269(3):1575–1578

Lefkowitz RJ (1993) G protein-coupled receptor kinases. Cell 74:409–412

Lefkowitz RJ (1998) G protein-coupled receptors. III. New roles for receptor kinases and beta-arrestins in receptor signaling and desensitization (review, 47 refs). J Biol Chem 273(30):18677–18680

Lowe JB (1997) Selectin ligands, leukocyte trafficking, and fucosyltransferase genes (review, 113 refs). Kidney Int 51(5):1418–1426

Manzella SM, Dharmesh SM, Beranek MC, Swanson P, Baenziger JU (1995) Evolutionary conservation of the sulfated oligosaccharides on vertebrate glycoprotein hormones that control circulatory half-life. J Biol Chem 270(37):21665–21671

Manzella SM, Hooper LV, Baenziger JU (1996) Oligosaccharides containing b1,4-linked N-acetylgalactosamine, a paradigm for protein-specific glycosylation (review). J Biol Chem 271(21):12117–12120

Manzella SM, Dharmesh SM, Cohick CB, Soares MJ, Baenziger JU (1997) Developmental regulation of a pregnancy-specific oligosaccharide structure, NeuAca2,6GalNAcb1,4GlcNAc, on select members of the rat placental prolactin family. J Biol Chem 272(8):4775–4782

Martinez-Pomares L, Gordon S (1999) Potential role of the mannose receptor in antigen transport. Immunol Lett 65:9–13

Martinez-Pomares L, Kosco-Vilbois M, Darley E, Tree P, Herren S, Bonnefoy JY, Gordon S (1996) Fc chimeric protein containing the cysteine-rich domain of the murine mannose receptor binds to macrophages from splenic marginal zone and lymph node subcapsular sinus and to germinal centers. J Exp Med 184(5):1927–1937

Martinez-Pomares L, Mahoney JA, Kaposzta R, Linehan SA, Stahl PD, Gordon S (1998) A functional soluble form of the murine mannose receptor is produced by macrophages in vitro and is present in mouse serum. J Biol Chem 273(36):23376–23380

McEver RP, Cummings RD (1997) Role of PSGL-1 binding to selectins in leukocyte recruitment. J Clin Invest 100(3):485–491

McEver RP, Moore KL, Cummings RD (1995) Leukocyte trafficking mediated by selectin-carbohydrate interactions (review). J Biol Chem 270(19):11025–11028

Metzler M, Gertz A, Sarkar M, Schachter H, Schrader JW, Marth JD (1994) Complex asparagine-linked oligosaccharides are required for morphogenic events during post-implantation development. EMBO J 13:2056–2065

Mukherjee S, Palczewski K, Gurevich V, Benovic JL, Banga JP, Hunzicker-Dunn M (1999) A direct role for arrestins in desensitization of the luteinizing hormone/choriogonadotropin receptor in porcine ovarian follicular membranes. Proc Natl Acad Sci USA 96(2):493–498

Nicolas JP, Lambeau G, Lazdunski M (1995) Identification of the binding domain for secretory phospholipases A2 on their M-type 180-kDa membrane receptor. J Biol Chem 270(48):28869–28873

Powell LD, Varki A (1995) I-type lectins (review). J Biol Chem 270(24):14243–14246

Rosen SD, Bertozzi CR (1994) The selectins and their ligands. Curr Opin Cell Biol 6(5):663–673

Rosenberg RD, Shworak NW, Liu J, Schwartz JJ, Zhang L (1997) Heparan sulfate proteoglycans of the cardiovascular system. Specific structures emerge but how is synthesis regulated? (Review, 88 refs.) J Clin Invest 99(9):2062–2070

Segaloff DL, Ascoli M (1993) The lutropin/choriogonadotropin receptor...4 years later. Endocr Rev 14:324–347

Siciliano RA, Morris HR, Bennett HPJ, Dell A (1994) O-glycosylation mimics N-glycosylation in the 16-kDa fragment of bovine pro-opiomelanocortin. The major O-glycan attached to Thr-45 carries SO₄-4GalNAcb1–4GlcNAcb1-, which is the archetypal non-reducing epitope in the N-glycans of pituitary glycohormones. J Biol Chem 269:910–920

Skelton TP, Kumar S, Smith PL, Beranek MC, Baenziger JU (1992) Pro-opiomelanocortin synthesized by corticotrophs bears asparagine-linked oligosaccharides terminating with SO₄-4GalNAcb1,4GlcNAcb1,2Mana. J Biol Chem 267:12998–3006

Smith PL, Kaetzel D, Nilson J, Baenziger JU (1990) The sialylated oligosaccharides of recombinant bovine lutropin modulate hormone bioactivity. J Biol Chem 265:874–881

Smith PL, Skelton TP, Fiete D, Dharmesh SM, Beranek MC, MacPhail L, Broze GJ Jr, Baenziger JU (1992) The asparagine-linked oligosaccharides on tissue factor pathway inhibitor terminate with SO₄-4GalNAcb1,4GlcNAcb1, 2Mana. J Biol Chem 267:19140–19146

Smith PL, Bousfield GR, Kumar S, Fiete D, Baenziger JU (1993) Equine lutropin and chorionic gonadotropin bear oligosaccharides terminating with SO₄-4-GalNAc and sialic acida2,3Gal, respectively. J Biol Chem 268:795–802

Stahl PD, Ezekowitz RA (1998) The mannose receptor is a pattern recognition receptor involved in host defense. (Review, 81 refs.) Curr Opin Immunol 10(1):50–55

Stockell Hartree A, Renwick AGC (1992) Molecular structures of glycoprotein hormones and functions of their carbohydrate components. Biochem J 287:665–679

Szkudlinski MW, Thotakura NR, Tropea JE, Grossmann M, Weintraub BD (1995) Asparagine-linked oligosaccharide structures determine clearance and organ distribution of pituitary and recombinant thyrotropin. Endocrinology 136(8):3325–3330

Taylor ME (1997) Evolution of a family of receptors containing multiple C-type carbohydrate-recognition domains. Glycobiology 7(3):V–XV

Taylor ME, Drickamer K (1993) Structural requirements for high affinity binding of complex ligands by the macrophage mannose receptor. J Biol Chem 268:399–404

Taylor ME, Conary JT, Lennartz MR, Stahl PD, Drickamer K (1990) Primary structure of the mannose receptor contains multiple motifs resembling carbohydrate-recognition domains. J Biol Chem 265:12156–12162

Taylor ME, Bezouska K, Drickamer K (1992) Contribution to ligand binding by multiple carbohydrate- recognition domains in the macrophage mannose receptor. J Biol Chem 267:1719–1726

Thotakura NR, Blithe DL (1995) Glycoprotein hormones: glycobiology of gonadotropins, thyrotropin, and the free a subunit. Glycobiology 5:3–10

Trombetta ES, Helenius A (1998) Lectins as chaperones in glycoprotein folding (review, 48 refs). Curr Opin Struct Biol 8(5):587–592

Van den Eijnden DH, Neeleman AP, Van der Knaap WP, Bakker H, Agterberg, M, Van Die I (1995) Novel glycosylation routes for glycoproteins: the lacdiNAc pathway (review, 49 refs). Biochem Soc Trans 23(1):175–179

Vanrooijen JJM, Kamerling JP, Vliegenthart JFG (1998) Sulfated di-, tri- and tetraantennary N-glycans in human Tamm-horsfall glycoprotein. Eur J Biochem 256(2):471–487

Williams DB (1995) Calnexin: a molecular chaperone with a taste for carbohydrate (review, 50 refs). Biochem Cell Biol 73:123–132

Wu K, Yuan J, Lasky LA (1996) Characterization of a novel member of the macrophage mannose receptor type C lectin family. J Biol Chem 271(35):21323–21330

Sialoadhesin Structure

Andrew P. May[1] and E. Yvonne Jones[2,3]

1 Introduction

Members of the immunoglobulin superfamily (IgSF) make up the most widely distributed group of cell surface receptors (Barclay et al. 1997). The IgSF comprises proteins containing domains of approximately 100 amino acids folded into a structure (known as the Ig fold) consisting of two antiparallel β-sheets sandwiched together. Cell adhesion processes involving the Ig-fold are usually mediated by either heterotypic or homotypic protein-protein interactions. The siglec (sialic-acid-binding *Ig*-like *lec*tins; Crocker et al. 1998) family comprises a group of cell surface molecules, classified by sequence similarity as members of the IgSF, which exhibit functional protein-carbohydrate recognition. The siglecs mediate cellular interactions through the recognition of specific sialylated glycoconjugates as their counter-receptors (Kelm et al. 1994; Freeman et al. 1995; Cornish et al. 1998; Patel et al. 1999; Falco et al. 1999; Nicoll et al. 1999; Angata and Varki 2000a; Floyd et al. 2000; Zhang et al. 2000; Angata and Varki 2000b; Kikly et al. 2000). This group of molecules currently consists of sialoadhesin, CD22, the myelin-associated glycoprotein (MAG), CD33, Siglec-5, Siglec-6, Siglec-7, Siglec-8 and Siglec-9.

Outside of a central core of 5 β-strands the peripheral β-strands and loops of Ig-folds exhibit a high degree of structural variability that has lead to their classification into subsets (Williams 1987; Harpaz and Chothia 1994; Tan et al. 1998). Members of the siglec family have similar domain organizations consisting of an N-terminal V-set Ig domain followed by differing numbers of C2-set domains, the number varying from 1 in CD33 and Siglec-7 to 16 in sialoadhesin (Fig. 1). The highest level of sequence similarity is found in the N-terminal two domains (Crocker et al. 1994; Freeman et al. 1995; Cornish et al. 1998; Floyd et al. 2000; Angata and Varki 2000b). These two domains contain an unusual conserved pattern of cysteine residues. Each domain contains three cysteine residues: in the V-set domain, two of these cysteines are predicted to

[1] Department of Structural Biology, Stanford University School of Medicine, Fairchild Building, Stanford, CA 94305, USA
[2] Structural Biology, Wellcome Trust Centre for Human Genetics, Roosevelt Drive, Headington, Oxford, OX3 7BN. UK
[3] Oxford Centre for Molecular Sciences, New Chemistry Building, South Parks Road, Oxford, OX1 3QT, UK

Results and Problems in Cell Differentiation, Vol. 33
Paul R. Crocker (Ed.): Mammalian Carbohydrate Recognition Systems
© Springer-Verlag Berlin Heidelberg 2001

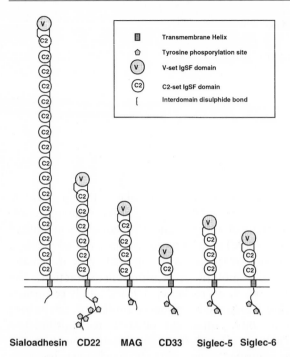

| Transmembrane Helix |
| Tyrosine phosporylation site |
| V-set IgSF domain |
| C2-set IgSF domain |
| Interdomain disulphide bond |

Sialoadhesin   CD22   MAG   CD33   Siglec-5 Siglec-6

Fig. 1. The siglec family. Members of the siglec family have a similar domain organisation, and consist of an N-terminal V-set domain, followed by differing numbers of C2-set domains, a transmembrane helix, and a short cytoplasmic tail. The line drawn between the N-terminal V- and C2-set domains indicates the presence of a predicted interdomain disulphide bond. All siglecs except sialoadhesin and Siglec-8 contain tyrosine residues within potential signalling motifs in their cytoplasmic tails

form a distinctive intra-β-sheet disulphide bond (Williams et al. 1989) whilst the third forms a disulphide bond with one of the three cysteine residues in domain 2 (Pedraza et al. 1990). The remaining two cysteine residues in domain 2 form an inter-β-sheet disulphide bond found typically in Ig domains.

The prototypic siglec sialoadhesin is restricted to macrophages and under non-inflammatory conditions the highest level of expression is on resident macrophages of haemopoeitic and secondary lymphoid tissues (Crocker et al. 1991). Molecular cloning of murine sialoadhesin showed that it consisted of 17 extracellular Ig-like domains, a transmembrane helix, and a short cytoplasmic tail containing a number of potential serine/threonine phosphorylation sites (Crocker et al. 1994). As mentioned above, the N-terminal two domains show closest sequence similarity to other siglecs. Of the remaining 15 domains in sialoadhesin, the 'stem' region (domains 4–17) is made up of alternating shorter and longer C2-set domains (Crocker et al. 1994). This suggests that this region is likely to have arisen through duplication of an ancestral module consisting of one short and one long domain (Vinson et al. 1997). Domain 3 is most similar to the longer domains.

With 17 extracellular domains, sialoadhesin is the largest cell surface member of the IgSF known to date. Furthermore, this number of domains is conserved in the human and rat homologues of sialoadhesin (A. Hartnell et al., manuscript in prep.; T. van den Berg, unpubl. data), suggesting an important functional role. It has been proposed that the large number of domains project the sialic acid-binding surface away from the macrophage glycocalyx. This would minimise cis-interactions and allow adhesive interactions between macrophages and cells bearing the appropriate counterreceptors (Crocker et al. 1997; P.R. Crocker et al., unpubl. data).

Domain deletion experiments showed that for sialoadhesin, the N-terminal domain was both necessary and sufficient for sialic acid mediated adhesion (Nath et al. 1995), a conclusion further supported by subsequent site directed mutagenesis studies (Vinson et al. 1996). A soluble recombinant form of the N-terminal domain of sialoadhesin (SnD1), defined as residues 1–119, contains no glycosylation sites and remains competent to bind sialic acids. As a result, this represents an ideal system in which to probe the structural basis of sialic acid-mediated recognition in the siglecs by crystallographic methods. Crystallographic studies have provided the first insight into adaptations of the Ig fold utilised by sialoadhesin and, by analogy, other siglec molecules. The details of the mode of sialic-acid binding described here were determined directly from the crystal structures of complexes of SnD1 with sialic acid-containing ligands (May et al. 1998; A.P. May and E.Y. Jones, unpubl. data).

2 Sialoadhesin Carbohydrate-Binding Domain Structure and Function

As expected from sequence analysis, structural comparisons place SnD1 within the subgroup of IgSF V-set domains. The structure, illustrated in Fig. 2 in complex with 3'-sialyllactose (Neu5Acα2,3Galβ1,4Glc), comprises two β sheets formed from the ABED and A'GFCC' β-strands (May et al. 1998). The sialic acid lies along strand G, making interactions with side chains from the A, G and F strands, and main chain atoms of strand G (Figs. 2, 3, 5 and Table 1). Inspection of the SnD1 structure indicates two notable differences from the standard V-set Ig framework that appear linked to its ability to bind sialic acid. Firstly, the substitution of the intra-sheet disulphide between residues from strands B and E, in place of the canonical Ig domain inter-sheet disulphide, widens the space between the β-sheets. This exposes aromatic residues W2 and W106 for interaction with the acetamido methyl group and the glycerol side chain of the sialic acid. Secondly, the G-strand, distinctively split in SnD1 into two shorter strands, contributes key hydrogen bonds to the glycan interaction.

The carbohydrate binding interactions of SnD1 are dominated by those made to sialic acid. The details of the sialic acid interaction are identical for complexes of SnD1 with 3'-sialyllactose and αMeNeuAc (May et al. 1998; APM and EYJ unpublished data) confirming that the mode of interaction between

Table 1. Interactions at the sialic acid-binding sites of sialoadhesin (SnD1) and influenza virus haemagglutinin (influenza HA). Distances are given in Angstroms (Å)

Type of interaction	Sugar	SnD1	Distance/Å	Influenza HA	Distance/Å
H-bond	Sia O1A	Arg 97 Nη2	2.95	Asn 137 N	2.90
	Sia O1B	Arg 97 Nη1	2.74	Ser 136 Oγ	2.87
	Sia N5	Arg 105 O	2.81	Gly 135 O	3.15
	Sia O4	Ser 103 O	2.74		
		Wat O1	2.81		
	Sia O7	Wat O1	2.87		
	Sia O8	Leu 107 N	2.83	Tyr 98 OH	2.75
	Sia O9	Leu 107 O	2.77	Tyr 98 OH	2.73
		Wat O1	2.83	Ser 228 Oγ	2.69
		Wat O1	3.37	Glu 190 Oε2	2.79
	Gal O6	Tyr 44 OH	2.70	Gly 225 O	2.83
		Wat O1	2.77		
		Wat O1	2.53		
	Gal O2	Wat O1	2.83		
Hydrophobic	Sia C11	Trp 2 Cε2	3.40	Trp 153 Cε2	3.86
		Trp 2 Nε1	3.32	Trp 153 Nε1	
		Trp 2 Cζ2	3.99	Trp 153 Cζ2	3.93
		Trp 2 Cδ1	3.52	Trp 153 Cδ1	
		Trp 2 Cη2	4.66	Trp 153 Cη2	4.01
		Trp 2 Cζ3	4.84	Trp 153 Cζ3	4.05
		Trp 2 Cε3	4.40	Trp 153 Cε3	4.02
		Trp 2 Cδ2	3.64	Trp 153 Cδ2	3.92
		Trp 2 Cγ	3.73	Trp 153 Cγ	4.50
	Sia C9	Trp 106 Cζ3	3.81		
		Trp 106 Cε3	3.83		
		Trp 106 Cδ2	3.93		
		Trp 106 Cε2	4.02		
		Trp 106 Cζ2	4.03		
		Trp 106 Cη2	3.91		
	Gal C5	Leu 107 Cδ4	4.47	Leu 107 Cδ4	4.20

sialoadhesin and sialic acid observed in the SnD1/3′-sialyllactose complex is independent of the extended oligosaccharide. Many of the interactions are mediated by atoms from the SnD1 main chain but three residues W2, R97 and W106, stand out as a specific sialic acid-binding template contributing key interactions through their side chains (May et al. 1998). Sequence comparisons of all current members of the siglec family (Fig. 4) reinforce the template hypothesis with total conservation of R97 and mainly conservative substitutions of W2 (to phenylalanine in CD33 and Siglec-6; to tyrosine in Siglecs 5, 7 and 8) and W106 (to tyrosine in CD33, MAG, Siglecs 5 and 6). This structure-based observation is also consistent with functional studies. Mutations of R97 in sialoadhesin (Vinson et al. 1996) and the equivalent residue in CD22 (van der Merwe et al. 1996), MAG (Tang et al. 1997), CD33 (Taylor et al. 1999) and Siglecs 7 and 9 (Angata and Varki 2000a,b) show that the salt

Fig. 2. The structure of the N-terminal domain of sialoadhesin in complex with 3'-sialyllactose. The structure is represented as a ribbon diagram. Each strand is labelled. The 3'-sialyllactose lies along strand G and makes interactions with residues from the A, G and F strands

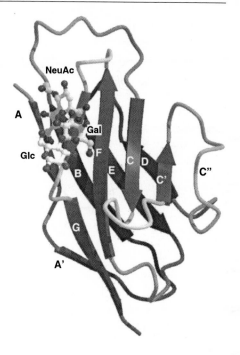

bridge this residue forms with the carboxylate of sialic acid is a critical interaction. For both sialoadhesin and CD22 even a conservative substitution to lysine resulted in a significant reduction in binding. In the case of CD22 (van der Merwe et al. 1996) the mutation of W138 (equivalent to W106 in sialoadhesin) also abrogated sialic acid-dependent binding. This result is again consistent with the importance of the hydrophobic contact between this aromatic side chain and the terminal carbon of the glycerol side chain (C9) of sialic acid. Similarly, the mutation of W2 in sialoadhesin (May et al. 1998) abolished sialic acid-dependent adhesion in line with the loss of the van der Waals contact between this second aromatic side chain and the acetamido methyl group. However, in Siglec-9, which remains fully capable of supporting sialic acid binding, W2 is replaced by a lysine residue, suggesting that this interaction may not be as critical in other siglecs (Zhang et al. 2000; Angata and Varki 2000b). The importance of the interactions of SnD1 with the chemical substituents of sialic acid has also been demonstrated in a number of studies in which sialic acid-bearing glycoconjugates have been subjected to chemical modification (see the chapter by Kelm). Removal of any of the sialic acid side chains renders the carbohydrate unrecognisable by sialoadhesin. NMR studies on sialic acid have also confirmed the validity of the interactions observed in the crystal structure, demonstrating essential roles for R97 and the interactions of the glycerol and amide side chains with aromatic amino acids in sialoadhesin (Crocker et al. 1999).

Since the remainder of the interactions made with sialic acid are mediated by main chain groups it is difficult to assess what effect substitutions at these

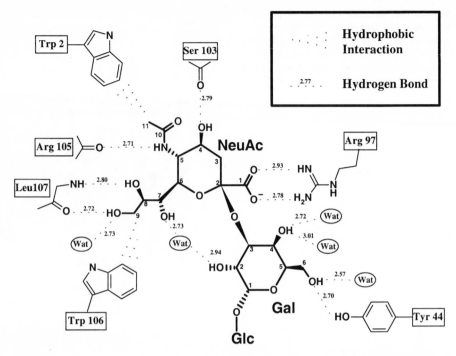

Fig. 3. Interactions at the 3′-sialyllactose binding site. A schematic view of the carbohydrate-binding site of SnD1. 3′-sialyllactose makes a number of interactions with sialoadhesin. The carboxylate forms a salt bridge with the side chain of R97. Hydrophobic interactions are made between the acetamido methyl group and the side chain of W2; C9 of the glycerol side chain and the side chain of W106; C6 of galactose and the side chain of L107. Hydrogen bonds (shown as *broken lines*) are made between sialic acid and main chain atoms of S103, L107 and R105, and between galactose and the side chain of Y44. A number of water molecules are also involved in the SnD1/3′-sialyllactose interaction. In the crystal, the glucose does not interact with the same SnD1 molecule to which the sialic acid is bound, and is not shown in the schematic as a result

positions in SnD1, and the other siglecs, would have on the affinity of ligand binding. The G strand main chain plays a prominent role in these interactions but there are clearly significant differences between siglecs in the distinctive G strand insert (Fig. 4). There is little sequence similarity between family members in this region, so the detailed structure and the magnitude of the contribution to ligand affinity and specificity of this segment seems likely to differ between them.

Contacts with the galactose portion of the sialyllactose are through a hydrogen bond with Y44 and a long-range hydrophobic contact with L107. However, unlike the amino acids interacting with sialic acid, these residues are not well conserved in other siglecs (Fig. 4). Furthermore, the sialoadhesin mutation Y44 A does not affect adhesion. Although Y44 forms a hydrogen bond with the 6-hydroxyl of galactose, this result indicates that removal of this hydrogen bond

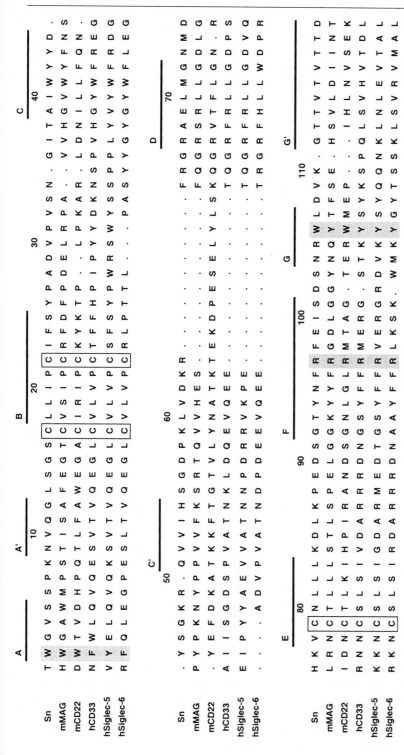

Fig. 4. Sequence alignment of the N-terminal domains of current members of the siglec family. Only one species orthologue is shown for each siglec. The three cysteine residues are *boxed* and the three residues which form the sialic acid-binding template, W2, R97 and W106, are *highlighted*. Secondary structure elements are shown as *lines* above the sequences, and every tenth residue of sialoadhesin is numbered. Sequences are: *Sn* mouse sialoadhesin, *mMAG* mouse myelin-associated glycoprotein, *mCD22* mouse CD22, *hCD33* human CD33, *hSiglec-5* human siglec-5, *hSiglec-6* human siglec-6, *hSiglec-7* human siglec-7, *hSiglec-8* human siglec-8, *hSiglec-9* human siglec-9

A

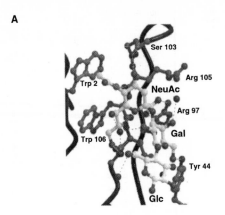

B

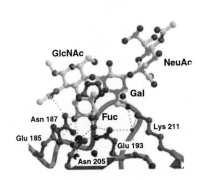

Fig. 5. A comparison of the carbohydrate binding sites in (**A**) sialoadhesin and (**B**) an E-selectin-like mutant of rat mannose binding protein (MBP). In the complex of sialyl-Lewis[x] with the E-selectin-like mutant of MBP, the majority of the protein-carbohydrate interactions are mediated by the fucose of sialyl-Lewis[x]. The sialic acid does not make any contact with the protein surface. In direct contrast to this, in sialoadhesin, the majority of protein-carbohydrate contacts are made via sialic acid, and few contacts are made by the rest of the oligosaccharide

is insufficient to reduce the level of interaction as measured by the solid-phase assay used to assess sialic acid-dependent adhesion.

3 Comparison with Other Sialic Acid-Binding Proteins

A number of structures exist for lectins in complex with sialic acid-containing ligands, namely influenza haemagglutinin (HA; Weis et al. 1988; Sauter et al. 1992), wheat germ agglutinin (WGA; Wright 1984, 1990), cholera toxin (CT; Merrit et al. 1994), pertussis toxin (PT; Stein et al. 1994) VP1 from polyoma virus (PV; Stehle et al. 1994) and an E-selectin-like mutant of rat mannose binding protein (E-MBP; Ng and Weis 1997). In all of these except the E-MBP structure, which will be discussed later, the sialic acid is involved in a number of contacts both directly with the protein, and through ordered water molecules. A comparison of the structural basis of sialic acid binding in sialoadhesin and in these other complexes reveals enormous diversity in the details of the binding but does highlight several key themes. The binding site of SnD1 is most closely related to that of HA, to which it has been previously compared on the basis of its adhesive properties (Crocker et al. 1991); this structural comparison is detailed in Table 1.

Sialic acid binding is dominated by interactions with its acetamido, carboxylate and glycerol side chain substituents. At the atomic level only two of the interactions with the sialic acid are found in all of the known complexes, these are hydrogen bonds to the acetamido N5 and to the carboxylate, although the acceptors and donors for these respective interactions vary between structures. The only structure other than SnD1 in which the carboxylate forms a salt bridge is the 3.65 Å structure of PV, where again, an arginine residue is used. In all other complexes, the carboxylate moeity hydrogen bonds to main chain amide groups, polar side chains and ordered water molecules. However, the use of a charged group in interaction with the carboxylate of sialic acid may not be as unusual as indicated by currently available structural data. A number of bacterial sialic acid-binding lectins such as the *Escherichia coli* adhesins Sfas, K99, CFA/I and LT1-B have a highly conserved motif containing basic amino acids (lysine or arginine) which have been implicated in ligand binding (reviewed in Kelm and Schauer 1997).

The involvement of the hydrophobic portions of the glycerol and acetamido substituents in the present examples of sialic acid / protein interactions is not universal, but does also represent a significant general theme. In HA, the glycerol side chain is buried, and makes extensive hydrogen bonds to the protein through the 8- and 9-hydroxyls, with the non-polar face stacked against a complementary hydrophobic surface. The glycerol O8 and O9 atoms are bound via structural water molecules in CT, and in WGA, the glycerol side chain carbon backbone is in van der Waals contact with a tyrosine in a similar fashion to the SnD1 interaction between C9 and W106. The stacking of the acetamido group onto the ring of W2 is closely resembled in HA and WGA by interactions with a tryptophan residue and a tyrosine residue respectively.

Specific interactions with the 4-hydroxyl of sialic acid appear to be somewhat optional, however, for complexes with PV and WGA; as in SnD1, this polar group is involved in a hydrogen bond. It is also interesting to note that in all of the known structures of sialyllactose complexes (SnD1, HA and WGA) the only conserved interaction observed with the lactose group is through a hydrogen bond to the 6-hydroxyl of galactose. In HA the side chain of L226 is in van der Waals contact with galactose (C5), and structural superposition of the binding sites of HA and SnD1 shows that the side chain of L107 adopts a similar position to L226 relative to galactose.

4 Sialic Acid Mediated Cell Adhesion

There are a limited number of mammalian proteins known to utilise sialic acid in cellular recognition events, the best characterised being the siglecs and the selectins (E-, P- and L-selectin). The structure of the sialic-acid-binding domain of sialoadhesin described here reveals the mode of interaction for the siglecs is with the major substituent groups of the sialic acid. The selectins recognise oligosaccharides containing sulphated and sialyllated Lewis[X] and Lewis[A], but as yet there is no direct structural information on their mode of binding. However, the structure of a mutant of rat mannose binding protein with E-selectin-like specificity (E-MBP; Blanck et al. 1996) has been determined in the presence of both 3'-sialyl-Lewis[X] and 3'- and 4'-sulpho-Lewis[X] (Ng and Weis 1997). This form of carbohydrate-recognition complex is compared with that of SnD1 in Fig. 5. The majority of contacts to the carbohydrate in the E-MBP structure are via the fucose of Lewis[X] with further interactions being made to the GlcNAc and galactose. Surprisingly, these studies show no direct contacts are made to either the 3'sulpho-group, or the sialic acid, despite observations that both the binding of E-selectin to neutrophils and L-selectin to endothelial cells can be abrogated by pre-treatment with sialidase (Yednock and Rosen 1989; Corrall et al. 1990). It is therefore noteworthy that replacement of sialic acid with a sulphate ester results in retention of binding by all selectins. Since in the E-MBP complex the sialic acid is located in the vicinity of a patch of lysine residues (implicated by mutagenesis as being important in neutrophil adhesion; Graves et al. 1994) one plausible interpretation is that its contribution may be indirect, possibly through electrostatic interactions guiding the initial binding. Chemical modification studies demonstrate that only alterations to the carboxylate group of sialic acid affect adhesion (Tyrrell et al. 1991; Brandley et al. 1993). Furthermore, studies on P-selectin showed, through a combination of mutagenesis and chemical modification of one of these lysine residues, that it is the charge which is critical for interaction with sialyl-Lewis[x] (Hollenbough et al. 1995). This strongly contrasts with the binding of sialic acid in the sialoadhesin family where it constitutes a highly specific component of the final complex. Therefore, despite both families of mammalian proteins having a critical requirement for sialic acid as a recognition determinant in their carbohydrate-dependent interactions, it is clear that the role that the sialic acid plays is very different.

5 Conclusion

The structure of a functional fragment of sialoadhesin complexed with 3'-sialyllactose (May et al. 1998) has allowed some reappraisal and rationalisation of the currently available functional data for protein-carbohydrate mediated cell–cell recognition. In particular, as discussed above, the contribution of the sialic acid moiety in this type of recognition is now clear. Although the current structural data provides some important pointers, further structural and functional data will be required to elucidate the basis of differences in oligosaccharide affinity and sialic acid-galactose linkage specificity between members of the siglec family. The existing data do, however, represent a template on which to plan manipulation of this form of sialic acid-based cell–cell recognition.

References

Angata T, Varki A (2000a) Siglec-7: a sialic acid-binding lectin of the immunoglobulin super-family. Glycobiology 10:431–438

Angata T, Varki A (2000b) Cloning, characterization, and phylogenetic analysis of siglec-9, a new member of the CD33-related group of siglecs. Evidence for co-evolution with sialic acid synthesis pathways. J Biol Chem 275:22127–22135

Barclay AN, Brown MH, Law SKA, Mcknight AJ, Tomlinson MG, van der Merwe PA (1997) The leucocyte antigen facts book, 2nd edn. Academic Press, London

Blanck O, Iobst ST, Gabel C, Drickamer K (1996) Introduction of selectin-like binding specificity into a homologous mannose-binding protein. J Biol Chem 271:7289–7292

Brandley BK, Kiso M, Abbas S, Nikrad P, Srivasatava O, Foxall C, Oda Y, Hasegawa A (1993) Structure-function studies on selectin carbohydrate ligands – modifications to fucose, sialic-acid and sulfate as a sialic-acid replacement. Glycobiology 3:633–641

Cornish AL, Freeman S, Forbes G, Ni J, Zhang M, Cepeda M, Gentz R, Augustus M, Carter KC, Crocker PR (1998) Characterization of siglec-5, a novel glycoprotein expressed on myeloid cells related to CD33. Blood 92:2123–2132

Corrall L, Singer MS, Macher BA, Rosen SD (1990) Requirement for sialic-acid on neutrophils in gmp-140 (padgem) mediated adhesive interaction with activated platelets. Biochem Biophys Res Commun 172:1349–1356

Crocker PR, Kelm S, Dubois C, Martin B, McWilliam AS, Shotton DM, Paulson JC, Gordon S (1991) Purification and properties of sialoadhesin, a sialic acid-binding receptor of murine tissue macrophages. EMBO J 10:1661–1669

Crocker PR, Mucklow S, Bouckson V, McWilliam A, Willis AC, Gordon S, Milon G, Kelm S, Bradfield P (1994) Sialoadhesin, a macrophage sialic acid binding receptor for haemopoietic cells with 17 immunoglobulin-like domains. EMBO J 13:4490–4503

Crocker PR, Hartnell A, Munday J, Nath D (1997) The potential role of sialoadhesin as a macrophage recognition molecule in health and disease. Glycoconj J 14:601–609

Crocker PR, Clark EA, Filbin M, Gordon S, Jones Y, Kehrl JH, Kelm S, Le Douarin N, Powell L, Roder J, Schnaar RL, Sgroi DC, Stamenkovic K, Schauer R, Schachner M, van den Berg TK, van der Merwe PA, Watt SM, Varki, A. (1998) Siglecs: a family of sialic-acid binding lectins. Glycobiology 8:v

Crocker PR, Vinson M, Kelm S, Drickamer K (1999) Molecular analysis of sialoside binding to sialoadhesin by NMR and site-directed mutagenesis. Biochem J 341:355–361

Falco M, Biassoni R, Bottino C, Vitale M, Sivori S, Augugliaro R, Moretta L, Moretta A (1999) Identification and molecular cloning of p75/AIRM1, a novel member of the sialoadhesin family that functions as an inhibitory receptor in human natural killer cells. J Exp Med 190:793–802

Floyd H, Ni J, Cornish AL, Zeng Z, Liu D, Carter KC, Steel J, Crocker PR (2000) Siglec-8. A novel eosinophil-specific member of the immunoglobulin superfamily. J Biol Chem 275:861–6

Freeman SD, Kelm S, Barber EK, Crocker PR (1995) Characterization of CD33 as a new member of the sialoadhesin family of cellular interaction molecules. Blood 85:2005–2012

Graves BJ, Crowther RL, Chandran C, Rumberger JM, Li S, Huang KS, Presky DH, Familletti PC, Wolitzky BA, Burns DK (1994) Insight into E-selectin/ligand interaction from the crystal structure and mutagenesis of the lec/EGF domains. Nature 367:532–538

Harpaz Y, Chothia C (1994) Many of the immunoglobulin superfamily domains in cell adhesion molecules and surface receptors belong to a new structural set which is close to that containing variable domains. J Mol Biol 238:528–539

Hollenbaugh D, Aruffo A, Senter PD (1995) Effects of chemical modification on the binding activities of p-selectin mutants. Biochemistry 34:5678–5684

Kelm S, Schauer R (1997) Sialic acids in molecular and cellular interactions. Int Rev Cytol 175:137–240

Kelm S, Pelz A, Schauer R, Filbin MT, Tang S, de Bellard ME, Schnaar RL, Mahoney JA, Hartnell A, Bradfield P, Crocker P (1994) Sialoadhesin, myelin-associated glycoprotein and CD22 define a new family of sialic acid-dependent adhesion molecules of the immunoglobulin superfamily. Curr Biol 4:965–972

Kikly KK, Bochner BS, Freeman SD, Tan KB, Gallagher KT, D'alessio KJ, Holmes SD, Abrahamson, JA, Erickson-Miller CL, Murdock PR, Tachimoto H, Schleimer RP, White JR (2000) Identification of SAF-2, a novel siglec expressed on eosinophils, mast cells, and basophils. J Allergy Clin Immunol 105:1093–1100

May AP, Robinson RC, Vinson M, Crocker PR, Jones EY (1998) Crystal structure of the N-terminal domain of sialoadhesin in complex with 3'sialyllactose at 1.85 Å resolution. Mol Cell 1:719–728

Merrit EA, Sarfaty S, Vandenakker F, Lhoir C, Martial JA, Hol WGJ (1994) Crystal-structure of cholera-toxin b-pentamer bound to receptor G_{M1} pentasaccharide. Protein Sci 3:166–175

Nath D, van der Merwe PA, Kelm S, Bradfield P, Crocker PR (1995) The amino-terminal immunoglobulin-like domain of sialoadhesin contains the sialic acid binding site. Comparison with CD22. J Biol Chem 270:26184–26191

Ng KK, Weis WI (1997) Structure of a selectin-like mutant of mannose-binding protein complexed with sialylated and sulfated Lewis(x) oligosaccharides. Biochemistry 36:979–988

Nicoll G, Ni J, Liu D, Klenerman P, Munday J, Dubock S, Mattei MG, Crocker PR (1999) Identification and characterization of a novel siglec, siglec-7, expressed by human natural killer cells and monocytes. J Biol Chem 274:34089–34095

Patel N, Brinkman-Van der Linden EC, Altmann SW, Gish K, Balasubramanian S, Timans JC, Peterson D, Bell MP, Bazan JF, Varki A, Kastelein RA (1999) OB-BP1/Siglec-6. a leptin- and sialic acid-binding protein of the immunoglobulin superfamily. J Biol Chem 31:22729–22738

Pedraza L, Owens GC, Green LA, Salzer JL (1990) The myelin-associated glycoproteins: membrane disposition, evidence of a novel disulfide linkage between immunoglobulin-like domains, and posttranslational palmitylation. J Cell Biol 111:2651–2661

Sauter NK, Hanson JE, Glick GD, Brown JH, Crowther RL, Park SJ, Skehel JJ, Wiley DC (1992) Binding of influenza-virus hemagglutinin to analogs of its cell-surface receptor, sialic-acid – analysis by proton nuclear-magnetic-resonance spectroscopy and X-ray crystallography. Biochemistry 31:9609–9621

Stehle T, Yan Y, Benjamin TL, Harrison SC (1994) Structure of murine polyoma virus complexed with an oligosaccharide receptor fragment. Nature 369:160–163

Stein PE, Boodhoo A, Armstrong GD, Heerze LD, Cockle SA, Klein MH, Read RJ (1994) Structure of a pertussis toxin sugar complex as a model for receptor-binding. Nat Struct Biol 1:591–596

Takei Y, Sasaki S, Fujiwara T, Takahashi E, Muto T, Nakamura Y (1997) Molecular cloning of a novel gene similar to myeloid antigen CD33 and its specific expression in placenta. Cytogenet Cell Genet 78:295–300

Tan K, Casasnovas JM, Liu J, Briskin MJ, Springer TA, Wang J (1998) The structure of immunoglobulin superfamily domains 1 and 2 of MAdCAM-1 reveals novel features important for integrin recognition. Structure 6:793–801

Tang S, Shen YJ, DeBellard ME, Mukhopadhyay G, Salzer JL, Crocker PR, Filbin MT (1997) Myelin-associated glycoprotein interacts with neurons via a sialic acid binding site at ARG118 and a distinct neurite inhibition site. J Cell Biol 138:1355–1366

Taylor VC, Buckley CD, Douglas M, Cody AJ, Simmons DL, Freeman SD (1999) The myeloid-specific sialic acid-binding receptor, CD33, associates with the protein-tyrosine phosphatases, SHP-1 and SHP-2. J Biol Chem 274:11505–11512

Tyrrell D, James P, Rao N, Foxall C, Abbas S, Dasgupta F, Nashed M, Hasegawa A, Kiso M, Asa D, Kidd J, Brandley BK (1991) Structural requirements for the carbohydrate ligand of E-selectin. Proc Natl Acad Sci USA 88:10372–10376

van der Merwe PA, Crocker PR, Vinson M, Barclay AN, Schauer R, Kelm S (1996) Localization of the putative sialic acid-binding site on the immunoglobuin superfamily cell-surface molecule CD22. J Biol Chem 271:9273–9280

Vinson M, van der Merwe PA, Kelm S, May A, Jones EY, Crocker PR (1996) Characterization of the sialic acid-binding site in sialoadhesin by site-directed mutagenesis. J Biol Chem 271:9267–9272

Vinson M, Mucklow S, May AP, Jones EY, Kelm S, Crocker PR (1997) Sialic acid recognition by sialoadhesin and related lectins. Trends Glycosci Glycotech 9:283–297

Weis W, Brown JH, Cusack S, Paulson JC, Skehel JJ, Wiley DC (1988) Structure of the influenza-virus hemagglutinin complexed with its receptor, sialic-acid. Nature 333:426–431

Williams AF (1987) A year in the life of the immunoglobulin superfamily. Immunol Today 8:298–302

Williams AF, Davis SJ, He Q, Barclay AN (1989) Structural diversity in domains of the immunoglobulin superfamily. Cold Spring Harb Symp Quant Biol 54:637–647

Wright CS (1984) Structural comparison of the 2 distinct sugar binding-sites in wheat-germ-agglutinin isolectin-I. J Mol Biol 178:91–104

Wright CS (1990) 2.2 Å resolution structure analysis of 2 refined N-acetylneuraminyl-lactose-wheat-germ-agglutinin isolectin complexes. J Mol Biol 215:635–651

Yednock TA, Rosen SD (1989) Lymphocyte homing. Adv Immunol 44:313–378

Zhang JQ, Nicoll G, Jones C, Crocker PR (2000) Siglec-9, a novel sialic acid binding member of the immunoglobulin superfamily expressed broadly on human blood leukocytes. J Biol Chem 275:22121–22126

Ligands for Siglecs

Soerge Kelm[1]

1 Introduction

1.1 Sialic Acids in Cell Recognition

Sialic acids (Sia) are derivatives of the nine-carbon neuraminic acid mono-saccharide (Fig. 1). To date, about 40 of these neuraminic acid modifications have been found to occur in nature (Table 1). Their structure, occurrence and metabolic pathways have been reviewed in detail (Rosenberg 1995; Kelm and Schauer 1997; Schauer and Kamerling 1997; Varki 1997).

Sia are ubiquitous components of cell surface glycoconjugates in all animals from the echinoderms onward. As they are usually terminal sugars, they are well positioned to play important parts in cellular interactions. For several decades (Hirst 1941, 1942; McClelland and Hare 1941) it has been known that pathogens like viruses and bacteria use Sia-containing glycoconjugates to attach to the surfaces of their host cells (Kelm and Schauer 1997). It even seems possible that the variability of Sia modifications is a result of the ongoing strug-gle between pathogens and their hosts, in which the latter try to prevent infec-tions by modifications of the pathogen recognition determinant Sia, which in turn leads to pathogens with new binding specificities (Gagneux and Varki 1999). in contrast, functions of Sia in endogenous cellular interactions have been under debate, since no Sia-binding proteins had been described in higher animals until about 10 years ago, when the binding of selectins (Lowe et al. 1990; Phillips et al. 1990; Walz et al. 1990; Tiemeyer et al. 1991) and siaload-hesin (Sn; Crocker et al. 1991) to sialylated cell surface glycoconjugates were discovered.

Other well-described roles for Sia are their influence on the physical prop-erties of molecules like the mucins, the protection of carrier molecules from enzymatic digestion, and the masking effect which prevents the recognition of underlying structures by lectins or antibodies (Kelm and Schauer 1997; Schauer and Kamerling 1997).

---

[1] Institute of Biochemistry, University of Kiel, Olshausenstrasse 40, 24098 Kiel, Germany

Results and Problems in Cell Differentiation, Vol. 33
Paul R. Crocker (Ed.): Mammalian Carbohydrate Recognition Systems
© Springer-Verlag Berlin Heidelberg 2001

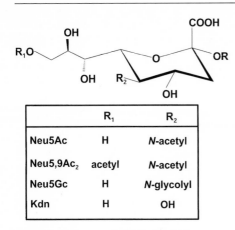

Fig. 1. Structure of common sialic acids. *N*-acetyl neuraminic acid (Neu5Ac) is the most common Sia. *O*-acetylation of the hydroxyl at C-9 (Neu5,9Ac$_2$) and hydroxylation of the *N*-acetyl at C-5 (Neu5Gc) are frequently found modifications of Neu5Ac. More recently, the wide distribution of the de-*N*-acetylated derivative Kdn has been recognised. More examples for naturally occurring sialic acids are given in Table 1

Table 1. Examples of naturally occurring sialic acids. Several Sia occurring in mammals with their names and recommended abbreviations are shown. A list of all Sia and citations for their occurrence can be found in Schauer and Kamerling (1997)

Name of sialic acid	Abbreviation
N-Acetylneuraminic acid	Neu5Ac
5-*N*-Acetyl-4-*O*-acetyl-neuraminic acid	Neu4,5Ac$_2$
5-*N*-Acetyl-7-*O*-acetyl-neuraminic acid	Neu5,7Ac$_2$
5-*N*-Acetyl-8-*O*-acetyl-neuraminic acid	Neu5,8Ac$_2$
5-*N*-Acetyl-9-*O*-acetyl-neuraminic acid	Neu5,9Ac$_2$
N-Glycolylneuraminic acid	Neu5Gc
4-*O*-Acetyl-5-*N*-glycolyl-neuraminic acid	Neu4Ac5Gc
7-*O*-Acetyl-5-*N*-glycolyl-neuraminic acid	Neu7Ac5Gc
8-*O*-Acetyl-5-*N*-glycolyl-neuraminic acid	Neu8Ac5Gc
9-*O*-Acetyl-5-*N*-glycolyl-neuraminic acid	Neu9Ac5Gc
N-(*O*-Acetyl)glycolylneuraminic acid	Neu5GcAc
2-Keto-3-deoxynononic acid	Kdn
9-*O*-Acetyl-2-keto-3-deoxy-nononic acid	Kdn9Ac

1.2 Selectins

About 10 years ago it was discovered that a group of proteins involved in the adherence of white blood cells to endothelia bind to cell surface glycans containing terminal Sia. Subsequently, these proteins were called selectins (Bevilacqua et al. 1991). Now it is clear that although naturally occurring binding partners of the selectins often contain Sia, it is not necessary for binding, and can usually be replaced by many other negatively charged moieties (Lasky 1995; Crocker and Feizi 1996; Kelm and Schauer 1997). Therefore, the selectins cannot be considered as truly Sia-recognising proteins.

1.3 Siglecs

At the same time as the binding specificity of the selectins was described, it was shown that an adhesion protein found on macrophage subpopulations specifically recognised sialylated glycans. This portein was therefore called sialoadhesin (Crocker et al. 1991).

The primary structure of Sn, derived from its cDNA sequence, revealed that it is a member of the immunoglobulin superfamily (IgSF) and, moreover, that it shares several structural similarities with some other members of the IgSF, namely CD22, CD33 and MAG as described below (Crocker et al. 1994). For all these proteins it has been demonstrated that they bind specifically to sialylated glycans (Powell et al. 1993; Sgroi et al. 1993; Crocker et al. 1994; Kelm et al. 1994; Freeman et al. 1995). Therefore, this group was called the Sialoadhesin family (Kelm et al. 1994), and later the siglecs (Crocker et al. 1998), an acronym of sialic acid-binding immunoglobulin-like lectins. By definition, the following properties have to be fulfilled by a protein to be a siglec: (1) capability to bind sialylated glycans and (2) significant sequence similarity with other siglecs within the N-terminal V-set and adjacent C2-set domains. Since 1998, three additional siglecs, designated siglec-5 (Cornish et al. 1998), siglec-6 (Patel et al. 1999) and siglec-7 (Falco et al. 1999; Nicoll et al. 1999) have been described. All these are more similar to CD33 than to any other siglec and have been identified on the basis of their sequence similarity with the known siglecs from either random sequencing of cDNA libraries (siglec-5 and siglec-7) or expression cloning of leptin-binding proteins (siglec-6).

Seven designated siglecs have been described in the literature up to now, but it is very likely that this family contains further members which are likely to be identified after finishing the Human Genome Project. Interestingly, the expression of all siglecs is relatively restricted to specific cell types (Table 2). With the exception of the myelin-associated glycoprotein (MAG, siglec 4-a) and the Schwann cell myelin protein (SMP, siglec 4-b), all siglecs found so far are only expressed by cells of the haematopoietic system.

2 Structures of Siglecs

The siglecs are typical type I transmembrane proteins with N-terminal extracellular parts containing several (2– 17) Ig-like domains, a single transmembrane domain and a cytoplasmic tail at the C-terminus which is likely to be involved in signal transduction, since it contains potential phosphorylation sites including so-called immune receptor tyrosine based inhibitory motifs

Table 2. Occurrence and potential function of siglecs

Name	Occurrence	Potential function
Sialoadhesin (Sn, siglec-1)	Macrophage subpopulations	Cellular interactions of macrophages
CD22 (siglec-2)	B lymphocytes	Modulation of B cell-dependent immune response; bone marrow homing
CD33 (siglec-3)	Myeloid precursor cells	Unknown
Myelin-associated glycoprotein (MAG, siglec-4a); Schwann cell myelin protein (SMP, siglec-4b)	Myelinating cells of the central and peripheral nervous systems (oligodendrocytes and Schwann cells)	Maintenance of myelin structure and function; regulation of neurite outgrowth
OB-BP2 (siglec-5)	Neutrophils, monocytes	Unknown; signalling?
OB-BP1 (siglec-6)	Placenta (cyto- and syncytiotrophoblasts); B lymphocytes	Sia-independent high affinity leptin binding; Sia-binding function unknown
p75/AIRM1 (siglec-7)	Natural killer cells	Inhibitory receptor of natural killer cells

(ITIM) (Cornish et al. 1998; Patel et al. 1999). At least for CD22 (Schulte et al. 1992; Campbell and Klinman 1995; Law et al. 1996; Wu et al. 1998; Yohannan et al. 1999) and MAG (Jaramillo et al. 1994; Umemori et al. 1994), phosphorylation of tyrosine have been demonstrated.

Typical Ig-like domains are characterised by two opposing β-sheets held together by an intersheet disulfide bridge leading to an overall relatively stable structure. Based on characteristic compositions of these sheets, they can be classified in different sets (Barclay and Brown 1997).

2.1 Sialic Acid-Binding Domain

It was demonstrated, using domain deletion and site-directed mutagenesis, that the GFC-face of the N-terminal domain 1 contains the Sia binding site at least for Sn and CD22 (Nath et al. 1995; van der Merwe et al. 1996; Vinson et al. 1996; Tang et al. 1997; May et al. 1998; Crocker et al. 1999). This has been confirmed and elucidated in detail by X-ray crystallography of the N-terminal domain of Sn complexed with 2,3-sialylactose (May et al. 1998), which has been described in detail in the chapter by May and Jones.

The N-terminal domains of all siglecs share several similarities in their amino acid sequences (Fig. 2). In contrast to the regular pattern of cysteins found in typical members of the IgSF, which form an intersheet disulfide bridge between the B and F strands, siglecs contain an intrasheet bridge between the B and E strands. This allows a somewhat larger distance between the two β-sheets which appears to be important for the binding of Sia residues (May et al. 1998). One additional cystein residue occurs at the C-terminal end of the B strand which builds an interdomain bridge to the following domain 2.

Several amino acids are highly conserved among the siglecs (Fig. 2). Whereas most of these are critical for the overall structure of the Ig-fold, some of these amino acids are important for the interaction with the bound Sia, especially an arginine (Arg^{97} in Sn) towards the C-terminal end of the F-strand, as discussed in detail in the chapter by May and Jones and below.

2.2 Glycosylation

The extracellular parts of all siglecs contain several potential N-glycosylation sites. Only in a few cases has their use been characterised to some extent. With the exceptions of Sn and SMP, all siglecs described contain N-glycans also in domain 1, the Sia binding domain. One of these sites occurs at corresponding positions in CD22 (Asn^{111}), all CD33-like siglecs (Asn^{98}) and MAG (Asn^{99}; Fig. 2) From site-directed mutagenesis experiments it was concluded that glycosylation at this position is required for binding activity (in CD22), can prevent binding to Sia if this glycan is sialylated (in CD33; Sgroi et al. 1996), or does not influence binding (in MAG; Tropak and Roder 1997). However, this asparagine residue is located at the N-terminal end of the E-strand, on the opposite site with respect to the Sia binding site. Therefore, it is difficult to understand how this glycosylation site can have an effect on the binding site of the same molecule. The effect observed with CD33 could be explained by occupation of the other binding site in the dimeric Fc-chimeras used in these experiments. Furthermore, this asparagine is right next to the cysteine which forms the intrasheet disulfide to the B-strand. The loss of binding detected for CD22 may be due to an aberrant folding of the domain also affecting the Sia binding site.

3 Carbohydrate Recognition

3.1 Methodology

Several methods have been used to investigate the binding specificity of siglecs for their sialylated binding partners (Crocker and Kelm 1996; Kelm et al. 1998). Like most other carbohydrate-binding proteins, the affinity for single glycans is relatively low and an oligovalent presentation of both the lectin and the sugar is required to obtain strong binding. Therefore, it is not surprising that many of the methods applied in such experiments involve multivalent glycan presentations created on cell or plastic surfaces or on synthetic neoglyconjugates.

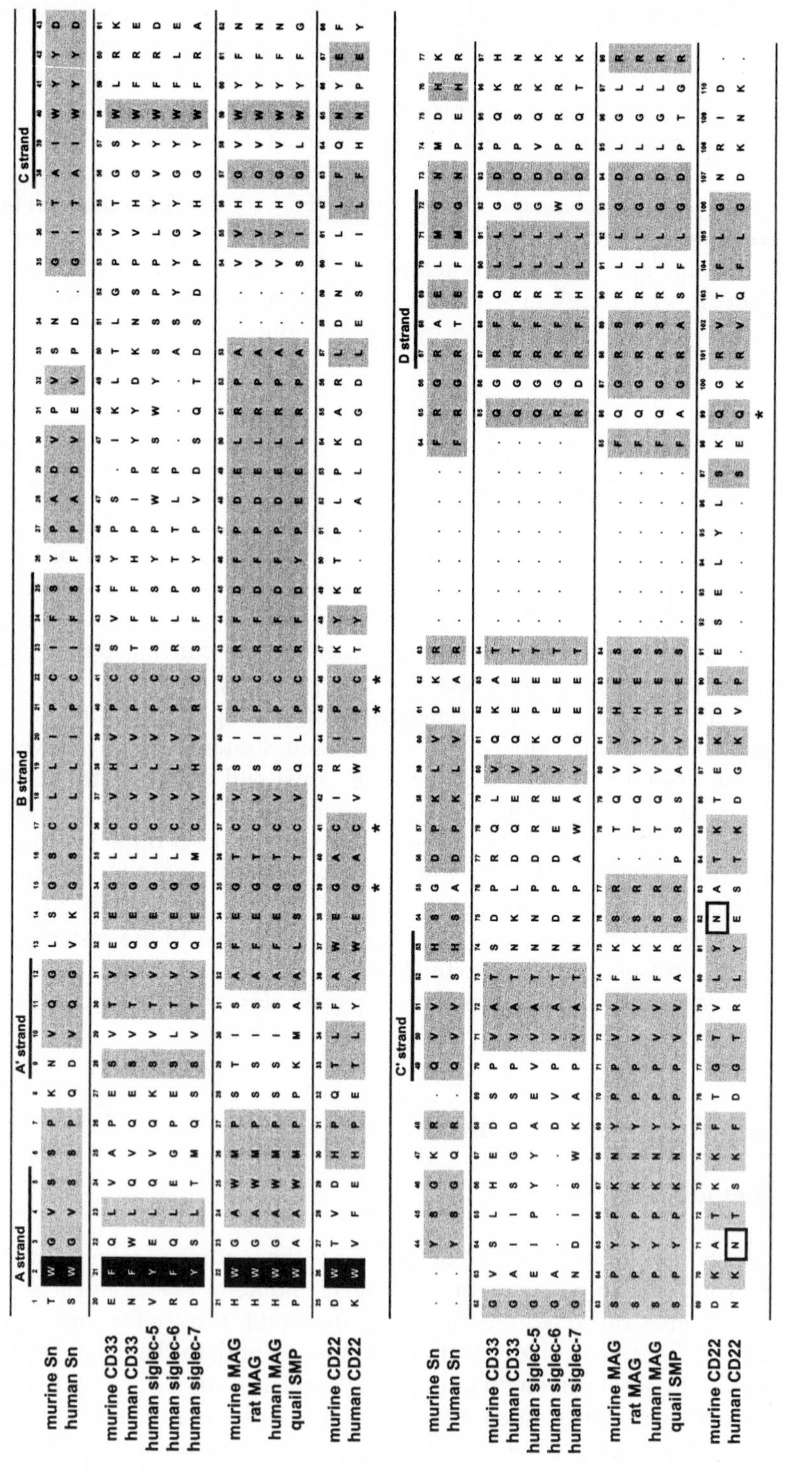
A strand
A' strand
B strand
C strand
C' strand
D strand
murine Sn
human Sn
murine CD33
human CD33
human siglec-5
human siglec-6
human siglec-7
murine MAG
rat MAG
human MAG
quail SMP
murine CD22
human CD22

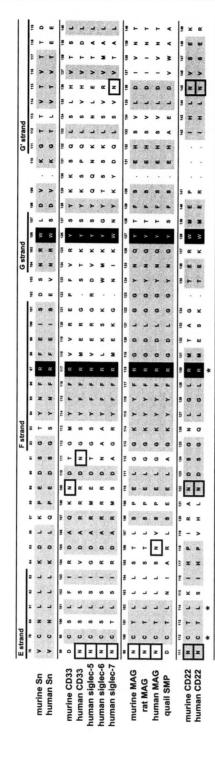

Fig. 2. Alignment of the N-terminal domains of siglecs described to date. The N-terminal domains of all known siglec sequences are grouped according to sequence similarities. Amino acid residues identical in each group are on a *grey* background, amino acids with side chains involved in Sia-binding (deduced from the Sn structure) are printed *white on a black background*, amino acids identical in all siglecs are marked with an *asterix* under the CD22 sequence. The amino acid numbering for the top sequence in each group is given *above* that sequence. Positions of potential *N*-glycans are *boxed*. The strand assignment is based on the crystal structure of Sn (May et al. 1998)

3.1.1 Cell Surface Resialylation and Neoglycoconjugates

Human erythrocytes have been widely used as cellular models for the presentation of sialylated glycans. In order to dissect the specificity for types and linkages of Sia recognised, the glycans on these cells can be modified enzymatically by first removing all Sia with sialidases and then resialylating them with glycan and linkage-specific sialyltransferases and CMP-Sia (Kelm et al. 1994, 1998; Crocker and Kelm 1996). Due to the relatively wide specificity of sialyltransferases for the type of Sia in the donor substrate, naturally occurring or synthetic Sia modifications can be transferred to the erythrocytes. Such resialylated cells have been useful for demonstrating the specificity of siglecs. Although this experimental approach can also be used for other cells, the continuous glycosylation process will replace the artificial glycans with the naturally occurring structures in a relatively short time. Another approach for obtaining unusual Sia on cell surfaces is to feed the cells with precursors, namely mannosamine derivatives, which then are converted into the corresponding Sia (Kayser et al. 1992; Keppler et al. 1995; Herrmann et al. 1997). Obviously, this approach is restricted to modifications of the N-acyl residue at C-5 of Sia.

The presentation of glycans on synthetic glycoconjugates (neoglycoconjugates) has become an important tool for investigations of lectin biochemistry and have also been used in studies on siglecs. Carrier molecules used today are synthetic polymers like polyacrylamide (Cornish et al. 1998; Strenge et al. 1998; Patel et al. 1999) and proteins like bovine serum albumin (BSA; Kelm et al. 1994) or streptavidine (Hashimoto et al. 1998). On these, basically all chemically accessible carbohydrate structures can be presented in an oligomeric fashion. In addition, they can be labelled with radioactive, chromogenic or fluorescent tracers for quantification or histochemical analysis.

3.1.2 Cell Binding Assays

Most assays with siglecs involve cells to present the siglec or the glycans on their surface (Crocker and Kelm 1996). Molecular cloning of cDNAs coding for siglecs have been instrumental for their characterisation. The easiest and most direct approach is to express these transiently as full-length constructs in COS cells (monkey kidney fibroblast cell line, stably expressing the large T antigen from SV40 virus). Therefore, it is not surprising that transfected COS cells have been used in all initial experiments demonstrating the Sia-binding activity of siglecs. The simplest way to do this is by cell binding assays with erythrocytes and/or other cells, since untransfected COS cells are not adhesive for most cell types. Sia-dependent binding can be shown with sialidase-treated cells which should not be bound.

3.1.3 Fc-Chimeras

Recombinant soluble siglecs fused to the Fc part of human IgG (Fc-chimeras) have been very useful tools in further characterisations of siglecs (Sgroi et al. 1993; Kelm et al. 1994; Powell and Varki 1994; Freeman et al. 1995; Crocker and Kelm 1996). Where possible, these Fc-chimeras contain only those N-terminal domains of siglecs which are necessary for binding. Fc-chimeras are easy to purify on protein-A-agarose and can be used in many different assays. In addition, these have provided useful tools for site-directed mutagenesis (Nath et al. 1995; van der Merwe et al. 1996; Vinson et al. 1996; May et al. 1998).

Fc-chimeras can easily be immobilised on plastic surfaces coated with anti-Fc antibodies. To such captured Fc-chimeras of siglecs, cells or glycoproteins bind in a Sia-dependent manner (Crocker and Kelm 1996). It is important to note that for binding, a threshold density of captured Fc-chimeras is essential. Interestingly, this is tenfold higher for glycoprotein than for cell binding (Kelm et al. 1994). This assay has been particularly useful for identifying the cell types binding to Sn or CD22 from cell mixtures such as bone marrow or blood cells (Crocker et al. 1995).

Since this solid phase assay gives no quantitative data regarding the amount of siglecs bound to cells, a radioactive fluid phase assay with soluble complexes of Fc-chimeras with anti-Fc antibodies has been developed (Kelm et al. 1994; Crocker and Kelm 1996). This assay has not only been used to quantify the amount of siglecs bound to cells but also as a basis for hapten inhibition assays. Direct binding of small monovalent sialosides is difficult to detect, since they bind only with low affinity (0.1 – 10 mM Kd). However, the inhibitory potential of such molecules as competitors for Sia-dependent binding is a useful way to determine their affinity for siglecs (Kelm et al. 1998; Strenge et al. 1998). Such assays have been used to elucidate the contributions of structural elements of synthetic and naturally occurring sialosides and oligosaccharides for their interaction with siglecs as discussed below.

3.2 Sialic Acid Interactions with Siglecs

Over the last few years, contributions from several laboratories using a variety of different methods have led to a relatively clear picture for the interactions of siglecs with Sia on the atomic level. The three-dimensional structure of the Sia-binding domain from Sn with its bound cognate ligand 2,3-sialyllactose has been described in detail in the chapter by May and Jones. It also provides the basis for model structures of the other siglecs. The crystal structure shows that most of the Sia residue is in close contact with the protein, forming a bidentate salt bridge, several hydrogen bonds and hydrophobic contacts as discussed below. This demonstrates that during evolution the binding site of Sn and probably also the other siglecs has been adapted very well to the specific recognition of sialylated glycans, suggesting that Sia recognition is important for their biological functions. In contrast, the penultimate monosaccharides have almost no contacts with the protein, but are more or less completely surrounded by the solvent.

3.2.1 Functional Groups of Sialic Acids and Amino Acids

Several characteristic elements can be found in the structure of Sia (Fig. 1) which could contribute to specific recognition by proteins. These are the carboxyl group at C-2, the glycerol side chain (C-7 through C-9), the different substituents at C-5 and the missing hydroxyl at C-3. It is interesting to elucidate which of these structural features are important for the binding of siglecs to Sia.

Crystallography gives a very detailed three-dimensional structure, and distances between the atoms of amino acids involved and the bound carbohydrate can be determined. Although this allows the assignment of hydrogen bonds, hydrophobic contacts and salt bridges, from this information it cannot be deduced which of these contacts really contribute to the binding strength (Janin 1997). Synthetic Sia analogues with modifications of the functional groups have therefore been used to evaluate such contributions.

3.2.1.1 Carboxyl Group and Ring Substitutents

Elimination of the carboxylate of Sia, e.g. by esterification, abolishes binding of Sn, MAG or SMP (Collins et al. 1997a,b). On glycoconjugates, Sia occur in the α-linkage placing the carboxylate in an axial position. This is necessary for binding to siglecs, since β-linked glycosides of Sia are not bound by Sn or MAG in hapten inhibition assays (Table 3). Furthermore, NMR experiments with the N-terminal domain of Sn demonstrated concentration-dependent binding of α-methyl sialoside, whereas no binding could be detected with the β-methyl isomer (Crocker et al. 1999). In the crystal structure of Sn the carboxylate of Sia is in close proximity to an arginine residue (Arg[97]) of the F-strand. This residue is highly conserved in all siglecs identified (Fig. 2) and seems to be essential for Sia binding, since even a conservative change to lysine causes a dramatic reduction in affinity as shown for Sn (Vinson et al. 1996; Crocker et al. 1999), CD22 (Arg[130]) (van der Merwe et al. 1996) and MAG (Arg[118]) (Tang et al. 1997; own unpubl. obs.).

The crystal structure of Sn revealed a hydrogen bond between the C-4 hydroxyl and the side chain of Ser[103] (May et al. 1998). However, inhibition experiments with 4-deoxy-Sia provided evidence that this interaction is not essential for binding (Strenge et al. 1998).

Table 3. Relative inhibitory potencies of synthetic and naturally occurring sialosides. The rIPs of each compound was calculated by dividing the IC_{50} of the reference substance Neu5AcαMe (for monosaccharides) or 2,3-sialyllactose (for oligosaccharides) by the IC_{50} of the compound of interest. This results in rIPs above 1.0 for Sia-derivatives binding better than the reference and rIPs lower than 1 for structures binding more weakly (Kelm et al. 1998; Strenge et al. 1998)

Compound	rIP for Sn	rIP for MAG
Neu5AcαMe	1.00	1.00
Neu5AcβMe	n.a.[a]	n.a.[a]
Neu5propαMe	1.56	1.56
Neu5NH$_2$AcαMe	0.14	n.a.
Neu5FAcαMe	1.67	16.94
Neu5ClAcαMe	0.78	n.a.
Neu5F$_3$AcαMe	1.40	4.04
Neu5thioAcαMe	0.87	3.85
9-deoxy-Neu5Ac	0.03	n.a.
9-Cl-Neu5Ac	0.04	n.a.
9-NH$_2$-Neu5Ac	1.56	2.98
Neu5Acα2,3Galβ1,4Glc	1.00	1.00
Neu5Acα2,6Galβ1,4Glc	0.21	0.12
Neu5Acα2,3Galβ1,4AllNAcβ1,3Galβ1,4Glcβ-SE	1.3	6.3
Neu5Acα2,6Galβ1,3GlcNAcβ1,3Galβ1,4Glcβ-SE	0.52	0.82

[a] n.a., not applicable, less than 50% inhibition at the highest concentration tested (20 mM)

3.2.1.2 N-*acyl group*

The finding that the murine orthologues of Sn and CD22 distinguish between Neu5Ac and Neu5Gc already suggested that the *N*-acyl residues of Sia contribute significantly to the binding specificity (Kelm et al. 1994). The crystal structure of Sn complexed with 2,3-sialyllactose then showed that the methyl group of the acetyl substituent in Neu5Ac is in contact with the Trp[2] residue of Sn (May et al. 1998). Furthermore, H[1]-NMR binding experiments clearly demonstrated a strong interaction leading to a line broadening and an upfield shift of the signal for the methyl group (Crocker et al. 1999). The importance of this amino acid is further supported by the observation that a mutation to glutamine abolished any detectable Sia binding (May et al. 1998; Crocker et al. 1999). Experiments with Sia containing modified *N*-acyl residues also demonstrated its relevance for the recognition by siglecs. As mentioned above, Neu5Gc is recognised much more poorly by Sn (Kelm et al. 1994, 1998). This could be explained by an interference of the additional hydroxyl group in this Sia with the van der Waals contact. It is unlikely to be caused by steric hindrance, since other modifications, like an extension of the acetyl residue to a propionyl or the introduction of halogen atoms, are well tolerated.

All siglecs described so far contain an aromatic residue at the position analogous to Trp[2] in Sn (Fig. 2), suggesting that this hydrophobic interaction is also important for Sia-binding by other members of the family. However, several observations provide evidence that its contribution is less pronounced for other siglecs or at least occurs in a different context. For example, in H^1-NMR experiments with Fc-chimeras of murine CD22, CD33 and MAG, the changes of signal for the methyl group of the N-acetyl was much smaller compared with that obtained with Sn (Crocker et al. 1999). The same modification of the N-acetyl residue at C-5 of Neu5Ac can enhance or prevent binding by siglecs. For example, in contrast to Sn, CD22 binds Neu5Gc very well (Kelm et al. 1994, 1998). In addition, murine CD22 strongly prefers Neu5Gc over Neu5Ac, whereas human CD22 binds to both Neu5Gc and Neu5Ac about equally well. This would suggest that the hydrophobic interaction with Trp^2 in Sn discussed above does not play a role in Sia binding by CD22. Nevertheless, the tryptophane residue also seems to be important in murine CD22, as shown by site-directed mutagenesis (K. Piperi and P.A. van der Merwe, pers. comm.). Apparently, the aromatic residue found in these siglecs does not occupy exactly the same position relative to Sia in the binding site.

One interesting possibility is that of designing Sia analogues with appropriate modifications of the substituent at C-5 which would bind preferentially with one siglec. Examples of this have been shown for Sn, C22 and MAG (Kelm et al. 1998). Whereas Sia containing a halogenated N-acetyl group at C-5 bind to MAG better than Neu5Ac (Table 3), no enhanced binding to Sn or CD22 has been found. Replacing the oxygen of the acetyl residue with sulphur also caused enhanced binding to MAG but not to Sn. Even changing the entire N-acetyl substituent to a hydroxyl group as in Kdn improves binding, whereas binding to Sn is drastically reduced (Strenge et al. 1998). All these data lead to the conclusion that for the binding of Sia by MAG, a hydrogen bond from the substituent at C-5 is possibly important for binding, which more than compensates the loss of a potential hydrophobic interaction of the N-acetyl substituent. This could come either from the nitrogen of the N-acetyl group in Neu5Ac, which is enhanced by the halogen or sulphfur in the Sia analogues described above, or from the hydroxyl residue in Kdn. The crystal structure of Sn reveals a potential hydrogen bond of the amido group in Neu5Ac to the backbone carbonyl of Arg^{105} (May et al. 1998). A comparison of the primary amino acid sequences does not reveal an obvious explanation for these different binding properties and a final clarification can only come from a crystal structure of MAG complexes with the appropriate Sia.

3.2.1.3 Extracyclic Side Chain

The simplest experiment to demonstrate the importance of the extracyclic glycerol side chain for binding is its cleavage by periodate oxidation under specific mild conditions followed by reduction (Schauer 1982). Only the hydroxyl at C-7 remains, whereas those at C-8 and C-9 are removed. This modification can be performed easily on both soluble glycoconjugates and on cells. For Sn, CD22 and MAG, it has been demonstrated that such mild periodate oxidation abrogates binding (Sgroi et al. 1993; Powell and Varki 1994; Hanasaki et al. 1995; Powell et al. 1995; Collins et al. 1997a,b; Razi and Varki 1998; Sawada et al. 1999). Although these experiments provided evidence that either the hydroxyl at C-8 or at C-9, or both, are important, their individual contributions or those of hydrophobic interactions can not be deduced. According to the crystal structure, both hydroxyl groups form hydrogen bonds with the peptide backbone (Leu[107] in Sn). Sia analogues were useful tools to demonstrate that both hydroxyl groups at C-8 and at C-9 significantly strengthen binding of the Sia residue, since the removal of one of these substituents was sufficient to lose binding to a large extent, whereas the hydroxyl at C-7 does not appear to be necessary (Kelm et al. 1998; Strenge et al. 1998).

Masking of Sia hydroxyl groups by O-acetylation is commonly found in nature (Table 1). Particularly interesting is the 9-O-acetylation, since it prevents binding by Sn, CD22 and MAG (Kelm et al. 1945; Sjoberg et al. 1994; Shi et al. 1996; Kelm and Schauer 1997). In principle, this could be caused by steric hindrance and/or by loss of a hydrogen bond. The crystal structure of Sn shows additional space around the C-9 hydroxyl along the G-strand, suggesting that steric hindrance is unlikely to be relevant for the O-acetylation effect. In agreement with the crystal structure, the hydroxyl at C-9 appears to donate a hydrogen bond to the carbonyl of Leu[107], since it can be replaced by an amino group (Kelm et al. 1998). In contrast to a hydroxyl group, an amino function can also donate a hydrogen bond after acetylation. Therefore, Sia analogues with aminoacyl substituents at C-9 were used in inhibition assays. These experiments demonstrated that acetyl groups as well as even bulkier residues such as a hexanoyl group were tolerated (Fig. 3).

In addition to the hydrogen bonds discussed, the glycerol side chain is placed over Trp[106,] allowing a van der Waals contact (May et al. 1998).

In summary, the most significant contributions to the recognition of Sia by siglecs come from (1) a bidentate salt bridge between a conserved arginine residue in the F-strand and the carboxylate, (2) hydrogen bonds between the peptide backbone of the G-strand with hydroxyls at C-8 and C-9, and (3) interactions with the N-acyl substituent at C-5 with the N-terminal parts of the A- and G-strands (Fig. 4). Whereas the first two interactions are common for all siglecs investigated so far, the types of interactions with the N-acyl group and their contributions are different between siglecs, leading to specific recognition of naturally occurring and synthetic Sia.

3.3 Glycan Interactions with Siglecs

3.3.1 Linkage Specificity

In common mammalian glycans, Sia are glycosidically linked to galactose (2,3 or 2,6), N-acetylgalactosamine (2,6), Sia (2,8) or, less commonly, to N-acetyl-glucosamine (2,6-linked) units (Schauer and Kamerling 1997). Most siglecs

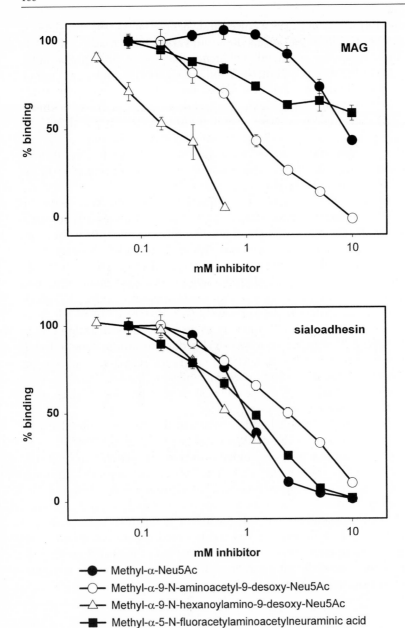

Fig. 3. Hapten inhibition assays with sialic acid analogues containing 9-aminoacyl residues. Hapten inhibition assays were performed with Fc-chimeras of MAG and Sn with human erythrocytes as target cells as described (Kelm et al. 1998)

Fig. 4. Interactions of the sialic acid *N*-acetyl and glycerol side chain with amino acids of the Sn and MAG binding site. Only the bound 2,3-sialyllactose and selected amino acids from the A- and G-strand are shown. The position of amino acids from Sn are from the crystal structure (May et al. 1998), the MAG model has been constructed on the basis of homology to Sn

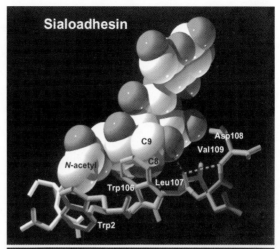

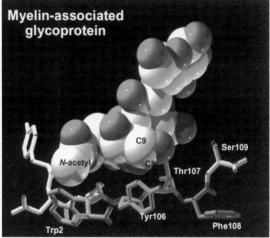

bind better to one type of linkage (Table 4). An obvious question is how this specificity relates to the structures of the binding sites. In principle, the specificity could be due to additional contacts of the undelying glycan or due to steric restraints which prevent the Sia residue from binding optimally.

The only contact between the subterminal galactose and glucose units and the binding site of Sn observed in the crystal structure is between the hydroxyl at C-6 of galactose and Tyr[44]. However, mutation of this amino acid to alanine (May et al. 1998) or removal of the hydroxyl group of galactose (Strenge et al. 1998) did not lead to a significant loss of binding. Therefore, it appears that this hydrogen bond is not relevant for binding. Binding assays with Sn and a series of modified glycans provided evidence that the underlying glycan

Table 4. Specificity of siglecs for the linkage and type of sialic acid

Name	Linkage specificity	Specificity for Sia modification
Sialoadhesin (Sn, siglec-1)	2,3>>2,6	Neu5Ac>>Neu5Gc, Masking by 9-O-acetylation
CD22 (siglec-2)	2,6>>2,3	Neu5Gc>>Neu5Ac (murine) Neu5Gc = Neu5Ac (human), Masking by 9-O-acetylation
CD33 (siglec-3)	2,3>>2,6	Unknown
Myelin-associated glycoprotein (MAG, siglec-4a); Schwann cell myelin protein (SMP, siglec-4b)	2,3>>2,6	Neu5Ac>>Neu5Gc, Masking by 9-O-acetylation
OB-BP2 (siglec-5)	2,3=2,6	Unknown
OB-BP1 (siglec-6)	2,6GalNAc	Unknown
p75/AIRM1 (siglec-7)	2,3=2,6	Unknown

only slightly enhances binding, since 2,3-sialyllactose and even longer oligosaccharides were bound only marginally better than the methyl glycoside of Neu5Ac.

MAG apparently binds better to longer oligosaccharides (Table 3), since pentasaccharides were almost tenfold better inhibitors than the corresponding trisaccharides. It is important to note that this enhancement was also observed if the oligosaccharides contained 2,6 linked Sia (Strenge et al. 1998). Similar conclusions can be drawn from H^1-NMR experiments with Sn and the 2,3 and 2,6 isomers of sialyllactose (Crocker et al. 1999). Therefore, steric hindrance appears to have only a small effect on Sn or MAG binding, at least for these simple oligosaccharides.

Binding of oligosaccharides to CD22 is significantly enhanced if the Sia is 2,6-linked to galactose, N-acetylgalactosamine, or N-acetylglucosamine (Powell et al. 1995). These experiments suggest that CD22 interacts at least with the subterminal monosaccharide moiety as well as the Sia residue. Such contacts could be to the CC′ area of the GFC face which shows a considerable difference in the amino acid sequence between CD22 and Sn. Interestingly, Sia-dependent binding of murine CD22 was lost if Tyr[66] was mutated to aspartic acid. Although this is a drastic change in the amino acid, it may be related to additional contacts between the glycan and the protein. Only an X-ray crystallographic analysis and further mutagenesis experiments can provide conclusive answers to these questions.

In general it appears that only in the context of a multivalent presentation, as on cell surfaces or complex glycoconjugates, do these small differences add up to the linkage specificity generally found in cell binding assays. Since such multivalent interactions are also likely to occur in vivo, it seems reasonable that the linkage specificity of siglecs is of biological significance.

3.4 Ligands

3.4.1 Cell Surface Glycoproteins

The glycans to which siglecs bind occur on a large variety of cell surface glycoproteins and glycolipids. The first characterisation of Sn demonstrated that it can bind to both types of glycoconjugates on Western blots and thin layer chromatograms (Crocker et al. 1991). Also, MAG binding to glycoproteins and glycolipids can be demonstrated in vitro (Kelm et al. 1994; Collins et al. 1997b; Sawada et al. 1999; Strenge et al. 1999). An obvious question is which of these potential ligands are biologically relevant in vivo. Using Fc-chimeras of CD22, Sn or MAG, several glycoproteins could be identified as potential binding partners by affinity precipitations of cell extracts or serum (Sgroi et al. 1993; Hanasaki et al. 1995; Strenge et al. 1999). Interestingly, the proteins precipitated are specific for the siglec used. Even siglecs with very similar glycan specificity to Sn and MAG precipitate different proteins (Fig. 5). This is evidence that additional structural elements contribute to recognition by siglecs. However, further characterisation of these potential ligands, of both the protein and the carbohydrate parts, are necessary for a better understanding of their biological significance.

3.4.2 Extracellular Glycoproteins

Binding partners for siglecs can occur on cell surfaces or the extracellular space. In fact, two serum glycoproteins, IgM and haptoglobin bind to CD22 with high affinity (Hanasaki et al. 1995). The extracellular matrix contains several glycoproteins which can bind to siglecs, if they carry the appropriate glycans.

3.4.3 Glycolipids

Glycolipids have been proposed as important binding partners for siglecs, especially for MAG (Collins et al. 1997a,b; Sawada et al. 1999). However, several lines of evidence argue in favour of glycoproteins as ligands for MAG, at least for net binding in vitro: (1) trypsin treatment of neurons abrogates binding of MAG to these cells; (2) cultivation of neuroblastoma cells in the presence of swainsonine, an inhibitor of N-glycan processing, reduces MAG-binding to background levels (Strenge et al. 1999); (3) binding of MAG to neuroblastoma cells is not enhanced if the cells have been cultivated in the presence of the sialidase inhibitor 2,3-dideoxy-N-acetylneuraminic acid, which leads to a higher expression level of complex glycolipids (Kpoitz et al. 1996) identified as potential MAG ligands. Although these studies provide strong evidence that the majority of MAG binding to neuronal cells in vitro is to glycoproteins car-

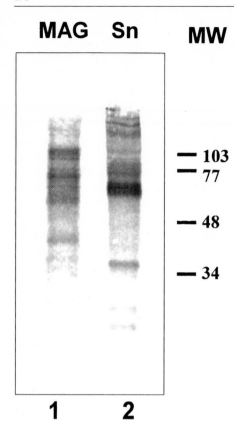

Fig. 5. Glycoproteins from neuroblastoma cells precipitated with Fc-chimeras of MAG and Sn. Glycoproteins from N₂A neuroblastoma cells labelled with ³H-glucosamine were affinity precipitated with either Fc-MAG or Fc-Sn and run on an SDS-PAGE, blotted on nitrocellulose and visualised with a phospho imager (Strenge et al. 1998). Molecular weight standards are indicated on the *right*

rying *N*-glycans, important regulatory roles of glycolipids interacting with MAG are quite possible.

3.5 Potential Regulation by *cis*-Interactions or Soluble Competitors

As discussed above, potential binding partners for siglecs occur on cell surfaces and in the extracellular space. Besides glycoconjugates on the opposing cell, molecules on the same cell can also interact with siglecs (Fig. 6). Such *cis*-interactions can actually mask the binding activity of siglecs expressed on cells carrying the sialylated glycans recognised by the siglec. For example, the Sia-binding activity of CD33 was only detected after sialidase treatment of the COS cells used as the expression system (Freeman et al. 1995). Similar observations were made with MAG and SMP (Tropak and Roder 1997) or siglec-5 (Cornish et al. 1998) if expressed in CHO cells which are rich in 2,3-linked Sia, the structure recognised by these siglecs. Obviously, the glycosyltransferases, i.e. sialyltransferases, expressed by a cell control the pattern of glycans available for siglec binding and therefore also control the potential for *cis*-interactions. For

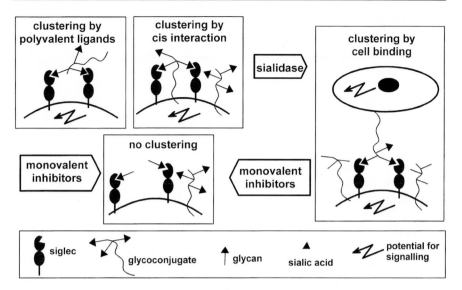

Fig. 6. Potential binding partners of siglecs and their potential role in signal transduction

example, the Sia-binding activity of CD22 is easily detected if it is expressed in COS cells which only contain 2,3-linked Sia, but is masked if the COS cells are co-transfected with cDNA for ST6GalI – the enzyme synthesising the glycan recognised by CD22 (Braisch-Andersen and Stamenkovic 1994; Hanasaki et al. 1995). All siglecs described are glycoproteins and therefore have the potential for *cis*-interactions. As described above such a mechanism has been proposed for CD22 and CD33 (Sgroi et al. 1996).

Of course, *cis*-interactions can only occur if the glycans can spatially reach the siglec binding site. With 17 Ig-domains, Sn is the largest member of the family and it has been proposed that during evolution it reached this size to position the Sia-binding site far enough from the plasma membrane outside of the glycocalyx, since Sn constructs containing less than six Ig-domains cannot mediate cell-cell binding even at high levels of expression (P.R. Crocker, pers. comm.). But even the potential of full-length Sn to mediate cell adhesion seems to be regulated by the level of 2,3-sialyltransferase activity (Barnes et al. 1999).

cis-Interactions of siglecs can have different biological consequences. Besides competing with glycoconjugates in the environment, binding to *cis*-ligands can also have specific functions on the cell itself, such as enabling or supporting the association of a siglec with other receptors, and through this mechanism regulating their biological activity.

In conclusion, binding of carbohydrate ligands to siglecs can regulate cellular responses as summarised in Fig. 6. Siglecs can be clustered on cell surfaces by interacting with multivalent binding partners such as cell surfaces, extracellular matrix or soluble glycoproteins. This then may lead to intracellular signalling, for example through the ITIM motifs found in several siglecs. As mentioned above, for CD22 and MAG it has been demonstrated that clustering of these siglecs leads to an increased phosphorylation of tyrosine residues within their cytoplasmic domain. However, although it can be expected, it has not been shown whether this phosphorylation can also be induced by their glycoconjugate ligands. Another effect of siglecs binding to *trans* ligands could be that the siglec molecules are dissociated from their *cis*-binding ligands and influence the signalling properties of these ligands. For example, CD22 could be withdrawn from the B-cell receptor complex after binding to sialylated glycans.

Also, the binding partners of siglecs on the opposing cell could be clustered through a sialic acid-mediated interaction and then could start a signal cascade in that cell. For example, the effects of MAG on neurite outgrowth could follow this route.

All the scenarios mentioned depend on clustering of siglecs, which generally have a much higher affinity for multivalent ligands. However, monovalent sialosides can also bind to siglecs and inhibit the interaction with multivalent ligands.

Since the glycans to which siglecs can bind are relatively common, it is likely that cellular interactions of siglecs are controlled by the relative distribution of these glycans in each situation. Furthermore, this distribution is controlled by the expression levels of glycosyltransferases, i.e. the sialyltransferases, and by the enzymes controlling the modifications of sialic acids. For example, the CMP-Neu5Ac hydroxylase regulates the relative concentration of CMP-Neu5Ac and CMP-Neu5Gc, donor substrates for sialyltransferases. As mentioned above, in the mouse CD22 will bind with high preference only to cells carrying Neu5Gc, whereas Sn does not bind well to this Sia. Therefore, high levels of CMP-Neu5Ac hydroxylase activity are necessary for CD22 binding but will prevent Sn binding, even if these cells contain high levels of 2,3-linked Sia (Kelm et al. 1998). Similarly, expression of the appropriate sialate-9-*O*-acetyl-transferase leading to 9-*O*-acetylated Sia can therefore reduce binding to siglecs. This mechanism prevents binding of CD22 to many cell types, since de-*O*-acetylation of Sia unmasks these binding sites (Sjoberg et al. 1994).

Also, extracellular glycosidases, i.e. sialidases, can regulate the availability of binding sites. This has been shown for CD22. Whereas on resting B cells CD22 binding activity is masked by *cis*-binding, in vitro activation of these cells unmasks the binding activity at least partly, possibly by sialidase activity (Razi and Varki 1998).

4 Perspectives

The siglecs are a group of probably more than ten Ig-like proteins which bind sialylated glycans. Most of them show a selectivity for the type of Sia, both the linkage to the penultimate sugar and the modification of Sia. In general, they are found only on restricted cell populations. Although this is evidence that they have specific biological functions, these are still unclear in most cases. For two siglecs, CD22 and MAG, the effects of gene disruption have been described. From these it could be concluded that they are important for the regulation of the B cell-dependent immune response and the maintenance of functional myelin, respectively. However, it is still unclear what role the Sia-binding activities have.

Answering this question will certainly be an important task for the next few years. Understanding the protein-carbohydrate interaction as discussed here will be helpful not only for a better understanding of the molecular mechanism of Sia-recognition but also for the development of potent monovalent inhibitors, which in turn will be useful tools for biological studies and possibly for medicine, if it is required to modulate these siglec-mediated processes.

References

Barclay AN, Brown MH (1997) Heterogeneity of interactions mediated by membrane glycoproteins of lymphocytes. Biochem Soc Trans 25:224–228

Barnes YC, Skelton TP, Stamenkovic I, Sgroi DC (1999) Sialylation of the sialic acid binding lectin sialoadhesin regulates its ability to mediate cell adhesion. Blood 93:1245–1252

Bevilacqua M, Butcher E, Furie B, Gallatin M, Gimbrone M, Harlan J, Kishimoto K, Lasky L, McEver R, Paulson J, Rosen S, Seed B, Siegelman M, Springer T, Stoolman L, Tedder T, Varki A, Wagner D, Weissman I, Zimmerman G (1991) Selectins – a family of adhesion receptors. Cell 67: 233–233

Braesch-Andersen S, Stamenkovic I (1994) Sialylation of the B lymphocyte molecule CD22 by a 2,6-sialyltransferase is implicated in the regulation of CD22-mediated adhesion. J Biol Chem 269:11783–11786

Campbell MA, Klinman NR (1995) Phosphotyrosine-dependent association between CD22 and protein tyrosine phosphatase 1C. Eur J Immunol 25:1573–1579

Collins BE, Kiso M, Hasegawa A, Tropak MB, Roder JC, Crocker PR, Schnaar RL (1997a) Binding specificities of the sialoadhesin family of I-type lectins – sialic acid linkage and substructure requirements for binding of myelin-associated glycoprotein, Schwann cell myelin protein, and sialoadhesin. J Biol Chem 272:16889–16895

Collins BE, Yang LJ, Mukhopadhyay G, Filbin MT, Kiso M, Hasegawa A, Schnaar RL (1997b) Sialic acid specificity of myelin-associated glycoprotein binding. J Biol Chem 272:1248–1255

Cornish AL, Freeman S, Forbes G, Ni J, Zhang M, Cepeda M, Gentz R, Augustus M, Carter KC, Crocker PR (1998) Characterization of siglec-5, a novel glycoprotein expressed on myeloid cells related to CD33. Blood 92:2123–2132

Crocker PR, Feizi T (1996) Carbohydrate recognition systems: functional triads in cell-cell interactions. Curr Opin Struct Biol 6:679–691

Crocker PR, Kelm S (1996) Methods for studying the cellular binding properties of lectin-like receptors. In: Herzenberg LA, Weir DM, Blackwell C (eds) Weir's handbook of experimental immunology. Blackwell Science, Cambridge, pp 166.1–166.11

Crocker PR, Kelm S, Dubois C, Martin B, McWilliam AS, Shotton DM, Paulson JC, Gordon S (1991) Purification and properties of sialoadhesin, a sialic acid-binding receptor of murine tissue macrophages. EMBO J 10:1661–1669

Crocker PR, Mucklow S, Bouckson V, McWilliam A, Willis AC, Gordon S, Milon G, Kelm S, Bradfield P (1994) Sialoadhesin, a macrophage sialic acid binding receptor for haemopoietic cells with 17 immunoglobulin-like domains. EMBO J 13:4490–4503

Crocker PR, Freeman S, Gordon S, Kelm S (1995) Sialoadhesin binds preferentially to cells of the granulocytic lineage. J Clin Invest 95:635–643

Crocker PR, Clark EA, Filbin MT, Gordon S, Jones Y, Kehrl JH, Kelm S, Le Douarin NM, Powell L, Roder J, Schnaar R, Sgroi D, Stamenkovic I, Schauer R, Schachner M, Tedder T, van den Berg TK, van der Merwe PA, Watt SM, Varki A (1998) Siglecs – a family of sialic acid-binding lectins. Glycobiology 8 (Glycoforum 2):v–vi

Crocker PR, Vinson M, Kelm S, Drickamer K (1999) Molecular analysis of sialoside binding to sialoadhesin by NMR and site-directed mutagenesis. Biochem J 341:355–361

Falco M, Biassoni R, Bottino C, Vitale M, Sivori S, Augugliaro R, Moretta L, Moretta A (1999) Identification and molecular cloning of p75/AIRM1, a novel member of the sialoadhesin family that functions as an inhibitory receptor in human natural killer cells. J Exp Med 190:793–801

Freeman SD, Kelm S, Barber EK, Crocker PR (1995) Characterization of CD33 as a new member of the sialoadhesin family of cellular interaction molecules. Blood 85:2005–2012

Gagneux P, Varki A (1999) Evolutionary considerations in relating oligosaccharide diversity to biological function. Glycobiology 9:747–755

Hanasaki K, Powell LD, Varki A (1995) Binding of human plasma sialoglycoproteins by the B cell-specific lectin CD22 – selective recognition of immunoglobulin M and haptoglobin. J Biol Chem 270:7543–7550

Hanasaki K, Varki A, Powell LD (1995) CD22-mediated cell adhesion to cytokine-activated human endothelial cells. Positive and negative regulation by a2–6-sialylation of cellular glycoproteins. J Biol Chem 270:7533–7542

Hashimoto Y, Suzuki M, Crocker PR, Suzuki A (1998) A streptavidin-based neoglycoprotein carrying more than 140 GT1b oligosaccharides: quantitative estimation of the binding specificity of murine sialoadhesin expressed on CHO cells. J Biochem Tokyo 123:468–478

Herrmann M, von der Lieth C-W, Stehling P, Reutter W, Pawlita M (1997) Consequences of a subtle sialic acid modification on the murine polyoma virus receptor. J Virol 71:5922–5931

Hirst GK (1941) Agglutination of red cells by allantoic fluid of chick embryos infected with influenza virus. Science 94:22–23

Hirst GK (1942) Adsorption of influenza virus hemagglutinins and virus by red blood cells. J Exp Med 76:195–209

Janin J (1997) Ångströms and calories. Structure 5:473–479

Jaramillo ML, Afar DEH, Almazan G, Bell JC (1994) Identification of tyrosine 620 as the major phosphorylation site of myelin associated glycoprotein and its implication in interacting with signaling molecules. J Biol Chem 269:27240–27245

Kayser H, Zeitler R, Kannicht C, Grunow D, Nuck R, Reutter W (1992) Biosynthesis of a non-physiological sialic acid in different rat organs, using N-propanoyl-D-hexosamines as precursors. J Biol Chem 267:16934–16938

Kelm S, Pelz A, Schauer R, Filbin MT, Tang S, de Bellard ME, Schnaar RL, Mahoney JA, Hartnell A, Bradfield P, Crocker PR (1994) Sialoadhesin, myelin-associated glycoprotein and CD22 define a new family of sialic acid-dependent adhesion molecules of the immunoglobulin superfamily. Curr Biol 4:965–972

Kelm S, Brossmer R, Gross HJ, Strenge K, Schauer R (1998) Functional groups of sialic acids involved in binding to sialoadhesins defined by synthetic analogues. Eur J Biochem 255: 663–672

Kelm S, Schauer R (1997) Sialic acids in molecular and cellular interactions. Int Rev Cytol 175:137–240

Kelm S, Schauer R, Manuguerra JC, Gross HJ, Crocker PR (1994) Modifications of cell surface sialic acids modulate cell adhesion mediated by sialoadhesin and CD22. Glycoconjugate J 11:576–585

Keppler OT, Stehling P, Herrmann M, Kayser H, Grunow D, Reutter W, Pawlita M (1995) Biosynthetic modulation of sialic acid-dependent virus-receptor interactions of two primate polyoma viruses. J Biol Chem 270:1308–1314

Kopitz J, von Reitzenstein C, Sinz K, Cantz M (1996) Selective ganglioside desialylation in the plasma membrane of human neuroblastoma cells. Glycobiology 6:367–376

Lasky LA (1995) Selectin-carbohydrate interactions and the initiation of the inflammatory response. Annu Rev Biochem 64:113–139

Law CL, Sidorenko SP, Chandran KA, Zhao ZH, Shen SH, Fischer EH, Clark EA (1996) CD22 associates with protein tyrosine phosphatase 1 C, Syk, and phospholipase C-1γ upon B cell activation. J Exp Med 183:547–560

Lowe JB, Stoolman LM, Nair RP, Larsen RD, Berhend TL, Marks RM (1990) ELAM-1-dependent cell adhesion to vascular endothelium determined by a transfected human fucosyltransferase cDNA. Cell 63:475–484

May AP, Robinson RC, Vinson M, Crocker PR, Jones EY (1998) Crystal structure of the N-terminal domain of sialoadhesin in complex with 3′ sialyllactose at 1.85 Aangström resolution. Mol Cell 1:719–728

McClelland L, Hare R (1941) The adsorption of influenza virus by red cells and a new in vitro method of measuring antibodies for influenza virus in the embryonated egg. Can J Public Health 32:530–538

Nath D, van der Merwe PA, Kelm S, Bradfield P, Crocker PR (1995) The amino-terminal immunoglobulin-like domain of sialoadhesin contains the sialic acid binding site – comparison with CD22. J Biol Chem 270:26184–26191

Nicoll G, Ni J, Liu D, Klenerman P, Munday J, Dubock S, Mattei M-G, Crocker PR (1999) Identification and characterisation of a novel siglec, siglec-7, expressed by human natural killer cells and monocytes. J Biol Chem274: 34086–34095

Patel N, Brinkman-Van der Linden ECM, Altmann SW, Balasubramanian S, Timans CS, Peterson D, Bazan JF, Varki A, Kastelein RA (1999) OB-BP1/Siglec-6 – a leptin- and sialic acid-binding protein of the immunoglobulin superfamily. J Biol Chem 427:22729–22738

Phillips ML, Nudelman E, Gaeta FCAPM, Singhal AK, Hakomori S-I, Paulson JC (1990) ELAM-1 mediates cell adhesion by recognition of a carbohydrate ligand, Sialyl-Le^x. Science 250: 1130–1132

Powell LD, Varki A (1994) The oligosaccharide binding specificities of CD22 b, a sialic acid-specific lectin of B cells. J Biol Chem 269:10628–10636

Powell LD, Sgroi D, Sjoberg ER, Stamenkovic I, Varki A (1993) Natural ligands of the B cell adhesion molecule CD22 b carry N-linked oligosaccharides with a-2,6-linked sialic acids that are required for recognition. J Biol Chem 268:7019–7027

Powell LD, Jain RK, Matta KL, Sabesan S, Varki A (1995) Characterization of sialyloligosaccharide binding by recombinant soluble and native cell-associated CD22 – evidence for a minimal structural recognition motif and the potential importance of multisite binding. J Biol Chem 270:7523–7532

Razi N, Varki A (1998) Masking and unmasking of the sialic acid-binding lectin activity of CD22 (Siglec-2) on B lymphocytes. Proc Natl Acad Sci USA 95:7469–7474

Rosenberg A (1995) Biology of the sialic acids. Plenum, New York

Sawada N, Ishida H, Collins BE, Schnaar RL, Kiso M (1999) Ganglioside GD1a analogues as high-affinity ligands for myelin-associated glycoprotein (MAG). Carbohydr Res 316:1–5

Schauer R (1982) Sialic acids. Adv Carbohydr Chem Biochem 40:131–234

Schauer R, Kamerling JP (1997) Chemistry, biochemistry and biology of sialic acids. In: Montreuil J, Vliegenthart JFG, Schachter H (eds) Glycoproteins II. Elsevier, Amsterdam, pp 243–402

Schulte RJ, Campbell MA, Fischer WH, Sefton BM (1992) Tyrosine phosphorylation of CD22 during B cell activation. Science 258:1001–1004

Sgroi D, Varki A, Braesch-Andersen S, Stamenkovic I (1993) CD22, a B cell-specific immunoglobulin superfamily member, is a sialic acid-binding lectin. J Biol Chem 268:7011–7018

Sgroi D, Nocks A, Stamenkovic I (1996) A single N-linked glycosylation site is implicated in the regulation of ligand recognition by the I-type lectins CD22 and CD33. J Biol Chem 271: 18803–18809

Shi WX, Chammas R, Varki NM, Powell L, Varki A (1996) Sialic acid 9-O-acetylation on murine erythroleukemia cells affects complement activation, binding to I-type lectins, and tissue homing. J Biol Chem 271:31526–31532

Sjoberg ER, Powell LD, Klein A, Varki A (1994) Natural ligands of the B cell adhesion molecule CD22 b can be masked by 9-*O*-acetylation of sialic acids. J Cell Biol 126:549–562

Strenge K, Schauer R, Bovin N, Hasegawa A, Ishida H, Kiso M, Kelm S (1998) Glycan specificity of myelin-associated glycoprotein and sialoadhesin defined by synthetic oligosaccharides. Eur J Biochem 258:677–685

Strenge K, Schauer R, Kelm S (1999) Binding partners for the myelin-associated glycoprotein of N(2)A neuroblastoma cells. FEBS Lett 444:59–64

Tang S, Shen YJ, deBellard ME, Mukhopadhyay G, Salzer JL, Crocker PR, Filbin MT (1997) Myelin-associated glycoprotein interacts with neurons via a sialic acid binding site at ARG118 and a distinct neurite inhibition site. J Cell Biol 138:1355–1366

Tiemeyer M, Swiedler SJ, Ishihara M, Moreland M, Schweingruber H, Hirtzer P, Brandley BK (1991) Carbohydrate ligands for endothelial leukocyte adhesion molecule. Proc Natl Acad Sci USA 88:1138–1142

Tropak MB, Roder JC (1997) Regulation of myelin-associated glycoprotein binding by sialylated *cis*-ligands. J Neurochem 68:1753–1763

Umemori H, Sato S, Yagi T, Aizawa S, Yamamoto T (1994) Initial events of myelination involve Fyn tyrosine kinase signalling. Nature 367:572–576

van der Merwe PA, Crocker PR, Vinson M, Barclay AN, Schauer R, Kelm S (1996) Localization of the putative sialic acid-binding site on the immunoglobulin superfamily cell-surface molecule CD22. J Biol Chem 271:9273–9280

Varki A (1997) Sialic acids as ligands in recognition phenomena. FASEB J 11:248–255

Vinson M, van der Merwe PA, Kelm S, May A, Jones EY, Crocker PR (1996) Characterization of the sialic acid-binding site in sialoadhesin by site-directed mutagenesis. J Biol Chem 271:9267–9272

Walz G, Aruffo A, Kolanus W, Bevilacqua M, Seed B (1990) Recognition by ELAM-1 of the sialyl-Lex determinant on myeloid and tumor cells. Science 250:1132–1135

Wu YJ, Nadler MJS, Brennan LA, Gish GD, Timms JF, Fusaki N, Jongstrabilen J, Tada N, Pawson T, Wither J, Neel BG, Hozumi N (1998) The B-cell transmembrane protein CD72 binds to and is an in vivo substrate of the protein tyrosine phosphatase SHP-1. Curr Biol 8:1009–1017

Yohannan J, Wienands J, Coggeshall KM, Justement LB (1999) Analysis of tyrosine phosphorylation-dependent interactions between stimulatory effector proteins and the B cell co-receptor CD22. J Biol Chem 274:18769–18776

Functions of Selectins

Klaus Ley[1]

1 Selectin Structure and Expression

1.1 Selectin Structure

The selectins encompass a family of three type I transmembrane glycoproteins exclusively expressed on blood cells and vascular cells. All known functions of the selectins require the N-terminal C-type lectin domain. The lectin domains show the highest degree of homology among different species (about 72% identity) and considerable homology among the three selectins within the same species (about 52% for human selectins; Kansas 1996). All three selectins strictly require the presence of free calcium ions for ligand binding, and incubation with a chelator eliminates all selectin binding. One EGF domain follows the lectin domain, and two or more short consensus repeats are between the EGF domain and the transmembrane domain. Each domain is encoded by a separate exon, except the cytoplasmic domain, which is encoded by two exons. Complementary DNAs for all three selectins were cloned in 1989 (Bevilacqua et al. 1989; Camerini et al. 1989; Johnston et al. 1989; Siegelman et al. 1989; Tedder et al. 1989), and no new selectins have been found in the past 10 years, suggesting that this protein family consists of only three members.

1.2 Expression of E-selectin

E-selectin (CD62E, ELAM-1) is exclusively expressed on activated endothelial cells and represents one of the few truly endothelial-specific markers. A recombinant, truncated E-selectin containing the lectin and EGF domains was successfully crystallized and subjected to X-ray crystallography (Graves et al. 1994). The structural analysis showed a likely ligand binding site associated with a divalent cation binding site in the lectin domain (Graves et al. 1994). Although the physiological ligands for E-selectin are not known, a molecular model of E-selectin with the ligand mimetic tetrasaccharide, sialyl-Lewis^x, revealed a binding site at the 'top' face of the lectin domain (away from the EGF domain). Human E-selectin has six SCR domains with unknown function. The E-selectin

---

[1] Department of Biomedical Engineering, University of Virginia, Charlottesville, Virginia 22908, USA

Results and Problems in Cell Differentiation, Vol. 33
Paul R. Crocker (Ed.): Mammalian Carbohydrate Recognition Systems
© Springer-Verlag Berlin Heidelberg 2001

promoter contains three nuclear factor kappa-B (NFκB) sites and an ATF-2 site, which are required for cytokine-induced expression (Collins et al. 1995).

1.3 Expression of P-selectin

P-selectin (CD62P, PADGEM, GMP-140) is stored in secretory granules of endothelial cells (Weibel-Palade bodies) and platelets (α-granules). It is rapidly surface-expressed after stimulation with secretagogues or phorbol esters (Hattori et al. 1989; Geng et al. 1990), and expression is enhanced by hydrogen peroxide (Patel et al. 1991). P-selectin in murine endothelial cells is also expressed in response to cytokine stimulation (Sanders et al. 1992; Weller et al. 1992). Consistent with this observation, the murine P-selectin promoter contains NFκB sites (Pan et al. 1998b). However, human P-selectin does not appear to be regulated at the transcriptional level by inflammatory cytokines (Pan et al. 1998a). P-selectin contains nine SCR domains, and its extracellular domain is about 47 nm long (Ushiyama et al. 1993), making it the longest selectin. P-selectin expression is reduced by re-internalization and degradation (Subramaniam et al. 1993), and by shedding from the cell surface (Dunlop et al. 1992).

1.4 Expression of L-selectin

L-selectin (CD62L, LAM-1, gp90^{MEL}, MEL-14 antigen) is constitutively expressed on the surface of most leukocytes (Gallatin et al. 1983; Kansas et al. 1985; Lewinsohn et al. 1987). Human L-selectin contains only two SCR domains, making it the shortest selectin. Analysis of the subcellular distribution of L-selectin shows preferential expression on the tips of leukocyte microvilli (Picker et al. 1991), and this location is important for some of the functions of L-selectin, including the initiation of adhesion under flow (capture or tethering; von Andrian et al. 1995). L-selectin expression is transcriptionally regulated during hematopoietic differentiation and in mature lymphocytes. In all white blood cells, L-selectin is shed from the cell surface after cell activation (Jutila et al. 1990; Kishimoto et al. 1990). Prior to being shed, L-selectin may undergo a transient increase in ligand binding avidity in response to cellular stimulation (Spertini et al. 1991a), probably caused by post-receptor events like oligomerization (Li et al. 1998).

2 Functions Common to All Selectins

2.1 Adhesion Under Flow

All three selectins have been shown to support cell adhesion in the presence of shear forces. Blood flow through microvessels causes a force on each cell in the flow field. This force is largest right next to the vessel wall and directly proportional to the rate of change of velocity, or wall shear rate, the viscosity of the fluid, and the surface area of the cell exposed to the flow. Since the velocity of a fluid particle is maximal in the center of the vessel and equal to zero at the vessel wall, the wall shear rate can be estimated based on the centerline velocity and the fluid properties of blood (Reneman et al. 1992). Forces on adhering leukocytes in vivo range between 10 pN and 10 nN (10^{-11} to 10^{-8} N; House and Lipowsky 1988), resulting from wall shear rates between less than 10 and more than 1000 s^{-1} (Damiano et al. 1996) and wall shear stresses between 0.01 and 10 Pa, or between 0.1 and 100 dyn/cm^2. Although the forces on an individual cell are exceedingly small, it is important to realize that these forces are balanced by the binding of one or a handful of adhesion molecules. Also, these bonds must form extremely rapidly, because the time during which an adhesion molecule and its ligand are close enough to form a bond (nanometers) is of the order of microseconds. The selectins appear to have evolved to accommodate these kinetic constraints (Fig. 1).

2.2 Leukocyte Adhesion Cascade

Adhesion under flow takes at least four clearly distinguishable steps; capture, rolling, and slow rolling leading to firm adhesion (Jung et al. 1998a) (Fig. 1). First, a bond must form between the flowing leukocyte and the stationary endothelial cell. The formation of this first bond is called capture or tethering. The forward reaction rate, or on-rate, of an adhesion molecule-ligand system is thought to determine how efficiently this first bond can form. Once a single bond has formed, the flow forces will swing the leukocyte around until the adhesion molecule-ligand pair is pulled taut. The resulting torque pushes the leukocyte towards the endothelium, thereby facilitating the formation of more bonds (Lawrence et al. 1997). L-selectin and P-selectin glycoprotein ligand-1 (PSGL-1), the two most important molecules for capture, are strategically positioned on the tips of microvilli of neutrophils and other leukocytes (Picker et al. 1991; Moore et al. 1995; von Andrian et al. 1995). Therefore, one or a few bonds and the microvillus on which they reside form a functional unit with spring-like properties (Shao et al. 1998). The cell can continue to move downstream until the spring is maximally extended. If the bond breaks before a new bond has formed, a transient tethering or capture event will ensue. This can indeed be observed in flow chambers in vitro at low site densities of selectins and/or their ligands (Alon et al. 1995, 1997; Lawrence et al. 1997; Smith et al. 1999), and in vivo under conditions when the density of e.g. L-selectin ligand on the endothelium is low so the leukocytes 'hop' from capture to capture event (Ley et al. 1993; Jung et al. 1996). When new bonds form before the existing bond is broken, smooth rolling can ensue, keeping the leukocyte in continuous contact with the endothelium. Rolling is the most conspicuous property conferred by the selectins. All selectins can confer rolling, and very few other molecules can. Therefore, the selectins are rightfully called rolling molecules.

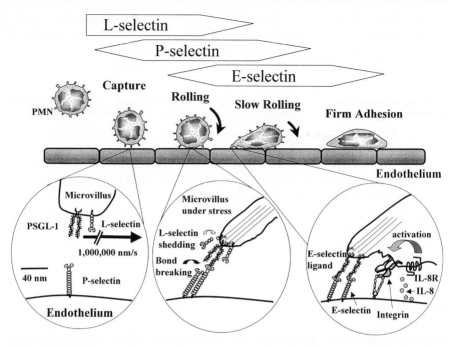

Fig. 1. Schematic representation of the first four steps of the leukocyte adhesion cascade. The series of *gray* boxes represents endothelial cells. A neutrophil (*PMN*) initiates capture from free flow (~1,000 µm/s), followed by rolling (~30 µm/s), slow rolling (~5 µm/s), and firm adhesion. Flow is from *left* to *right*. The contribution of each selectin is indicated by the *bands* on top of the respective processes, a *taper* indicates a partial contribution. The *circles* below show close-ups (field of view about 200 nm across) of the molecular interactions. *Left circle*: Capture is initiated by formation of a molecular bond between L-selectin and its endothelial ligand (not shown) or P-selectin and PSGL-1. Note that the velocity of a free-flowing leukocyte close to the wall is of the order of 1 mm/s, or 1,000,000 nm/s (*arrow*). Bond formation requires a proximity of about 1 nm, suggesting that the first selectin bond must form within about 1 µs. *Center circle*: Stable rolling ensues when new bonds engage before the load-bearing bonds at the rear end of the cell (shown here) break. Bond breakage happens at the receptor-ligand interface (shown for P-selectin binding PSGL-1, *black arrow*), through proteolytic shedding (shown for L-selectin, *white arrow*), and possibly through receptor uprooting from the membrane (not shown). Selectins or their ligands must be attached to the cytoskeleton (*thin lines in microvillus*) to mediate rolling (Kansas et al. 1993). *Right circle*: During slow rolling, mediated by E-selectin binding to unknown ligands on neutrophils, the rolling neutrophil is activated by chemokines, including interleukin-8 (Morgan et al. 1997), binding to their receptors (IL-8R) and causing activation (*arrow*) of β_2 (CD18) integrins (Jung et al. 1998a) and allowing them to bind to their ligands on the endothelium. Progressive integrin activation eventually leads to firm leukocyte adhesion (Kunkel et al. 2000)

Stable rolling can be mediated by each of the selectins, if the density of the selectin and its ligand(s) is high enough.

2.3 Selectin Bond Mechanics

The selectins bind to their ligands with moderate affinities, exhibiting equilibrium dissociation constants in the nanomolar range (Ushiyama et al. 1993). The selectins appear to be endowed with exceptionally rapid forward reaction rates (on-rates; Moore et al. 1991), although direct measurements have been difficult to obtain. Since the equilibrium binding dissociation constant is the ratio of the off-and on-rates, on-rates can indirectly be estimated by measuring the kinetics of individual bonds under flow. This idea was pioneered by Kaplanski and Bongrand (Kaplanski et al. 1993) for E-selectin and later applied to P-selectin (Alon et al. 1995) and L-selectin (Alon et al. 1997). These experiments suggest that E-selectin has a similar off-rate to P-selectin, and L-selectin may have a higher off-rate (Alon et al. 1995, 1997). In addition to the selectin dissociation rate under no-load conditions (Mehta et al. 1998), the bond response to mechanical load is very important, because the selectin bonds bear essentially all of the shear force exerted on the rolling leukocyte by the flowing blood. Modeling studies have predicted that the off-rate of selectins should increase with mechanical load (Tözeren and Ley 1992). This property was recently confirmed experimentally for P-selectin (Alon et al. 1995) and L-selectin (Alon et al. 1997). The increase in off-rate with shear stress, called reactive compliance or tensile strength, is much steeper for L-selectin than for E- or P-selectin (Smith et al. 1999).

3 Differential Functions of Selectins

Although all selectins can mediate attachment under flow, this ability is not equal among the three selectins. In vivo, it appears that L-selectin is most efficient at mediating capture, P-selectin is somewhat efficient, and E-selectin is not very efficient. This is based on observations of rolling flux, or the number of rolling leukocytes per unit of time, in the various selectin-deficient mice, and after antibody blockade of one or two selectins in wild-type mice (Kunkel and Ley 1996; Jung and Ley 1999). In vitro, a similar hierarchy of capture efficiency has been reported. Since the physiological ligands for the selectins are not known, with the exception of PSGL-1 for P-selectin, the in vitro data must rely on a cell bearing the physiological ligand and a purified selectin on a solid substrate (Smith et al. 1999). It is reassuring to note that the biophysical properties of recombinant PSGL-1 peptides coated on a bead and interacting with endothelial cells (Goetz et al. 1997), and sialyl-Lewis^x-coated beads interacting with an E-selectin surface (Brunk et al. 1996) are consistent with observations of selectin-mediated leukocyte attachment in vivo.

3.1 P-selectin

Intravital microscopic experiments in many laboratories have shown that P-selectin is responsible for 'spontaneous' leukocyte rolling (Dore et al. 1993; Mayadas et al. 1993; Nolte et al. 1994; Ley et al. 1995a,b). In most tissues, rolling is not truly spontaneous, because rolling is induced by the minimal trauma associated with careful preparation for intravital microscopy (Atherton and Born 1972; Fiebig et al. 1991; Ley 1994; Ley et al. 1995a,b; Jung and Ley 1997), and this induction is at least partially dependent on mast cell degranulation (Kubes et al. 1993; Kubes and Kanwar 1994; Ley 1994). Under these conditions, expression of P-selectin is limited to venules (Jung and Ley 1997), which coincides with the site of leukocyte rolling (Atherton and Born 1972; Ley and Gaehtgens 1991). In most venules, P-selectin produces rolling velocities between 20 and 50 µm/s. After cytokine stimulation with TNF-α, some P-selectin expression and leukocyte rolling is also seen on the endothelial lining of arterioles, which is completely absent in P-selectin-deficient mice (Kunkel et al. 1997). In P-selectin-deficient mice, leukocyte rolling is initially also absent in venules of all tissues studied (Mayadas et al. 1993; Ley et al. 1995a), but leukocyte rolling is induced to various degrees at later time points (Ley et al. 1995a), or after treatment with inflammatory cytokines like TNF-α (Johnson et al. 1995; Ley et al. 1995a; Kunkel and Ley 1996). In addition, endothelial P-selectin apparently binds to ligands on platelets to mediate platelet rolling in inflamed venules in vivo (Frenette et al. 1995; Romo et al 1999). Activated platelets can also bind to monocytes and other leukocytes through P-selectin (Larsen et al. 1990; Buttrum et al. 1993). This may direct monocytes to sites of vascular injury.

3.2 E-selectin

When E-selectin is expressed in models of acute inflammation, leukocytes roll very slowly (at less than 10 µm/s; Kunkel and Ley 1996). Careful side-by-side comparison in a reconstituted flow chamber system (Smith et al. 1999) has not revealed a major difference in the apparent off-rates of E-selectin and P-selectin. This suggests that the slower rolling velocity supported by E-selectin may be caused by a higher local site density of E-selectin than P-selectin on the endothelial surface, a higher density of E-selectin ligands than P-selectin ligands on the rolling leukocytes, or both. In addition, recent evidence suggests that E-selectin-, but not P-selectin-mediated leukocyte rolling can support the engagement of β_2 integrins during rolling, thus causing a further reduction in rolling velocity (Jung et al. 1998a). This phenomenon greatly facilitates leukocyte adhesion in response to chemoattractants (Ley et al. 1998). The importance of the interaction between E-selectin and CD18 integrins is illustrated in mice lacking both molecules. These mice have such a severe inflammatory defect that they die within weeks after birth (Forlow et al. 1999).

3.3 L-selectin

L-selectin mediates stable rolling on cytokine-stimulated endothelial cells in vitro (Abbassi et al. 1991; Spertini et al. 1992; Zakrzewicz et al. 1997) and in vivo (Jung et al. 1998b). Microvascular endothelium outside secondary lymphatic organs expresses ligand(s) for L-selectin with a peak at 6– 24 h after cytokine stimulation (Brady et al. 1992; Spertini et al. 1991b, 1992; Zakrzewicz et al. 1997; Jung et al. 1998b). The velocity of L-selectin-dependent rolling is modulated by proteolytic removal of L-selectin from the leukocyte surface during the rolling process (Walcheck et al. 1996; Hafezi-Moghadam and Ley 1999). The protease responsible for this removal or shedding appears to be a transmembrane protease acting on the L-selectin on the same, but not adjacent cells (Preece et al. 1996). Lymphocytes obtained from mouse embryos with a homozygous null mutation for TACE, the TNF-α cleaning enzyme, show impaired L-selectin shedding (Peschon et al. 1998). L-selectin shedding may be physiologically important for inflammatory cell recruitment. The concentration of soluble L-selectin in normal plasma is sufficient to competitively inhibit L-selectin-dependent adhesion (Schleiffenbaum et al. 1992).

In addition to mediating stable rolling after cytokine treatment, L-selectin also mediates transient interactions with partially activated endothelial cells (Ley et al. 1993; Jung et al. 1996), and with adherent leukocytes (Bargatze et al. 1994). By interacting with PSGL-1 (Guyer et al. 1996; Spertini et al. 1996) and other ligands (Alon et al. 1996; Ramos et al. 1998) on adherent leukocytes, L-selectin may amplify inflammatory cell recruitment. Although the contribution of this secondary capture phenomenon to neutrophil recruitment may be limited in vivo (Kunkel et al. 1998a), secondary capture could be important in recruiting other leukocytes, including T lymphocytes, to sites of inflammation. Beyond its role in inflammatory cell recruitment, L-selectin is strictly required for the homing of naïve lymphocytes to peripheral lymph nodes as shown by the absence of normally sized lymph nodes in L-selectin-deficient mice (Arbones et al. 1994). L-selectin also contributes to naïve lymphocyte homing to Peyer's patches (Arbones et al. 1994), but is not strictly required, because Peyer's patches in L-selectin-deficient mice show normal size and cellularity (Arbones et al. 1994; Kunkel et al. 1998b).

3.4 Homing of Bone Marrow Stem Cells

E- and P-selectins participate in the homing of hematopoietic stem cells to the bone marrow compartment. E- and P-selectin-deficient mice require somewhat higher numbers of intravenous bone marrow cells for successful engraftment (Frenette et al. 1998b). The contribution of E- and P-selectins to stem cell homing may be at the level of rolling in bone marrow sinusoids (Mazo et al. 1998), although the dominant rolling system in the bone marrow sinusoids appears to be $\alpha_4\beta_1$ integrin interacting with endothelial vascular cell adhesion molecule-1 (VCAM-1).

4 Signaling Through Selectins

Although many in vitro studies have shown that ligation of L-selectin (Waddell et al. 1994; Crockett-Torabi et al. 1995; Brenner et al. 1996; Steeber et al. 1997) or selectin ligands (Cooper et al. 1994; Elstad et al. 1995; Hidari et al. 1997) can activate leukocytes, there is no direct evidence that such signaling contributes to leukocyte recruitment. In most of these studies, antibody cross-linking was used as a method to engage and cluster L-selectin. In almost all protocols, at least one of the cross-linking antibodies used was an intact antibody containing an Fc portion able to interact with Fc receptors, which are abundantly expressed on neutrophils and monocytes. Engagement of Fc receptors either through attachment of a leukocyte to an antibody-coated surface or through ligation of Fc receptors with another cell surface molecule like a selectin or selectin ligand produces vigorous signaling (Edberg et al. 1992), but these signaling events cannot directly be attributed to selectins or selectin ligands. Importantly, neutrophils rolling over a substrate containing both P-selectin and intercellular adhesion molecule-1 (ICAM-1), a ligand for β_2 integrins, do not show signs of activation and do not adhere firmly, whereas introduction of a soluble chemoattractant into the system rapidly leads to firm adhesion (Lawrence and Springer 1991). Similarly, lymphocytes roll over a surface coated with a mixture of L-selectin ligands and ICAM-1 without becoming activated, unless a chemoattractant is also presented (Campbell et al. 1998). In vivo, rolling can be maintained for hours without leading to significant leukocyte attachment or transmigration (Ley and Gaehtgens 1991; Atherton and Born 1972). Taken together, it remains unclear whether engaging selectins during physiological rolling is insufficient to trigger activating signals. However, selectin-mediated activation may be important in other settings, including lymphocyte recruitment to secondary lymphoid organs. GlyCAM-1, a soluble ligand for L-selectin, has been proposed to selectively activate naïve lymphocytes (Hwang et al. 1996). Thus, L-selectin ligation may contribute to lymphocyte activation in high endothelial venules of peripheral lymph nodes.

5 Selectin Ligands

Selectin ligands are more extensively covered in Dr. Feizi's chapter of this book. She proposes a useful distinction between selectin ligands, encompassing the carbohydrate structures recognized by the selectins, and selectin counter-receptors, representing glycoproteins bearing these post-translational modifications and being able to support selectin-mediated adhesion under flow.

5.1 PSGL-1

Although many selectin ligands have been proposed, the only known molecule that fulfills all criteria of a biologically important selectin counter-receptor in vitro and in vivo is P-selectin Glycoprotein ligand-1 (PSGL-1) (Moore et al. 1992; Sako et al. 1993). PSGL-1 is a type 1 surface glycoprotein which is expressed on microprocesses of essentially all leukocytes (Moore et al. 1995). PSGL-1 requires post-translational modifications to function as a P-selectin ligand. Critical modifications include tyrosine sulfation (Sako et al. 1995; Pouyani and Seed 1995), sialylation (Norgard et al. 1993), decoration with core 2 oligosaccarides (Li et al. 1996) and, perhaps most importantly, fucosylation (Sako et al. 1993). PSGL-1 appears to be tyrosine-sulfated and sialylated in most or all leukocytes. Fucosylation is regulated by specific, inducible fucosyltransferases (FT). Fucosyltransferase VII (FTVII) plays a key role in producing functional PSGL-1 (Li et al. 1996; Snapp et al. 1997) and other functional selectin ligands (Knibbs et al. 1996; Smith et al. 1996; Snapp et al. 1997). The expression of FTVII appears to be regulated in lymphocytes, enabling effector T helper-1 cells to express functional P- and E-selectin ligands (Knibbs et al. 1996; Wagers et al. 1996; Austrup et al. 1997). FTVII is constitutively expressed in neutrophils, monocytes, and eosinophils (Natsuka et al. 1994; Sasaki et al. 1994). PSGL-1 is expressed on the cell surface as a homodimer, and dimerisation is necessary for optimal function (Snapp et al. 1998). Biochemical data have suggested that PSGL-1 can also serve as a ligand for E-selectin (Goetz et al. 1997; Jones et al. 1997), but functional data using transfected cell lines, various mutants (Kansas 1996; Snapp et al. 1997), and PSGL-1-deficient mice suggest that PSGL-1 is not relevant for mediating E-selectin dependent rolling under physiological conditions (Yang et al. 1999).

5.2 Other Selectin Ligands

Other proposed selectin counter-receptors include CD24 (Aigner et al. 1998) for P-selectin, glycosylation-dependent cell adhesion molecule-1 (GlyCAM-1; Lasky et al. 1992), mucosal addressin cell adhesion molecule-1 (MAdCAM-1; Berg et al. 1993), CD34 (Baumhueter et al. 1993), sgp200 (Hemmerich et al. 1994), sulfatides (Laudanna et al. 1994), and heparan sulfate proteoglycans (Nelson et al. 1993; Norgard-Sumnicht et al. 1993) for L-selectin, a 250-kDa molecule precipitated from bovine $\gamma\delta$T lymphocytes (Walcheck et al. 1993), E-selectin ligand-1 (ESL-1; Steegmaier et al. 1995), and neutrophil L-selectin (Picker et al. 1991; Zöllner et al. 1997) for E-selectin. Of all these candidate counter-receptors, only PSGL-1 has both been defined at the molecular level and shown to be relevant for leukocyte rolling and inflammation in vivo. Gene-targeted mice with null mutations for some of these candidate molecules have in fact failed to reveal significant inflammatory or immunological defects

(Cheng et al. 1996; Suzuki et al. 1996). These findings suggest that the true counter-receptors for E- and L-selectins have not been identified yet. Dr. Varki summarized this dilemma in a recent review article: 'Selectin ligands: will the real ones please stand up?' (Varki 1997).

6 Phenotype of Selectin-Deficient Mice

Null mutations have been induced by targeting the genes of each of the three selectins (Mayadas et al. 1993; Arbones et al. 1994; Labow et al. 1994), the E- and P-selectin combination (Bullard et al. 1996; Frenette et al. 1996), and all three selectins (Collins et al. 1999a). The two other combinations (L- and P-selectin or L- and E-selectin) have been produced by bone marrow transplantation (Jung and Ley 1999). All these mice are viable, but not all of them are healthy under vivarium conditions.

6.1 E- and P-Selectin Double Deficient Mice

The most severe defects are seen in E- and P-selectin double deficient mice (Bullard et al. 1996; Frenette et al. 1996). These mice develop spontaneous mucocutaneous infections with skin ulcerations, even in a specific pathogen-free environment. They have very high (10–20-fold elevated) circulating neutrophil counts, thought to result from an inability to recruit neutrophils to sites of inflammation, elevated circulating inflammatory cytokines (Bullard et al. 1996), or possibly alterations in the hematopoietic system (Frenette et al. 1996). These mice show no leukocyte rolling after tissue trauma or after short-term cytokine treatment (Bullard et al. 1996), but show a small number of rolling leukocytes resulting in significant leukocyte recruitment after prolonged (6–8 h) cytokine treatment (Jung et al. 1998b). The phenotype of E- and P-selectin double mutant mice is not cured by restoring P-selectin expression on platelets by bone marrow transplantation (Frenette et al. 1998a). The decreased inflammatory response in E- and P-selectin double mutant mice appears to be responsible for a reduction in edema and cell recruitment in delayed-type hypersensitivity (Staite et al. 1996), a reduced or delayed formation of atherosclerotic lesions (Dong et al. 1998), and an impaired wound healing response (Subramaniam et al. 1997).

6.2 Triple Selectin Deficient Mice

Interestingly, the E-, P- and L-selectin triple mutant mouse has a less severe phenotype than the E- and P-selectin-deficient mice. Skin lesions are not apparent, and the circulating neutrophil counts are lower (Collins et al. 1999a). These mice show no leukocyte rolling after surgical trauma or short-term

(2 h) stimulation with TNF-α, but show some α_4 integrin-dependent rolling 6–8 h after TNF-α (Jung and Ley 1999).

6.3 L-Selectin-Deficient Mice

The most severe of the single selectin-deficient mice is the one lacking L-selectin. This mouse not only has severely reduced inflammatory responses in response to chemical irritants (Arbones et al. 1994), and attenuated induction of delayed-type hypersensitivity (Tedder et al. 1995), but also shows prolonged survival of skin allografts (Tang et al. 1997) and reduced leukocyte rolling after trauma (Arbones et al. 1994) and in response to TNF-α (Ley et al. 1995a). The lymph nodes are small, lacking the normal number of naïve lymphocytes (Arbones et al. 1994), and lymphocyte homing to Peyer's patches is delayed, but not eliminated (Arbones et al. 1994; Kunkel et al. 1998b).

6.4 P-Selectin-Deficient Mice

P-selectin-deficient mice show a reduction of early neutrophil accumulation in the peritoneal cavity (Mayadas et al. 1993) and a lack of acute leukocyte rolling (Mayadas et al. 1993a; Ley et al. 1995a), but significant leukocyte rolling is restored after treatment with TNF-α (Ley et al. 1995a). Also, neutrophil recruitment in peritonitis becomes normal at later times, suggesting that absence of P-selectin delays, but does not eliminate the inflammatory response. P-selectin-deficient mice are significantly protected in models of ischemia and reperfusion (Connolly et al. 1997; Naka et al. 1997; Palazzo et al. 1998). They also show reduced formation of atherosclerotic lesions (Johnson et al. 1997; Nageh et al. 1997; Collins et al. 2000) and a dramatic reduction of the neointima formation in response to arterial injury in atherosclerosis-prone mice (Manka et al. 1999).

6.5 E-Selectin-Deficient Mice

E-selectin-deficient mice were originally described as showing normal leukocyte recruitment in at least some models of inflammation (Labow et al. 1994; Bullard et al. 1996). Interestingly, E-selectin-deficient mice succumb to systemic infection with *Streptococcus pneumoniae* (Munoz et al. 1997), suggesting that efficient elimination of some infectious organisms may require E-selectin. E-selectin-deficient mice cannot support slow (<10 μm/s) leukocyte rolling in response to cytokine stimulation (Kunkel and Ley 1996) and show reduced leukocyte adhesion in response to chemoattractants (Ley et al. 1998). E-selectin-deficient mice show marginally reduced lesion sizes in the apolipoprotein-E-deficient model of atherosclerosis (Collins et al. 2000).

6.6 L- and E- and L- and P-Selectin Double Deficient Mice

L- and E-selectin double deficient mice and L- and P-selectin double deficient mice have been investigated as bone marrow chimeras using intravital microscopy (Jung and Ley 1999). The combination of P- and L-selectin deficiency limits leukocyte rolling to a similar extent as the absence of E- and P-selectin, suggesting that P- and L-selectin are partially redundant and both involved in the same step of inflammatory cell recruitment. This common step is probably leukocyte capture (Ley and Tedder 1995). E- and L-selectin-deficient mice have leukocyte rolling and adhesion numbers similar to those found in the L-selectin single mutants, suggesting that E- and L-selectin mediate different steps in the inflammatory adhesion cascade and overlap with the function of other molecules. (These findings were recently confirmed).

7 Selectin Ligand Deficiencies

7.1 Selectin Ligand Deficiencies in Mice

Mice lacking selectin ligands have been produced by targeting the gene for fucosyl transferase VII (FTVII; Maly et al. 1996). This enzyme is responsible for attaching a fucose residue to the penultimate sugar in O-linked side chains of selectin ligands. In these mice, acute leukocyte rolling and neutrophil emigration into the peritoneal cavity are impaired (Maly et al. 1996). L-selectin ligand activity on high endothelial venules is also impaired, although not eliminated. Some of the residual leukocyte adhesion activity seen in these mice appears to depend on the activity of fucosyl transferase IV, which is also expressed in leukocytes. Interestingly, FTVII-deficient mice do not show spontaneous pathology as do the E- and P-selectin double mutant mice. Recently, mice deficient in core-2 acetyl-glucosaminyl-transferase have been generated (Ellies et al. 1998). This enzyme catalyzes the formation of biantennary structures on O-linked glycoproteins. Core-2 enzyme-deficient mice have a defect in early leukocyte recruitment to chemical peritonitis, and their leukocytes show reduced rolling on P-, E- and L-selectins. The phenotype is similar to, but milder than, that of the FTVII-deficient mice. Recently, mice lacking PSGL-1 have been produced by gene-targeting and homologous recombination. These mice have a phenotype similar to P-selectin-deficient mice (Yang et al. 1999).

7.2 Selectin Ligand Deficiency in Humans

The importance of selectin ligands is dramatically illustrated by a rare human disease, leukocyte adhesion deficiency-II (LAD-II). A fucosylation defect in these patients renders selectin ligands inactive, causing an inability of their leukocytes to roll (von Andrian et al. 1993). These patients have significant inflammatory defects (Etzioni et al. 1993). Although the exact molecular defect responsible for the inability to form functional selectin ligands is not known (Karsan et al. 1998), the pathology seen in these patients is consistent with that seen in FTVII null mice or triple selectin-deficient mice.

8 Selectins and Disease

8.1 Ischemia and Reperfusion

Selectins play an important role in ischemia-reperfusion injury. Interventions aimed at blocking P-selectin function have been shown to ameliorate reperfusion injury in models of lung (Carden et al. 1993; Naka et al. 1997), myocardial (Hayward et al. 1999), and renal (Takada et al. 1997) ischemia and reperfusion. P-selectin deficient mice are protected from renal injury induced by ischemia and reperfusion (Singbartl et al. 2000). It is important to bear in mind that many studies use rodent models, and the regulation of P-selectin expression is different in rodents and humans (Pan et al. 1998a,b).

L-selectin antibodies have been suggested to protect from myocardial ischemia-reperfusion injury (Ma et al. 1993). Based on gene-targeted mice, L-selectin seems to be involved in the exacerbation of hepatic ischemia-induced injury (Yadav et al. 1998), but not renal injury (Rabb et al. 1996). The role of L-selectin in ischemia-reperfusion injury seems to be less than that of P-selectin, but organ-specific differences exist.

E-selectin is expressed in several models, particularly those involving cutaneous ischemia (Stotland and Kerrigan 1997). Recently, protection from renal ischemia and reperfusion has been described in E-selectin-deficient mice (Singbartl and Ley 2000). Protection was seen also when a blocking antibody to E-selectin was administered, even after the onset of reperfusion (Singbartl and Ley 2000).

8.2 Cancer Metastasis

Many colon carcinoma cell lines express ligands for vascular selectins, suggesting that selectins may be involved in metastatic spread. Absence of P-selectin appears to limit the metastatic potential of human colon carcinoma cells transplanted into mice (Kim et al. 1998). Human colon carcinoma cells adhere to P-selectin in vitro, using a ligand other than PSGL-1 (Goetz et al. 1996). Breast carcinoma cells show a similar behavior and use CD24 to bind to P-selectin (Aigner et al. 1998). Small cell lung carcinoma can also attach to endothelial cells using P-selectin (Stone and Wagner 1993; Pottratz et al. 1996). P-selectin may influence the metastatic process by promoting platelet binding to tumor cells (Mannel and Grau 1997), which may protect them from the immune system while in transit in the circulation. For a review on the role of selectins in metastasis, see McEver (1997).

E-selectin also binds to a variety of tumor cell lines (Kojima et al. 1992; Iwai et al. 1993). Apparently, the site of metastasis formation can be influenced by the expression pattern of E-selectin in a transgenic mouse system (Biancone et al. 1996). Colon carcinoma cells expressing E-selectin ligands can be prevented from forming lung metastases by injecting mice with recombinant soluble E-selectin (Mannori et al. 1997). Similar results have been reported with an E-selectin antibody in a liver metastasis model (Brodt et al. 1997).

Interestingly, no correlation has been found between L-selectin and tumor metastasis (Jonas et al. 1998). However, in one study, high levels of soluble L-selectin were found to correlate with a negative outcome of acute myeloid leukemia (Extermann et al. 1998).

8.3 Autoimmune Diseases

There is some evidence that leukocyte adhesion molecules may be involved in autoimmune diseases, but the research in this area is at a relatively early stage. Patients with inflammatory bowel disease have elevated levels of P-selectin in the serum (Goke et al. 1997) and in tissue biopsies (Schürmann et al. 1995), but no experimental or interventional studies are available. E-selectin is also elevated in serum of patients with inflammatory bowel disease (Cellier et al. 1997; Goke et al. 1997). Little is known about the involvement of adhesion molecules in lupus, psoriasis, or rheumatoid arthritis. P-selectin appears to be involved in mediating cellular infiltration into arthritic joints in a rat model, based on antibody blocking studies (Issekutz 1998). At least one model of autoimmune glomerulonephritis is exacerbated in P-selectin-deficient mice (Rosenkranz et al. 1999).

8.4 Atherosclerosis

A number of studies have demonstrated localized expression of leukocyte adhesion molecules in atherosclerotic lesions and plaques. P-selectin expression is induced on vascular endothelium overlying human atherosclerotic plaques (Johnson-Tidey et al. 1994). Oxidized low density lipoprotein, a prominent component of fatty streaks in early atherosclerotic lesions, can enhance P-selectin surface expression by endothelium in vitro (Vora et al. 1997). Soluble selectins are found in the serum of healthy subjects and patients. Soluble P-selectin has been shown to correlate with atherosclerotic lesions in many studies (Blann et al. 1996, 1997a,b, 1998; Frijns et al. 1997; Saku et al. 1999) and may represent an independent risk factor for atherosclerosis.

There is good evidence that the development of atherosclerotic lesions in mice is dependent on P-selectin. LDL receptor-deficient mice develop smaller atherosclerotic lesions when they are also deficient in P-selectin (Johnson et al. 1997). However, lesion development in the double-deficient mice was similar to that for LDL receptor-deficient mice over longer time periods (Johnson et al. 1997), suggesting that P-selectin expression may primarily impact early lesion development. More recent data in apolipoprotein E-deficient mice show much more dramatic protection when P-selectin is absent (Collins et al. 2000). These mice are almost totally protected from neointima formation in response to arterial injury (Manka et al. 2000). Introducing the E- and P-selectin null mutation into LDLreceptor-deficient mice produces a 40% reduction in lesion size and less calcification (Dong et al. 1998), but these data are difficult to interpret because E- and P-selectin-deficient mice develop spontaneous disease under vivarium conditions which may influence the development of atherosclerotic lesions.

9 Summary

The selectins are cell surface lectins that have evolved to mediate the adhesion of white blood cells to endothelial cells and platelets under flow. They recognize fucosylated, sialylated and in some cases sulfated ligands expressed on scaffold glycoproteins serving as functional counter-receptors. Selectins are regulated at the transcriptional level, through proteolytic processing, through cellular sorting, and through regulated expression of glycosyl-transferases responsible for the formation of functional ligands. The selectins are physiologically important in inflammation, lymphocyte homing, immunological responses, and homing of bone marrow stem cells. They play a role in atherosclerosis, ischemia-reperfusion injury, inflammatory diseases, and metastatic spreading of some cancers.

References

Abbassi O, Lane CL, Krater S, Kishimoto TK, Anderson DC, McIntire LV, Smith CW (1991) Canine neutrophil margination mediated by lectin adhesion molecule-1 in vitro. J Immunol 147: 2107–2115

Aigner S, Ramos CL, Hafezi-Moghadam A, Lawrence MB, Altevogt P, Ley K (1998) CD24 mediates rolling of breast carcinoma cells on P-selectin. FASEB J 12:1241–1251

Alon R, Hammer DA, Springer TA (1995) Lifetime of the P-selectin-carbohydrate bond and its response to tensile force in hydrodynamic flow. Nature 374:539–542

Alon R, Fuhlbrigge RC, Finger EB, Springer TA (1996) Interactions through L-selectin between leukocytes and adherent leukocytes nucleate rolling adhesions on selectins and VCAM-1 in shear flow. J Cell Biol 135:849–865

Alon R, Chen SQ, Puri KD, Finger EB, Springer TA (1997) The kinetics of L-selectin tethers and the mechanics of selectin-mediated rolling. J Cell Biol 138:1169–1180

Arbones ML, Ord DC, Ley K, Ratech H, Maynard-Curry C, Otten G, Capon DJ, Tedder TF (1994) Lymphocyte homing and leukocyte rolling and migration are impaired in L-selectin-deficient mice. Immunity 1:247–260

Atherton A, Born GVR (1972) Quantitative investigations of the adhesiveness of circulating poly-morphonuclear leukocytes to blood vessels. J Physiol (Lond) 222:447–474

Austrup F, Vestweber D, Borges E, Lohning M, Brauer R, Herz U, Renz H, Hallmann R, Scheffold A, Radbruch A, Hamann A (1997) P- and E-selectin mediate recruitment of T-helper-1 but not T-helper-2 cells into inflamed tissues. Nature 385:81–83

Bargatze RF, Kurk S, Butcher EC, Jutila MA (1994) Neutrophils roll on adherent neutrophils bound to cytokine-induced endothelial cells via L-selectin on the rolling cells. J Exp Med 180: 1785–1792

Baumhueter S, Singer MS, Henzel W, Hemmerich S, Renz M, Rosen SD, Lasky LA (1993) Binding of L-selectin to the vascular sialomucin CD34. Science 262:436–438

Berg EL, McEvoy LM, Berlin C, Bargatze RF, Butcher EC (1993) L-selectin-mediated lymphocyte rolling on MAdCAM-1. Nature 366:695–698

Bevilacqua MP, Stengelin S, Gimbrone MA Jr, Seed B (1989) Endothelial leukocyte adhesion mol-ecule-1: an inducible receptor for neutrophils related to complement regulatory proteins and lectins. Science 243:1160–1165

Biancone L, Araki M, Araki K, Vassalli P, Stamenkovic I (1996) Redirection of tumor metastasis by expression of E-selectin in vivo. J Exp Med 183:581–587

Blann AD, Seigneur M, Boisseau MR, Taberner DA, McCollum CN (1996) Soluble P selectin in peripheral vascular disease – relationship to the location and extent of atherosclerotic disease and its risk factors. Blood Coagul Fibrinol 7:789–793

Blann AD, Faragher EB, McCollum CN (1997a) Increased soluble P-selectin following myocardial infarction – a new marker for the progression of atherosclerosis. Blood Coagul Fibrinol 8:383–390

Blann AD, Goode GK, Miller JP, McCollum CN (1997b) Soluble P-selectin in hyperlipidaemia with and without symptomatic vascular disease – relationship with von Willebrand factor. Blood Coagul Fibrinol 8:200–204

Blann AD, Kirkpatrick U, Devine C, Naser S, McCollum CN (1998) The influence of acute smoking on leucocytes, platelets and the endothelium. Atherosclerosis 141:133–139

Brady HR, Spertini O, Jimenez W, Brenner BM, Marsden PA, Tedder TF (1992) Neutrophils, mono-cytes, and lymphocytes bind to cytokine-activated kidney glomerular endothelial cells through L-selectin (LAM-1) in vitro. J Immunol 149:2437–2444

Brenner B, Gulbins E, Schlottmann K, Koppenhoefer U, Busch GL, Walzog B, Steinhausen M, Coggeshall KM, Linderkamp O, Lang F (1996) L-selectin activates the Ras pathway via the tyrosine kinase p56^{lck}. Proc Natl Acad Sci USA 93:15376–15381

Brodt P, Fallavollita L, Bresalier RS, Meterissian S, Norton CR, Wolitzky BA (1997) Liver endothe-lial E-selectin mediates carcinoma cell adhesion and promotes liver metastasis. Int J Cancer 71:612–619

Brunk DK, Goetz DJ, Hammer DA (1996) Sialyl Lewis(x)/E-selectin-mediated rolling in a cell-free system. Biophys J 71:2902–2907

Bullard DC, Kunkel EJ, Kubo H, Hicks MJ, Lorenzo I, Doyle NA, Doerschuk CM, Ley K, Beaudet AL (1996) Infectious susceptibility and severe deficiency of leukocyte rolling and recruitment in E-selectin and P-selectin double mutant mice. J Exp Med 183:2329–2336

Buttrum SM, Hatton R, Nash GB (1993) Selectin-mediated rolling of neutrophils on immoblized platelets. Blood 82:1165–1174

Camerini D, James SP, Stamenkovic I, Seed B (1989) Leu-8/TQ1 is the human equivalent of the Mel-14 lymph node homing receptor. Nature 342:78–82

Campbell JJ, Hedrick J, Zlotnik A, Siani MA, Thompson DA, Butcher EC (1998) Chemokines and the arrest of lymphocytes rolling under flow conditions. Science 279:381–384

Carden DL, Young JA, Granger DN (1993) Pulmonary microvascular injury after intestinal ischemia-reperfusion: role of P-selectin. J Appl Physiol 75:2529–2534

Cellier C, Patey N, Fromont-Hankard G, Cervoni JP, Leborgne M, Chaussade S, Barbier JP, Brousse N (1997) In-situ endothelial cell adhesion molecule expression in ulcerative colitis. E-selectin in-situ expression correlates with clinical, endoscopic and histological activity and outcome. Eur J Gastroenterol Hepatol 9:1197–1203

Cheng J, Baumhueter S, Cacalano G, Carver-Moore K, Thibodeaux H, Thomas R, Broxmeyer HE, Cooper S, Hague N, Moore M, Lasky LA (1996) Hematopoietic defects in mice lacking the sialomucin CD34. Blood 87:479–490

Collins RG, Jung U, Bullard DC, Hicks MJ, Ley K, Beaudet AL (1999a) Viable phenotype but impaired leukocyte rolling and peritoneal emigration in triple selectin (E, L and P) null mice. Keystone Symposia C4:39 (abstract)

Collins RG, Velji R, Guevara NV, Chan L, Beaudet AL (2000) P-selectin or ICAM-1 deficiency substantially protects against atherosclerosis in apo E-deficient mice. J Exp Med 191:189–194

Collins T, Read MA, Neish AS, Whitley MZ, Thanos D, Maniatis T (1995) Transcriptional regulation of endothelial cell adhesion molecules: NF-kappa B and cytokine-inducible enhancers. FASEB J 9:899–909

Connolly ES, Winfree CJ, Prestigiacomo CJ, Kim SC, Choudhri TF, Hoh BL, Naka Y, Solomon RA, Pinsky DJ (1997) Exacerbation of cerebral injury in mice that express the P- selectin gene-identification of P-selectin blockade as a new target for the treatment of stroke. Circ Res 81:304–310

Cooper D, Butcher CM, Berndt MC, Vadas MA (1994) P-selectin interacts with a b_2-integrin to enhance phagocytosis. J Immunol 153:3199–3209

Crockett-Torabi E, Sulenbarger B, Smith CW, Fantone JC (1995) Activation of human neutrophils through L-selectin and Mac-1 molecules. J Immunol 154:2291–2302

Damiano ER, Westheider J, Tözeren A, Ley K (1996) Variation in the velocity, deformation, and adhesion energy density of leukocytes rolling within venules. Circ Res 79:1122–1130

Dong ZM, Chapman SM, Brown AA, Frenette PS, Hynes RO, Wagner DD (1998) Combined role of P- and E-selectins in atherosclerosis. J Clin Invest 102:145–152

Dore M, Korthuis RJ, Granger DN, Entman ML, Smith CW (1993) P-selectin mediates spontaneous leukocyte rolling in vivo. Blood 82:1308–1316

Dunlop LC, Skinner MP, Bendall LJ, Favaloro EJ, Castaldi PA, Gorman JJ, Gamble JR, Vadas MA, Berndt MC (1992) Characterization of GMP-140 (P-selectin) as a circulating plasma protein. J Exp Med 175:1147–1150

Edberg JC, Salmon JE, Kimberly RP (1992) Functional capacity of Fc gamma receptor III (CD16) on human neutrophils. Immunol Res 11:239–251

Ellies LG, Tsuboi S, Petryniak B, Lowe JB, Fukuda M, Marth JD (1998) Core 2 oligosaccharide biosynthesis distinguishes between selectin ligands essential for leukocyte homing and inflammation. Immunity 9:881–890

Elstad MR, La Pine TR, Cowley FS, McEver RP, McIntyre TM, Prescott SM, Zimmerman GA (1995) P-selectin regulates platelet-activating factor synthesis and phagocytosis by monocytes. J Immunol 155:2109–2122

Etzioni A, Frydman M, Pollack S, Avidor I, Phillips ML, Paulson JC, Gershoni-Baruch R (1993) Recurrent severe infections caused by a novel leukocyte adhesion deficiency. N Engl J Med 327:1789–1792

Extermann M, Bacchi M, Monai N, Fopp M, Fey M, Tichelli A, Schapira M, Spertini O (1998) Relationship between cleaved L-selectin levels and the outcome of acute myeloid leukemia. Blood 92:3115–3122

Fiebig E, Ley K, Arfors K-E (1991) Rapid leukocyte accumulation by 'spontaneous' rolling and adhesion in the exteriorized rabbit mesentery. Int J Microcirc Clin Exp 10:127–144

Forlow SB, Bullard DC, Lu HF, Beaudet AL, Ley K (1999) Absence of slow leukocyte rolling and severe leukocyte recruitment defect in mice lacking E-selectin and CD18. FASEB J 13:A311 (abstract)

Frenette PS, Johnson RC, Hynes MR, Wagner DD (1995) Platelets roll on stimulated endothelium in vivo: an interaction mediated by endothelial P-selectin. Proc Natl Acad Sci USA 92: 7450–7454

Frenette PS, Mayadas TN, Rayburn H, Hynes RO, Wagner DD (1996) Susceptibility to infection and altered hematopoiesis in mice deficient in both P- and E-selectins. Cell 84:563–574

Frenette PS, Moyna C, Hartwell DW, Lowe JB, Hynes RO, Wagner DD (1998a) Platelet-endothelial interactions in inflamed mesenteric venules. Blood 91:1318–1324

Frenette PS, Subbarao S, Mazo IB, von Andrian UH, Wagner DD (1998b) Endothelial selectins and vascular cell adhesion molecule-1 promote hematopoietic progenitor homing to bone marrow. Proc Natl Acad Sci USA 95:14423–14428

Frijns CJM, Kappelle LJ, Vangijn J, Nieuwenhuis HK, Sixma JJ, Fijnheer R (1997) Soluble adhesion molecules reflect endothelial cell activation in ischemic stroke and in carotid atherosclerosis. Stroke 28:2214–2218

Gallatin WM, Weissman IL, Butcher EC (1983) A cell-surface molecule involved in organ-specific homing of lymphocytes. Nature 304:30–34

Geng JG, Bevilacqua MP, Moore KL, McIntyre TM, Prescott SM, Kim JM, Bliss GA, Zimmerman GA, McEver RP (1990) Rapid neutrophil adhesion to activated endothelium mediated by GMP-140. Nature 343:757–760

Goetz DJ, Ding H, Atkinson WJ, Vachino G, Camphausen RT, Cumming DA, Luscinskas FW (1996) A human colon carcinoma cell line exhibits adhesive interactions with P-selectin under fluid flow via a PSGL-1-independent mechanism. Am J Pathol 149:1661–1673

Goetz DJ, Greif DM, Ding H, Camphausen RT, Howes S, Comess KM, Snapp KR, Kansas GS, Luscinskas FW (1997) Isolated P-selectin glycoprotein ligand-1 mediates dynamic adhesion to P- and E-selectin. J Cell Biol 137:509–519

Goke M, Hoffmann JC, Evers J, Kruger H, Manns MP (1997) Elevated serum concentrations of soluble selectin and immunoglobulin type adhesion molecules in patients with inflammatory bowel disease. J Gastroenterol 32:480–486

Graves BJ, Crowther RL, Chandran C, Rumberger JM, Li S, Hunag KS, Presky DH, Familetti PC, Wolitzky BA, Burns DK (1994) Insight into E-selectin ligand interaction from the crystal structure and mutagenesis of the LEC and EGF domains. Nature 367:532–538

Guyer DA, Moore KL, Lynam EB, Schammel CMG, Rogelj S, McEver RP, Sklar LA (1996) P-selectin glycoprotein ligand-1 (PSGL-1) is a ligand for L-selectin in neutrophil aggregation. Blood 88:2415–2421

Hafezi-Moghadam A, Ley K (1999) Relevance of L-selectin shedding for leukocyte rolling in vivo. J Exp Med 189:939–948

Hattori R, Hamilton KK, Fugate RD, McEver RP, Sims PJ (1989) Stimulated secretion of endothelial von Willebrand factor is accompanied by rapid redistribution to the cell surface of the intracellular granule membrane protein GMP-140. J Biol Chem 264:7768–7771

Hayward R, Campbell B, Shin YK, Scalia R, Lefer AM (1999) Recombinant soluble P-selectin glycoprotein ligand-1 protects against myocardial ischemic reperfusion injury in cats. Cardiovasc Res 41:65–76

Hemmerich S, Butcher EC, Rosen SD (1994) Sulfation-dependent recognition of high endothelial venules (HEV)-ligands by L-selectin and MECA 79, an adhesion-blocking monoclonal antibody. J Exp Med 180:2219–2226

Hidari KIPJ, Weyrich AS, Zimmerman GA, McEver RP (1997) Engagement of P-selectin glycoprotein ligand-1 enhances tyrosine phosphorylation and activates mitogen-activated protein kinases in human neutrophils. J Biol Chem 272:28750–28756

House SD, Lipowsky HH (1988) In vivo determination of the force of leukocyte-endothelium adhesion in the mesenteric microvasculature of the cat. Circ Res 63:658–668

Hwang ST, Singer MS, Giblin PA, Yednock TA, Bacon KB, Simon SI, Rosen SD (1996) GlyCAM-1, a physiologic ligand for L-selectin, activates b_2 integrins on naive peripheral lymphocytes. J Exp Med 184:1343–1348

Issekutz AC (1998) Adhesion molecules mediating neutrophil migration to arthritis in vivo and across endothelium and connective tissue barriers in vitro. Inflamm Res 47:S123—S132

Iwai K, Ishikura H, Kaji M, Sugiura H, Ishizu A, Takahashi C, Kato H, Tanabe T, Yoshiki T (1993) Importance of E-selectin (ELAM-1) and sialyl Lewis[a] in the adhesion of pancreatic carcinoma cells to activated endothelium. Int J Cancer 54:972–977

Johnson RC, Mayadas TN, Frenette PS, Mebius RE, Subramaniam M, Lacasce A, Hynes RO, Wagner DD (1995) Blood cell dynamics in P-selectin-deficient mice. Blood 86:1106–1114

Johnson RC, Chapman SM, Dong ZM, Ordovas JM, Mayadas TN, Herz J, Hynes RO, Schaefer EJ, Wagner DD (1997) Absence of P-selectin delays fatty streak formation in mice. J Clin Invest 99:1037–1043

Johnson-Tidey RR, McGregor JL, Taylor PR, Poston RN (1994) Increase in the adhesion molecule P-selectin in endothelium overlying atherosclerotic plaques: coexpression with intercellular adhesion molecule-1. Am J Pathol 144:952–961

Johnston GI, Cook RG, McEver RP (1989) Cloning of GMP-140, a granule membrane protein of platelets and endothelium: sequence similarity to proteins involved in cell adhesion and inflammation. Cell 56:1033–1044

Jonas P, Holzmann B, Jablonskiwestrich D, Hamann A (1998) Dissemination capacity of murine lymphoma cells is not dependent on efficient homing. Int J Cancer 77:402–407

Jones WM, Watts GM, Robinson MK, Vestweber D, Jutila MA (1997) Comparison of E-selectin-binding glycoprotein ligands on human lymphocytes, neutrophils, and bovine gamma-delta T cells. J Immunol 159:3574–3583

Jung U, Ley K (1997) Regulation of E-selectin, P-selectin and ICAM-1 expression in mouse cremaster muscle vasculature. Microcirculation 4:311–319

Jung U, Ley K (1999) Mice lacking two or all three selectins demonstrate overlapping and distinct funtions of each selectin. J Immunol 162:6755–6762

Jung U, Bullard DC, Tedder TF, Ley K (1996) Velocity difference between L-selectin and P-selectin dependent neutrophil rolling in venules of the mouse cremaster muscle in vivo. Am J Physiol 271:H2740–H2747

Jung U, Norman KE, Ramos CL, Scharffetter-Kochanek K, Beaudet AL, Ley K (1998a) Transit time of leukocytes rolling through venules controls cytokine-induced inflammatory cell recruitment in vivo. J Clin Invest 102:1526–1533

Jung U, Ramos CL, Bullard DC, Ley K (1998b) Gene-targeted mice reveal the importance of L-selectin-dependent rolling for neutrophil adhesion. Am J Physiol 274:H1785–H1791

Jutila KL, Kishimoto TK, Butcher EC (1990) Regulation and lectin activity of the human neutrophil peripheral lymph node homing receptor. Blood 76:178–183

Kansas GS (1996) Selectins and their ligands: current concepts and controversies. Blood 88:3259–3287

Kansas GS, Wood GS, Fishwild DM, Engleman EG (1985) Functional characterization of human T lymphocyte subsets distinguished by monoclonal anti-Leu-8. J Immunol 134:2995–3002

Kansas GS, Ley K, Munro JM, Tedder TF (1993) Regulation of leukocyte rolling and adhesion to endothelium by the cytoplasmic domain of L-selectin. J Exp Med 177:833–838

Kaplanski G, Farnarier C, Tissot O, Pierres A, Benoliel A-M, Alessi M-C, Kaplanski S, Bongrand P (1993) Granulocyte-endothelium initial adhesion. Analysis of transient binding events mediated by E-selectin in a laminar shear flow. Biophys J 64:1922–1933

Karsan A, Cornejo CJ, Winn RK, Schwartz BR, Way W, Lannir N, Gershonibaruch R, Etzioni A, Ochs HD, Harlan JM (1998) Leukocyte adhesion deficiency type II is a generalized defect of de novo GDP-fucose biosynthesis – endothelial cell fucosylation is not required for neutrophil rolling on human nonlymphoid endothelium. J Clin Invest 101:2438–2445

Kim YJ, Borsig L, Varki NM, Varki A (1998) P-selectin deficiency attenuates tumor growth and metastasis. Proc Natl Acad Sci USA 95:9325–9330

Kishimoto TK, Jutila MA, Butcher EC (1990) Identification of a human peripheral lymph node homing receptor: a rapidly down-regulated adhesion molecule. Proc Natl Acad Sci USA 87:2244–2248

Knibbs RN, Craig RA, Natsuka S, Chang A, Cameron M, Lowe JB, Stoolman LM (1996) The fucosyltransferase FucT-VII regulates E-selectin ligand synthesis in human T cells. J Cell Biol 133:911–920

Kojima N, Handa K, Newman W, Hakomori S (1992) Inhibition of selectin-dependent tumor cell adhesion to endothelial cells and platelets by blocking O-glycosylation of these cells. Biochem Biophys Res Commun 182:1288–1295

Kubes P, Kanwar S (1994) Histamine induces leukocyte rolling in post-capillary venules: a P-selectin mediated event. J Immunol 152:3570–3577

Kubes P, Kanwar S, Niu X-F, Gaboury JP (1993) Nitric oxide synthesis inhibition induces leukocyte adhesion via superoxide and mast cells. FASEB J 7:1293–1299

Kunkel EJ, Ley K (1996) Distinct phenotype of E-selectin-deficient mice: E-selectin is required for slow leukocyte rolling in vivo. Circ Res 79:1196–1204

Kunkel EJ, Jung U, Ley K (1997) TNF-a induces selectin-dependent leukocyte rolling in mouse cremaster muscle arterioles. Am J Physiol 272:H1391–H1400

Kunkel EJ, Chomas JE, Ley K (1998a) Role of primary and secondary capture for leukocyte accumulation in vivo. Circ Res 82:30–38

Kunkel EJ, Ramos CL, Steeber DA, Muller W, Wagner N, Tedder TF, Ley K (1998b) The roles of L-selectin, b$_7$ integrins, and P-selectin in leukocyte rolling and adhesion in high endothelial venules of Peyers patches. J Immunol 161:2449–2456

Kunkel EJ, Dunne, JL, Ley K (2000) Leukocyte arrest during cytokine-dependent inflammation in vivo. J Immunol 164:3301–3308

Labow MA, Norton CR, Rumberger JM, Lombard-Gillooly KM, Shuster DJ, Hubbard J, Bertko R, Knaack PA, Terry RW, Harbison ML, Kontgen F, Stewart CL, McIntyre KW, Will PC, Burns DK, Wolitzky BA (1994) Characterization of E-selectin-deficient mice: demonstration of overlapping function of the endothelial selectins. Immunity 1:709–720

Larsen E, Palabrica T, Sajer S, Gilbert GE, Wagner DD, Furie BC (1990) PADGEM-dependent adhesion of platelets to monocytes and neutrophils is mediated by a lineage-specific carbohydrate, LNF III (CD15). Cell 63:467–474

Lasky LA, Singer MS, Dowbenko D, Imai Y, Henzel WJ, Grimley C, Fennie C, Gillett N, Watson SR, Rosen SD (1992) An endothelial ligand for L-selectin is a novel mucin-like molecule. Cell 69:927–938

Laudanna C, Constantin G, Baron P, Scardini E, Scarlato G, Cabrini G, Dechecchi C, Rossi F, Cassatella MA, Berton G (1994) Sulfatides trigger increase of cytosolic free calcium and enhanced expression of tumor necrosis factor a and interleukin-8 messenger RNA in human neutrophils – evidence for a role of L-selectin as a signaling molecule. J Biol Chem 269: 4021–4026

Lawrence MB, Springer TA (1991) Leukocytes roll on a selectin at physiologic flow rates: distinction from and prerequisite for adhesion through integrins. Cell 65:859–873

Lawrence MB, Kansas GS, Ghosh S, Kunkel EJ, Ley K (1997) Threshold levels of fluid shear promote leukocyte adhesion through selectins (CD62L,P,E). J Cell Biol 136:717–727

Lewinsohn DM, Bargatze RF, Butcher EC (1987) Leukocyte-endothelial cell recognition: evidence of a common molecular mechanism shared by neutrophils, lymphocytes, and other leukocytes. J Immunol 138:4313–4321

Ley K (1994) Histamine can induce leukocyte rolling in rat mesenteric venules. Am J Physiol 267:H1017–H1023

Ley K, Gaehtgens P (1991) Endothelial, not hemodynamic differences are responsible for preferential leukocyte rolling in venules. Circ Res 69:1034–1041

Ley K, Tedder TF (1995) Leukocyte interactions with vascular endothelium: new insights into selectin-mediated attachment and rolling. J Immunol 155:525–528

Ley K, Tedder TF, Kansas GS (1993) L-selectin can mediate leukocyte rolling in untreated mesenteric venules in vivo independent of E- or P-selectin. Blood 82:1632–1638

Ley K, Bullard DC, Arbones ML, Bosse R, Vestweber D, Tedder TF, Beaudet AL (1995a) Sequential contribution of L- and P-selectin to leukocyte rolling in vivo. J Exp Med 181:669–675

Ley K, Zakrzewicz A, Hanski C, Stoolman LM, Kansas GS (1995b) Sialylated O-glycans and L-selectin sequentially mediate myeloid cell rolling in vivo. Blood 85:3727–3735

Ley K, Allietta M, Bullard DC, Morgan SJ (1998) The importance of E-selectin for firm leukocyte adhesion in vivo. Circ Res 83:287–294

Li F, Wilkins PP, Crawley S, Weinstein J, Cummings RD, McEver RP (1996) Post-translational modifications of recombinant P-selectin glycoprotein ligand-1 required for binding to P- and E-selectin. J Biol Chem 271:3255–3264

Li X, Steeber DA, Tang MLK, Farrar MA, Perlmutter RM, Tedder TF (1998) Regulation of L-selectin-mediated rolling through receptor dimerization. J Exp Med 98:1385–1390

Ma X, Weyrich AS, Lefer DJ, Buerke M, Albertine KH, Kishimoto TK, Lefer AM (1993) Monoclonal antibody to L-selectin attenuates neutrophil accumulation and protects ischemic reperfused cat myocardium. Circulation 88:649–658

Maly P, Thall AD, Petryniak B, Rogers CE, Smith PL, Marks RM, Kelly RJ, Gersten KM, Cheng G, Saunders TL, Camper SA, Camphausen RT, Sullivan FX, Isogai Y, Hindsgaul O, von Andrian

UH, Lowe JB (1996) The a(1,3)fucosyltransferase Fuc-TVII controls leukocyte trafficking through an essential role in L-, E-, and P-selectin ligand biosynthesis. Cell 86:643–653

Manka DR, Collins RG, Ley K, Beaudet AL, Sarembock IJ (2000) Absence of P-selectin but not ICAM-1 limits neointimal growth after arterial injury in apolipoprotein-E (apoE)-deficient mice circulation, in press

Mannel DN, Grau GE (1997) Role of platelet adhesion in homeostasis and immunopathology. Mol Pathol 50:175–185

Mannori G, Santoro D, Carter L, Corless C, Nelson RM, Bevilacqua MP (1997) Inhibition of colon carcinoma cell lung colony formation by a soluble form of E-selectin. Am J Pathol 151:233–243

Mayadas TN, Johnson RC, Rayburn H, Hynes RO, Wagner DD (1993) Leukocyte rolling and extravasation are severely compromised in P selectin-deficient mice. Cell 74:541–554

Mazo IB, Gutiérrez-Ramos JC, Frenette PS, Hynes RO, Wagner DD, von Andrian UH (1998) Hematopoietic progenitor cell rolling in bone marrow microvessels – parallel contributions by endothelial selectins and vascular cell adhesion molecule 1. J Exp Med 188:465–474

McEver RP (1997) Selectin-carbohydrate interactions during inflammation and metastasis. Glycoconj J 14:585–591

Mehta P, Cummings RD, McEver RP (1998) Affinity and kinetic analysis of P-selectin binding to P-selectin glycoprotein ligand-1. J Biol Chem 273:32506–32513

Moore KL, Varki A, McEver RP (1991) GMP-140 binds to a glycoprotein receptor on human neutrophils: evidence for a lectin-like interaction. J Cell Biol 112:491–499

Moore KL, Stults NL, Diaz S, Smith DF, Cummings RD, Varki A, McEver RP (1992) Identification of a specific glycoprotein ligand for P-selectin (CD62) on myeloid cells. J Cell Biol 118:445–456

Moore KL, Patel KD, Breuhl RE, Fugang L, Johnson DA, Lichenstein HS, Cummings RD, Bainton DF, McEver RP (1995) P-selectin glycoprotein ligand-1 mediates rolling of human neutrophils on P-selectin. J Cell Biol 128:661–671

Morgan SJ, Moore MW, Cacalano G, Ley K (1997) Reduced leukocyte adhesion response and absence of slow leukocyte rolling in interleukin-8 (IL-8) receptor deficient mice. Microvasc Res 54:188–191

Munoz FM, Hawkins EP, Bullard DC, Beaudet AL, Kaplan SL (1997) Host defense against infection with S. pneumoniae is impaired in E-, P- and E-/P-selectin deficient mice. J Clin Invest 100:2099–2106

Nageh MF, Sandberg ET, Marotti KR, Lin AH, Melchior EP, Bullard DC, Beaudet AL (1997) Deficiency of inflammatory cell adhesion molecules protects against atherosclerosis in mice. Arterioscler Thromb Vasc Biol 17:1517–1520

Naka Y, Toda K, Kayano K, Oz MC, Pinsky DJ (1997) Failure to express the P-selectin gene or P-selectin blockade confers early pulmonary protection after lung ischemia or transplantation. Proc Natl Acad Sci USA 94:757–761

Natsuka S, Gersten KM, Zenita K, Kannagi R, Lowe JB (1994) Molecular cloning of a cDNA encoding a novel human leukocyte alpha-1,3-fucosyltransferase capable of synthesizing the sialyl Lewis x determinant. J Biol Chem 269:16789–16794

Nelson RM, Cecconi O, Roberts WG, Aruffo A, Linhardt RJ, Bevilacqua MP (1993) Heparin oligosaccharides bind L- and P-selectin and inhibit acute inflammation. Blood 82:3253–3258

Nolte D, Schmid P, Jäger U, Botzlar A, Roesken F, Hecht R, Uhl E, Messmer K, Vestweber D (1994) Leukocyte rolling in venules of striated muscle and skin is mediated by P-selectin, not by L-selectin. Am J Physiol 267:H1637–H1642

Norgard KE, Moore KL, Diaz S, Stults NL, Ushiyama S, McEver RP, Cummings RD, Varki A (1993) Characterization of a specific ligand for P-selectin on myeloid cells. A minor glycoprotein with sialylated O-linked oligosaccharides. J Biol Chem 268:12764–12774

Norgard-Sumnicht KE, Varki NM, Varki A (1993) Calcium-dependent heparin-like ligands for L-selectin in nonlymphoid endothelial cells. Science 261:480–483

Palazzo AJ, Jones SP, Anderson DC, Granger DN, Lefer DJ (1998) Coronary endothelial P-selectin in pathogenesis of myocardial ischemia-reperfusion injury. Am J Physiol Heart Circ Physiol 44:H1865–H1872

Pan JL, Xia LJ, McEver RP (1998a) Comparison of promoters for the murine and human P-selectin genes suggests species-specific and conserved mechanisms for transcriptional regulation in endothelial cells. J Biol Chem 273:10058–10067

Pan JL, Xia LJ, Yao LB, McEver RP (1998b) Tumor necrosis factor-alpha- or lipopolysaccharide-induced expression of the murine P-selectin gene in endothelial cells involves novel kappa-B sites and a variant activating transcription factor cAMP response element. J Biol Chem 273:10068–10077

Patel KD, Zimmerman GA, Prescott SM, McEver RP, McIntyre TM (1991) Oxygen radicals induce human endothelial cells to express GMP-140 and bind neutrophils. J Cell Biol 112:749–759

Peschon JJ, Slack JL, Reddy P, Stocking KL, Sunnarborg SW, Lee DC, Russell WE, Castner BJ, Johnson RS, Fitzner JN, Boyce RW, Nelson N, Kozlosky CJ, Wolfson MF, Rauch CT, Cerretti DP, Paxton RJ, March CJ, Black RA (1998) An essential role for ectodomain shedding in mammalian development. Science 282:1281–1284

Picker LJ, Warnock RA, Burns AR, Doerschuk CM, Berg EL, Butcher EC (1991) The neutrophil selectin LECAM-1 presents carbohydrate ligands to the vascular selectins ELAM-1 and GMP-140. Cell 66:921–933

Pottratz ST, Hall TD, Scribner WM, Jayaram HN, Natarajan V (1996) P-selectin-mediated attachment of small cell lung carcinoma to endothelial cells. Am J Physiol Lung Cell Mol Physiol 15:L918–L923

Pouyani T, Seed B (1995) PSGL-1 recognition of P-selectin is controlled by a tyrosine sulfation consensus at the PSGL-1 amino terminus. Cell 83:333–343

Preece G, Murphy G, Ager A (1996) Metalloproteinase-mediated regulation of L-selectin levels on leucocytes. J Biol Chem 271:11634–11640

Rabb H, Ramirez G, Saba SR, Reynolds D, Xu JC, Flavell R, Antonia S (1996) Renal ischemic-reperfusion injury in L-selectin-deficient mice. Am J Physiol 271:F408–F413

Ramos CL, Smith MJ, Snapp KR, Kansas GS, Stickney GW, Ley K, Lawrence MB (1998) Functional characterization of L-selectin ligands on human neutrophils and leukemia cell lines: evidence for mucin-like ligand activity distinct from P-selectin glycoprotein ligand-1. Blood 91:1067–1075

Reneman RS, Woldhuis B, oude Egbrink MGA, Slaaf DW, Tangelder GJ (1992) Concentration and velocity profiles of blood cells in the microcirculation. In: Hwang NHC, Turitto VT, Yen MRT (eds) Advances in cardiovascular engineering. Plenum Press, New York, pp 25–40

Robinson SD, Frenette PS, Rayburn H, Cummiskey M, Ullman-Cullere M, Wagner DD, Hynes RO (1999) Multiple, targeted deficiencies in selectins reveal a predominant role for P-selectin in leukocyte recruitment. Proc Natl Acad Sci USA 96:11452–11457

Romo GM, Dong JF, Schade AJ, Gardiner EE, Kansas GS, Li CQ, McIntire LV, Berndt MC, Lopez, JA (1999) The glycoprotein Ib-IX-V complex is a platelet counterreceptor for P-selectin. J Exp Med 190(6):803–813

Rosenkranz AR, Mendrick DL, Cotran RS, Mayadas TN (1999) P-selectin deficiency exacerbates experimental glomerulonephritis: a protective role for endothelial P-selectin in inflammation. J Clin Invest 103:649–659

Sako D, Chang X-J, Barone KM, Vachino G, White HM, Shaw G, Veldman GM, Bean KM, Ahern TJ, Furie B, Cumming DA, Larsen GR (1993) Expression cloning of a functional glycoprotein ligand for P-selectin. Cell 75:1179–1186

Sako D, Comess KM, Barone KM, Camphausen RT, Cumming DA, Shaw GD (1995) A sulfated peptide segment at the amino terminus of PSGL-1 is critical for P-selectin binding. Cell 83:323–331

Saku K, Zhang B, Ohta T, Shirai K, Tsuchiya Y, Arakawa K (1999) Levels of soluble cell adhesion molecules in patients with angiographically defined coronary atherosclerosis. Jpn Circ J 63:19–24

Sanders WE, Wilson RW, Ballantyne CM, Beaudet AL (1992) Molecular cloning and analysis of in vivo expression of murine P-selectin. Blood 80:795–800

Sasaki K, Kurata K, Funayama K, Nagata M, Watanabe E, Ohta S, Hanai N, Nishi T (1994) Expression cloning of a novel alpha 1,3-fucosyltransferase that is involved in biosynthesis of the sialyl Lewis^x carbohydrate determinants in leukocytes. J Biol Chem 269:14730–14737

Schleiffenbaum B, Spertini O, Tedder TF (1992) Soluble L-selectin is present in human plasma at high levels and retains functional activity. J Cell Biol 119:229–238

Schürmann GM, Bishop AE, Facer P, Vecchio M, Lee JCW, Rampton DS, Polak JM (1995) Increased expression of cell adhesion molecule P-selectin in active inflammatory bowel disease. Gut 36:411–418

Shao JY, Ting-Beall HP, Hochmuth RM (1998) Static and dynamic lengths of neutrophil microvilli. Proc Natl Acad Sci USA 95:6797–6802

Siegelman MH, van de Rijn M, Weissman IL (1989) Mouse lymph node homing receptor cDNA clone encodes a glycoprotein revealing tandem interaction domains. Science 243:1165–1172

Singbartl K, Ley K (2000) Protection from ischemia-reperfusion induced severe renal failure by blocking E-selectin. Crit Care Med 28:2507–2514

Singbartl K, Green SA, Ley K (2000) Blocking P-selectin protects from ischemia/reperfusion-induced acute renal failure. FASEB J 14:48–54

Smith MJ, Berg EL, Lawrence MB (1999) A direct comparison of selectin-mediated transient adhesive events using high temporal resolution Biophys J 77:3371–3383

Smith PL, Gersten KM, Petryniak B, Kelly RJ, Rogers C, Natsuka Y, Alford JA III, Scheidegger EP, Natsuka S, Lowe JB (1996) Expression of the a(1,3)fucosyltransferase Fuc-TVII in lymphoid aggregate high endothelial venules correlates with expression of L-selectin ligands. J Biol Chem 271:8250–8259

Snapp KR, Wagers AJ, Craig R, Stoolman LM, Kansas GS (1997) P-selectin glycoprotein ligand-1 (PSGL-1) is essential for adhesion to P-selectin but not E-selectin in stably transfected hematopoietic cell lines. Blood 89:896–901

Snapp KR, Craig R, Nelson RD, Stoolman LM, Kansas GS (1998) Dimerization of P-selectin glycoprotein ligand-1 (PSGL-1) required for optimal recognition of P-selectin. J Cell Biol 142:263–270

Spertini O, Kansas GS, Munro JM, Griffin JD, Tedder TF (1991a) Regulation of leukocyte migration by activation of the leukocyte adhesion molecule-1 (LAM-1) selectin. Nature 349:691–694

Spertini O, Luscinskas FW, Kansas GS, Munro JM, Griffin JD, Gimbrone MA Jr, Tedder TF (1991b) Leukocyte adhesion molecule-1 (LAM-1) interacts with an inducible endothelial cell ligand to support leukocyte adhesion. J Immunol 147:2565–2573

Spertini O, Luscinskas FW, Gimbrone MA Jr, Tedder TF (1992) Monocyte attachment to activated human vascular endothelium in vitro is mediated by leukocyte adhesion molecule-1 (L-selectin) under nonstatic conditions. J Exp Med 175:1789–1792

Spertini O, Cordey AS, Monai N, Giuffre L, Schapira M (1996) P-selectin glycoprotein ligand 1 is a ligand for L-selectin on neutrophils, monocytes, and CD34[+] hematopoietic progenitor cells. J Cell Biol 135:523–531

Staite ND, Justen JM, Sly LM, Beaudet AL, Bullard DC (1996) Inhibition of delayed-type contact hypersensitivity in mice deficient in both E-selectin and P-selectin. Blood 88:2973–2979

Steeber DA, Engel P, Miller AS, Sheetz MP, Tedder TF (1997) Ligation of L-selectin through conserved regions within the lectin domain activates signal transduction pathways and integrin function in human, mouse, and rat leukocytes. J Immunol 159:952–963

Steegmaier M, Levinovitz A, Isenmann S, Borges E, Lenter M, Kocher HP, Kleuser B, Vestweber D (1995) The E-selectin-ligand ESL-1 is a variant of a receptor for fibroblast growth factor. Nature 373:615–620

Stone JP, Wagner DD (1993) P-selectin mediates adhesion of platelets to neuroblastoma and small cell lung cancer. J Clin Invest 92:804–813

Stotland MA, Kerrigan CL (1997) E- and L-selectin adhesion molecules in musculocutaneous flap reperfusion injury. Plast Reconstr Surg 99:2010–2020

Subramaniam M, Koedam JA, Wagner DD (1993) Divergent fates of P- and E-selectins after their expression on the plasma membrane. Mol Biol Cell 4:791–801

Subramaniam M, Saffaripour S, Vandewater L, Frenette PS, Mayadas TN, Hynes RO, Wagner DD (1997) Role of endothelial selectins in wound repair. Am J Pathol 150:1701–1709

Suzuki A, Andrew DP, Gonzalo J-A, Fukumoto M, Spellberg J, Hashiyama M, Suda T, Takimoto H, Gerwin N, Webb J, Gutierrez-Ramos J-C, Molineux G, McNiece I, Ley K, Butcher EC, May WS,

Greaves MF, Amakawa R, Tada Y, Wakcham A, Mak TW (1996) CD34 deficient mice have reduced eosinophil accumulation after allergen exposure and reveal a novel crossreactive 90-kD protein. Blood 87:3550–3562

Takada M, Nadeau KC, Shaw GD, Marquette KA, Tilney NL (1997) The cytokine-adhesion molecule cascade in ischemia/reperfusion injury of the rat kidney – inhibition by a soluble P-selectin ligand. J Clin Invest 99:2682–2690

Tang MLK, Hale LP, Steeber DA, Tedder TF (1997) L-selectin is involved in lymphocyte migration to sites of inflammation in the skin – delayed rejection of allografts in L-selectin-deficient mice. J Immunol 158:5191–5199

Tedder TF, Isaacs CM, Ernst TJ, Demetri GD, Adler A, Disteche CM (1989) Isolation and chromosomal localization of cDNAs encoding a novel human lymphocyte cell surface molecule, LAM-1: homology with the mouse lymphocyte homing receptor and other human adhesion proteins. J Exp Med 170:123–133

Tedder TF, Steeber DA, Pizcueta P (1995) L-selectin deficient mice have impaired leukocyte recruitment into inflammatory sites. J Exp Med 181:2259–2264

Tözeren A, Ley K (1992) How do selectins mediate leukocyte rolling in venules? Biophys J 63:700–709

Ushiyama S, Laue TM, Moore KL, Erickson HP, McEver RP (1993) Structural and functional characterization of monomeric soluble P-selectin and comparison with membrane P-selectin. J Biol Chem 268:15229–15237

Varki A (1997) Selectin ligands – will the real ones please stand up? J Clin Invest 99:158–162

von Andrian UH, Berger EM, Ramezani L, Chambers JD, Ochs HD, Harlan JM, Paulson JC, Etzioni A, Arfors K-E (1993) In vivo behavior of neutrophils from two patients with distinct inherited leukocyte adhesion deficiency syndromes. J Clin Invest 91:2893–2897

von Andrian UH, Hasslen SR, Nelson RD, Erlandsen SL, Butcher EC (1995) A central role for microvillous receptor presentation in leukocyte adhesion under flow. Cell 82:989–999

Vora DK, Fang ZT, Liva SM, Tyner TR, Parhami F, Watson AD, Drake TA, Territo MC, Berliner JA (1997) Induction of P-selectin by oxidized lipoproteins – separate effects on synthesis and surface expression. Circ Res 80:810–818

Waddell TK, Fialkow L, Chan CK, Kishimoto TK, Downey GP (1994) Potentiation of the oxidative burst of human neutrophils. A signaling role for L-selectin. J Biol Chem 269:18485–18491

Wagers AJ, Lowe JB, Kansas GS (1996) An important role for the alpha-1,3 fucosyltransferase, FucT-VII, in leukocyte adhesion to E-selectin. Blood 88:2125–2132

Walcheck B, Watts G, Jutila MA (1993) Bovine gamma/d T cells bind E-selectin via a novel glycoprotein receptor: first characterization of a lymphocyte/E-selectin interaction in an animal model. J Exp Med 178:853–863

Walcheck B, Kahn J, Fisher JM, Wang BB, Fisk RS, Payan DG, Feehan C, Betageri R, Darlak K, Spatola AF, Kishimoto TK (1996) Neutrophil rolling altered by inhibition of L-selectin shedding in vitro. Nature 380:720–723

Weller A, Isenmann S, Vestweber D (1992) Cloning of the mouse endothelial selectins: expression of both E- and P-selectin is inducible by tumor necrosis factor-a. J Biol Chem 267:15176–15183

Yadav SS, Howell DN, Gao WS, Steeber DA, Harland RC, Clavien PA (1998) L-selectin and ICAM-1 mediate reperfusion injury and neutrophil adhesion in the warm ischemic mouse liver. Am J Physiol Gastrointest Liver Physiol 38:G1341–G1352

Yang J, Hirata T, Croce K, Merrill-Skoloff G, Tchernychev B, Williams E, Flaumenhaft R, Furie BC, Furie B (1999) Targeted gene disruption demonstrates that P-selectin glycoprotein ligand 1 (PSGL-1) is required for P-selectin–mediated but not E-selectin–mediated neutrophil rolling and migration. J Exp Med 190:1769–1782

Zakrzewicz A, Grafe M, Terbeek D, Bongrazio M, Auch-Schwelk W, Walzog B, Graf K, Fleck E, Ley K, Gaehtgens P (1997) L-selectin-dependent leukocyte adhesion to microvascular but not to macrovascular endothelial cells of the human coronary system. Blood 89:3228–3235

Zöllner O, Lenter MC, Blanks JE, Borges E, Steegmaier M, Zerwes HG, Vestweber D (1997) L-selectin from human, but not from mouse neutrophils binds directly to E-selectin. J Cell Biol 136:707–716

Carbohydrate Ligands for the Leukocyte-Endothelium Adhesion Molecules, Selectins

Ten Feizi[1]

1 Introduction

Much interest has been stimulated in carbohydrate biology as a result of selectin research. The three selectins, E-, L- and P-selectins are leukocyte-endothelium adhesion molecules (Brandley et al. 1990; Harlan et al. 1992; Bevilacqua et al. 1993). As discussed by Dr Klaus Ley (this volume), the three molecules are key players in mechanisms of innate immunity, for they have important roles at the initial stages of recruitment and emigration of leukocytes from the blood vascular compartment into sites of infection and injury. L-selectin, which is also involved in lymphocyte recirculation in and out of peripheral lymphoid tissues, is expressed constitutively on most types of leukocytes. E-selectin is expressed on activated endothelium, and P-selectin, which is in storage granules of platelets and endothelium, reaches the surface following cell activation. The selectins are long molecules that project from the cell surface. The special feature of these molecules is that they capture and mediate the rapid, but transient, attachment of leukocytes in the flowing blood stream to desired sites on the endothelium or to other leukocytes, and they tether and cause them to roll along the endothelial cell surface. This in turn allows cascades of other much tighter cell adhesive events to take place, mediated by other cell adhesion molecules, before the leukocytes can migrate through the endothelium and the underlying basement membrane. If the selectin-mediated rolling and tethering is inhibited, the subsequent steps are severely compromised. This knowledge has stimulated research into defining the ligands for the selectins in the hope of being able to design analogues that could serve as inhibitory, therapeutic substances for the treatment of disorders of inflammation.

Discovery of lectin-like domains at the membrane-distal tips of the selectins stimulated intense research into their carbohydrate ligands. The knowledge that human E- and P-selectins bind granulocytes and monocytes (Harlan et al. 1992) served to focus interest in carbohydrate differentiation antigens of these cell types, namely the 3'-fucosyl-N-acetyllactosamine (Le^x) and the related sialylated sequences which occur as capping groups on carbohydrate chains of

[1] The Glycosciences Laboratory, Imperial College School of Medicine, Northwick Park Campus, Watford Road, Harrow, Middlesex HA1 3UJ, UK

Results and Problems in Cell Differentiation, Vol. 33
Paul R. Crocker (Ed.): Mammalian Carbohydrate Recognition Systems
© Springer-Verlag Berlin Heidelberg 2001

Galβ1-4GlcNAcβ1-
    | 1,3
    Fucα

Le[x]

Galβ1-3GlcNAcβ1-
    | 1,4
    Fucα

Le[a]

Galβ1-4GlcNAcβ1-
    | 2,3 | 1,3
NeuAcα Fucα
3'-sialyl-Le[x]

Galβ1-3GlcNAcβ1-
    | 2,3 | 1,4
NeuAcα Fucα
3'-sialyl-Le[a]

HSO₃
  | 6
Galβ1-4GlcNAcβ1-
    | 2,3 | 1,3
NeuAcα Fucα
6'-sulfo-sialyl-Le[x]

        HSO₃
          | 6
Galβ1-4GlcNAcβ1-
    | 2, 3 | 1,3
NeuAcα Fucα
6-sulfo-sialyl-Le[x]

HSO₃   HSO₃
  | 6 | 6
Galβ1-4GlcNAcβ1-
    | 2,3 | 1,3
  NeuAcα Fucα
6',-sulfo-sialyl-Le[x]

Fig. 1. The Le[a] and Le[x] antigens and related capping sequences which occur naturally on the carbohydrate chains of glycoproteins or glycolipids

glycoproteins and glycolipids (Gooi et al. 1983; Thorpe and Feizi 1984; Feizi. 1985; Fukuda et al. 1999). Indeed, 3'-sialyl-Le[x] and the isomeric sequence 3'-sialyl-Le[a] (Fig. 1) were readily shown to be recognized by the three selectins (Bevilacqua et al. 1993; McEver et al. 1995; Rosen and Bertozzi 1996; Feizi and Galustian 1999). A picture has been emerging, however, of subtle differences in the binding specificities of the selectins, and there are examples of variant carbohydrate sequences related to the sialyl-Le[x] and sialyl-Le[a] sequences that are preferentially bound by one or other of these receptors. In addition, a feature of the L- and P-selectins, apparently not shared by E-selectin, is the ability to interact with a second class of ligands that present relatively small sulfated motifs.

Unlike the usual situation of receptor-ligand pairs (Liebecq 1992), the selectin recognition systems operate as triads: receptors, ligands and carriers (Crocker and Feizi 1996). Only when the oligosaccharide ligands are optimally presented on carriers (proteins or lipids), functional counter-receptors are formed. Information is accumulating on macromolecules that may serve as counter-receptors for the selectins. However, as critically discussed by Klaus Ley (this Vol.), only in the case of the membrane glycoprotein PSGL-1 is there certainty regarding the counter-receptor status for selectins. Most investigations of the counter-receptors are carried out with lysed cells, and it is not always possible to be certain as to whether oligosaccharides displayed on the isolated macromolecules (glycoproteins or glycolipids) are available on intact cell membranes for selectin binding.

This chapter is concerned predominantly with details of the sequences of the saccharide ligands, identified from natural sources, for the selectins. Developments in the field of glycosyltransferases are providing substantial insights into the biosynthesis and regulation of the ligands for the selectins. Selected aspects are discussed, but a detailed coverage of this important field is out of the scope of this review. The field of the selectin ligands has benefited tremendously from the availability of a rich array of ligand analogues made available through synthetic carbohydrate chemistry; many of these compounds have been reviewed by Simanek et al. (1998). A further contribution of synthetic carbohydrate chemistry has been to provide clues to the existence of novel saccharide ligands for the selectins, and this is discussed here in the context of the L-selectin ligands.

2 Carbohydrate Ligands for E-Selectin

2.1 The Initial Evaluations of the Le^x and Le^a Systems as Ligands

Early clues to the recognition of the sialyl-Le^x sequence as a ligand for E-selectin were observations of Lowe and colleagues on two spontaneously occurring variants of the HL60 cell line, designated A and B, which were found to differ in their binding to endothelial cells [human umbilical-vein endothelial cells (HUVECs)] that had been stimulated to express E-selectin by treatment with the cytokine tumor necrosis factor-α (Lowe et al. 1990): only HL60 B cells showed binding. Both cell lines express Le^x antigen (they bind anti-SSEA-1 (Gooi et al. 1981), but only HL60 B cells express the 3-sialylated form of Le^x, as demonstrated by their binding to the monoclonal antibody CSLEX-1 (Fukushima et al. 1984). Cultured COS-1 and Chinese hamster ovary (CHO) cells express the sialo-oligosaccharide sequence, 3'-sialyl-N-acteyllactosamine:

NeuAcα2–3Galβ1–4GlcNAc, but not the Fucα1–3-GlcNAc sequence, and they do not adhere to E-selectin-expressing HUVECs. Upon transfection of the COS-1 and CHO cell lines with a plasmid containing an α(1–3/1–4) fucosyl-transferase, which can transfer fucose in α1–3 linkage to GlcNAc of the above sialo-oligosaccharide sequence, adhesion to activated HUVECs was observed with both transfected cell lines.

Experiments using a soluble recombinant E-selectin IgG chimera produced in COS-7 m6 cells gave results in complete accord with the above conclusions. The myeloid cells HL60 and THP-1 were shown to bind to the immobilized protein and the binding was inhibited by anti-CSLEX-1 antibody but not by an anti-CD15 (anti-Le^x) monoclonal antibody (PM81) or by VIM-2 which is directed to the 3'-sialyl-poly-N-acetyllactosamine sequence with fucose α1–3-linked to the second (inner) rather than the outer N-acetyllactosamine (Macher et al. 1988), structure 6, Fig. 2. Binding mediated by E-selectin was abolished by sialidase treatment of the HL60 cells and was inhibited in the presence of glycoproteins known to contain the 3'-sialyl-3'-fucosyl-N-acetyl-lactosamine sequence.

Other studies raised the possibility that the sialylated and internally fuco-sylated sequences, as in the VIM-2 antigen, are bound by E-selectin. When CHO cells transfected to express the membrane form of E-selectin were overlaid onto acidic glycolipid fractions extracted from leukocytes and chromato-graphed on silica gel (Tiemeyer et al. 1991), binding was predominantly to long-chain sialo-glycolipids where the main components were identified by mass spectrometry as being 3-sialylated and internally 3-fucosylated, as in the case of VIM-2-active glycolipids. However, the presence of additional minor components with differing fucosylation patterns could not be ruled out. Obser-vation of a strong inhibition of HL60 cell adhesion to activated HUVECs, by liposomes containing a difucosylated glycolipid, related to VIM-2 but having an additional 3-linked fucose residue, on the outer N-acetylglucosamine (structure 7, Fig. 2), raised the possibility that one or both of these fucose residues are involved in E-selectin recognition (Phillips et al. 1990). These pos-sibilities are further considered, below.

Direct binding studies, in vitro, showed clearly that E-selectin recognizes sialyl-fuco-oligosaccharides with fucose 3-linked to outer N-acetylglu-cosamine, as in the 3'-sialyl-Le^x sequence (structure 1, Fig. 2), when these are presented linked to lipid (Tyrrell et al. 1991; Larkin et al. 1992; Yuen et al. 1992), or to protein (Berg et al. 1991).

E-selectin was also shown to bind to the 3'-sialyl-Le^a sequence terminating in the type 1, Galβ1–3GlcNAc, backbone sequence (as in structure 2, Fig. 2), when linked to lipid or to protein. The binding is stronger than to the 3-sialyl-Le^x analogue (Larkin et al. 1992; Yuen et al. 1992). Liposomes containing the lipid-linked 3'-sialyl-Le^a inhibited E-selectin binding to tumor cells (Takada et al. 1991). As 3'-sialyl-Le^a occurs as a cancer-associated antigen in some human epithelial tumors, this binding activity of E-selectin implicates the protein in

Galβ1-4GlcNAcβ1-3Galβ1-4Glc   ***Structure 1***
  |2,3 |1,3
NeuAcα Fucα

Galβ1-3GlcNAcβ1-3Galβ1-4Glc   ***Structure 2***
  |2,3 |1,4
NeuAcα Fucα

Galβ1-3/4GlcNAcβ1-3Gal   ***Structure 3***
  |3 |1,4/3
HSO₃   Fucα

Galβ-GlcNAc-Gal   ***Structure 4***
  | |
NeuAc Fuc

Galβ1-4GlcNAcβ1-3Galβ1-4GlcNAcβ1-3Galβ1-4Glc   ***Structure 5***
  |2,3 |1,3
NeuAcα Fucα

Galβ1-4GlcNAcβ1-3Galβ1-4GlcNAcβ1-3Galβ1-4Glc   ***Structure 6***
  |2,3 |1,3
NeuAcα   Fucα

Galβ1-4GlcNAcβ1-3Galβ1-4GlcNAcβ1-3Galβ1-4Glc   ***Structure 7***
  |2,3 |1,3 |1,3
NeuAcα Fucα   Fucα

Galβ1-4GlcNAcβ1-3Galβ1-4GlcNAcβ1-3Galβ1-4GlcNAcβ1-3Galβ1-4Glc   ***Structure 8***
  |2,3 |1,3
NeuAcα   Fucα

Galβ1-4GlcNAcβ1-3Galβ1-4GlcNAcβ1-3Galβ1-4GlcNAcβ1-3Galβ1-4GlcNAcβ1-3Galβ1-4Glc
  |2,3 |1,3
NeuAcα Fucα   ***Structure 9***

Galβ1-4GlcNAcβ1-3Galβ1-4GlcNAcβ1-3Galβ1-4GlcNAcβ1-3Galβ1-4GlcNAcβ1-3Galβ1-4Glc
  |2,3 |1,3   (|2,3)
NeuAcα   Fucα   (Fucα)   ***Structure 10***

       Gal β1-3\
             GalNAcβ1-3Galα1-4Galβ1-4Glc   ***Structure 11***
Galβ1-4GlcNAcβ1-6⁄
    |1,3
    Fucα

GlcUAβ1-3Galβ1-3GlcNAcβ1-3Galβ1-4Glc   ***Structure 12***
  |3
HSO₃

Fig. 2. Assignments for some Le^a- and Le^a-related oligosaccharide sequences for glycolipids, or for oligosaccharides isolated from myeloid and epithelial tissues, and investigated for binding to the selectins

tumor cell adhesion leading to metastasis, in addition to the pathogenesis of inflammation.

Co-operative effects of clustering both of E-selectin and its ligands were shown to be important for mediating adhesion. For example, a soluble, monomeric form of E-selectin gave no detectable binding signal with the immobilized clustered neoglycolipid ligands mentioned above, but could be induced to adhere when oligomerized in the form of a soluble complex using a non-neutralizing anti-E-selectin antibody and staphylococcal protein A (R.A. Childs and T. Feizi, unpubl.; cited by Feizi 1993). With CHO cells transfected to express E-selectin, adhesion intensity and even specificity were found to be dependent on the density of surface expression of the membrane-associated adhesion molecules (Larkin et al. 1992). There was a threshold of density of E-selectin required for binding to the lipid-linked acidic fuco-oligosaccharide ligands, and three types of adhesive specificity were observed (Larkin et al. 1992): (1) transfected cells with low levels of E-selectin expression showed no detectable carbohydrate-mediated adhesion; (2) cells with the highest density of E-selectin expression adhered not only to the lipid-linked acidic (3′-sialyl or 3′-sulfo) Le^a/Le^x sequences, but also to the non-acidic Le^a, and to a lesser extent Le^x-active analogues, although tenfold or higher levels of these latter neoglycolipids were required to give binding intensities equivalent to those observed with the acidic oligosaccharides; (3) cells with intermediate levels of E-selectin expression adhered only to the acidic oligosaccharides.

The in vivo significance of binding to the non-acidic Le^a and Le^x type sequences observed under static adhesion assay conditions, discussed above, is uncertain as there is considerable evidence for the importance of sialic acid in many in vivo and in vitro experiments investigating metastatic potential and cell adhesion. Interestingly, some evidence shows that under conditions of flow, in vitro, Le^x on the myeloid cell surface may play a greater role in E-selectin-mediated adhesion than sialyl-Le^x (Kojima et al. 1992). Amounts of Le^a antigen on human erythrocytes are too low to elicit E-selectin binding (Feizi. 1992) but the enormously dense expression of the Le^a and Le^x antigens on many adenocarcinoma cells may well be able to mediate adhesion. For these reasons, it is interesting to note that while the presence of the blood group H fucose $\alpha1$–2 linked to galactose of the sequence of Le^a (as in the Le^b antigen) does not hinder E-selectin adhesion, the additional presence of the blood group A or B monosaccharides, N-acetylgalactosamine $\alpha1$–3-linked as in ALe^b heptasaccharide (Larkin et al. 1992) or galactose $\alpha1$–3- linked to galactose, as in BLe^b heptasaccharide, hinders E-selectin binding (T. Feizi, E. Breimer, J. O'Brien, unpubl.; and cited by Feizi 1993). In adenocarcinomas of the distal colon, the blood group A, B, H, Le^b, ALe^b and BLe^b antigens are expressed as cancer-associated antigens of the distal colon, and we have raised the possibility (Larkin et al. 1992) that the variable display of these on the tumor cells in different individuals may be among factors that influence metastatic potential.

2.2 The Search for Epithelial and Myeloid Cell Ligands

A novel class of acidic selectin ligand, sulfate-containing, was identified by screening for E-selectin binding to lipid-linked oligosaccharide probes (neoglycolipids) generated from the carbohydrate chains (O-glycans) of a mucin-type glycoprotein that had been isolated from a human ovarian cystadenoma (Yuen et al. 1992). The oligosaccharide consisted of an equimolar mixture of 3'-sulfated tetrasaccharides of Le^a and Le^x type (structure 3, Fig. 2). The binding activity of the mixture was at least equal to that of the 3'-sialyl-Le^x analogue (Yuen et al. 1992). This was corroborated by subsequent work with the individual chemically synthesized Le^a- and Le^x-based sequences (Yuen et al. 1994).

Thus, in the E-selectin system (now also found with the L- and P-selectins; cited by Crocker and Feizi 1996) sulfate 3-linked to the terminal galactose can substitute for the carboxyl group of 3-linked sialic acid (Kogelberg et al. 1996). This suggests that the primary contribution of the sialic acid is the negatively charged group, and is in accord with evidence (Tyrrell et al. 1991) that other molecular features of sialic acid have no effect on E-selectin recognition.

Four papers (Osanai et al. 1996; Müthing et al. 1996; Stroud et al. 1996a,b) have served to rekindle interest in long chain glycolipid ligands for E-selectin. Both in the extracts of human (Müthing et al. 1996; Stroud et al. 1996a,b) and murine (Osanai et al. 1996) myeloid cells, and murine kidney (Osanai et al. 1996) examined by chromatogram binding assays, E-selectin binding was observed only to long chain, slow migrating glycolipids. These have R$_f$ values substantially less than those of the sialyl-Le^x hexasaccharide (ceramide) based on the tetrasaccharide backbone, as in structure 1, Fig. 2 (Fukushima et al. 1984) and of the VIM-2 octasaccharide-based on the hexasaccharide backbone (structure 6, Fig. 2; Macher et al. 1988). It must be emphasized that the bound components were highly heterogeneous, and extremely minor components among the bulk of glycolipids, as elaborated below.

The study by Stroud et al. (1996a,b) was very large-scale, starting with 1.2 l of packed HL60 cells. Using mass spectrometry and ^{1}H-NMR analysis, sequence information was obtained on monosialo-gangliosides based on backbones of di- to dodecasaccharides. Monofucosyl, sialyl sequences such as structure 9, Fig. 2, could not be detected among the glycolipids investigated. Rather, many sequences with fucose at one or more of the inner N-acetyglucosamines, as in structure 10, Fig. 2, were identified. Fucose at the outer N-acetylglucosamine was detected only in sequences that had one or two additional, inner, fucoses. The E-selectin binding observed in chromatogram-binding experiments was to regions of chromatograms containing these internally fucosylated glycolipids. The authors cited similar binding patterns with monosialo-gangliosides from normal human leukocytes, and suggested that E-selectin recognizes the internally fucosylated monosialo-gangliosides. They also stated that CSLEX-1 antibody may not be monospecific for the sialyl-Le^x sequence after all, and suggested that the antibody cross-reacts not only with internally fucosylated sialo-oligosaccharide sequences such as the VIM-2 sequence, but also with the non-fucosylated sialo-oligosaccharide backbones. The main questions raised by these data are whether E-selectin can recognize sialyl-poly-N-acetyllactosamine sequences that lack fucose at the outer N-acetylglucosamine, or whether the binding observed was to minor components that could not be detected by the physico-chemical techniques used.

The results of Müthing and colleagues (Müthing et al. 1996) were more in accord with the original notions regarding the antibody specificity and the sialyl-ligands for E-selectin. These were relatively small-scale experiments characterized by excellent resolution and fractionation of glycolipids from human granulocytes. Using a combination of immunochemical and mass spectrometric experiments, the authors clearly identified a sialyl-Le^x glycolipid based on a hexasaccharide backbone (structure 6, Fig. 2), which was bound by CSLEX-1 but not VIM-2 antibody. Two other mono-fucosyl gangliosides based on the octasaccharide backbone with fucose on the inner (second) N-acetylglucosamine as in structure 8, Fig. 2, and differing only in their ceramide moieties, were unreactive with CSLEX-1, but as expected, were bound by the VIM-2 antibody. Several other fucogangliosides based on the octa- or decasaccharide backbone, not fully sequenced, were bound by one or other antibody but not by both. According to these results, the specificity of CSLEX-1 is as originally described (Fukushima et al. 1984), and the sialyl-Le^x sequence, based on the hexasaccharide backbone [a predicted fucosylation product of Fuc-TVII (Maly et al. 1996)] which in the mouse is a key to the generation of selectin ligands, is to be found on human granulocytes. Müthing and colleagues also detected, in agreement with the findings of Stroud and colleagues, mono-sialyl-gangliosides based on octa- to decasaccharides, and with various internal fucosylation patterns.

Murine leukocytes, in sharp contrast to those of the human, do not express serologically detectable Le^x and sialyl-Le^x antigens (Smith et al. 1996) nor the Le^a and sialyl-Le^a antigens (Maly et al. 1996). Two approaches have been made to identifying carbohydrate ligands for E- selectin in the mouse (Osanai et al. 1996) by exploiting the neoglycolipid technology (Feizi et al. 1994). Having found evidence for the presence of poly-N-acetyllactosamine type sequences among the E-selectin-binding glycolipid population extracted from the murine neutrophilic cell line, 32D cl3, they used endo-β-galactosidase to release susceptible oligosaccharides of this series from the surface of these cells. Oligosaccharide fractions were converted into neoglycolipids, and the main E-selectin-binding component was identified, using TLC-liquid secondary ion mass spectrometry, as a sialo-fuco-oligosaccharide of the sialyl-Le^a/Le^x type (structure 4, Fig. 2), and it was bound by monoclonal anti-sialyl-Le^a (2D3) but not by CSLEX-1. We (J. Topping and T. Feizi, unpubl.) have subsequently found that CSLEX-1 does not give a binding signal with the sialyl-Le^x sequence based on the trisaccharide backbone Galβ1–4GlcNAcβ1–3Gal that was generated as a result of the endo-β-galactosidase digestion (rather, a tetrasaccharide or longer backbone is required for binding), whereas antibody 2D3 gives strong binding signals with the sialyl-Le^a analogue based on the trisaccharide backbone Galβ1–3GlcNAcβ1–3Gal. We now interpret the results as indicating that any sialyl-Le^x sequence (the predicted product of Fuc-TVII (Maly et al. 1996) within the fraction investigated would not have been detected. It is not known, however, why the sialyl-Le^a antigen is not detectable serologically on the murine leukocytes as well as the cell line investigated. A possible explanation is that the display of the saccharide epitope on these cells is suboptimal for antibody binding, possibly crowded out by other selectin ligands that are not recognized by the antibody. It is, moreover, possible that the products of another fucosyltransferase act in concert with those of FucT-VII to generate ligand densities above the threshold needed for selectin-mediated adhesion, but the densities of the individual sialyl-Le^x and Le^a determinants recognized by the antibodies are relatively low.

A second class of glycolipid bound by E-selectin was characterized in the kidney extracts of BALB/c strain mice (Osanai et al. 1996). Using chromatogram binding experiments and in situ mass spectrometry using neoglycolipids derived from oligosaccharides released with endoglycoceramidase, in conjunction with compositional and linkage analyses, the oligosaccharide moiety was identified as the Le^x-active, extended, branched globo sequence, structure 11, Fig. 2 (Sekine et al. 1988). Evidence was found for the presence of novel sialyl analogues of the branched globo sequence, raising the possibility that fucosyl (S-Le^x) analogues may also be present, though not detected. It will be interesting to determine the cellular distribution of the globo-Le^x type sequence, especially among epithelia and malignant cells derived from them, as its presence in the kidney is genetically determined (Sekine et al. 1988), and it may therefore contribute to the metastatic potential of epithelial tumors in different inbred strains of mice.

2.3 Conclusions and Questions Regarding the Carbohydrate Sequences Recognized by E-Selectin

The in vitro binding experiments with isolated molecules, taken together with the results of glycosyltransferase transfection experiments show that E-selectin recognizes the 3′-sialyl and 3′-sulfo-Le^x and Le^a sequences, and that it can also bind, albeit less strongly, to the asialo and the non-sulfated Le^a and Le^x sequences. Further investigations, particularly with intentionally synthesized oligosaccharides are required to resolve the issue of E-selectin recognition of the VIM-2 type, 3′-sialyl-poly-N-actyllactosamine sequences with fucose in the inner (second) rather than the outer N-acetyllactosamine.

3 Carbohydrate Ligands for L-Selectin

3.1 Initial Explorations of Carbohydrate Sequences Recognized by L-Selectin

Much evidence for a carbohydrate-mediated adhesive specificity for L-selectin, with involvement of sialic acid, fucose and sulfate was obtained before such evidence for the other two selectins, and before the discovery of a lectin-like protein motif on these proteins (see Rosen 1993 for a review of the elegant experiments that led to those conclusions). Research on the sequences of oligosaccharides recognized by L-selectin was boosted with the realization that the sialyl-Le^x sequence is recognized by the two related proteins E- and P-selectins. In vitro binding experiments using normal mouse lymphocytes (Berg et al. 1992) or a mouse L-selectin IgG chimera (Foxall et al. 1992) or mouse L1–2 pre-B cells transfected with human L-selectin cDNA (Berg et al. 1992), showed that both the murine and the human L-selectins can bind to 3′-sialyl-Le^x and -Le^a sequences (structures 1 and 2, Fig. 2), linked to protein or to lipid. These features, and the lack of recognition of the carboxyl group rather than the polyol tail of the sialic acid (Norgard et al. 1993) are shared with human E-selectin. It was apparent that the adhesive specificities of the two selectins are different. First, it was observed that L-selectin, but not E-selectin, adheres to the sulfate-containing glycosphingolipid sulfatide (Aruffo et al. 1991; Green et al. 1992; Foxall et al. 1992; Needham et al. 1993; Nelson et al. 1993; Suzuki et al. 1993). Second, L-selectin was shown to bind with a greater intensity than E-selectin to a glycoprotein 'PNAd' isolated from peripheral lymph nodes (Berg et al. 1992). Third, E-selectin but not L-selectin bound to glycoproteins expressing 'CLA', an antigen on cutaneous lymphocytes (Berg et al. 1992).

Work with structurally defined oligosaccharides of the Le^a, Le^x and sulfo-glucuronyl series (HNK-1, structure 12, Fig. 2) corroborated the importance of sulfate as a recognition element on oligosaccharide ligands for L-selectin (Feizi. 1993). From binding and inhibition experiments with the recombinant soluble protein, it became clear that there is an overall preference of the sulf-Le^a and -Le^x sequences over the sialyl analogues (Green et al. 1992; Green et al. 1995; Galustian et al. 1997a, 1999), to the extent that the non-fucosylated sulfated backbones are bound (albeit with lower intensities), but not the sialyl analogues (Green et al. 1992).

3.2 Novel Sulfated Sequences Detected on the Endothelial Glycoprotein GlyCAM-1

Considerable information is accumulating on sulfated oligosaccharide sequences of the type found on high endothelial venules, the key attachment sites for L-selectin. Among *O*-glycans released from the endothelial glycoprotein GlyCAM-1, evidence has been found for the occurrence of sulfated forms of sialyl-Le^x (Fig. 2) in the first, designated 6′-sulfo-sialyl-Le^x, the sulfate is at position 6 of the terminal galactose; in the second, 6-sulfo-sialyl-Le^x, sulfate is at position 6 of the penultimate *N*-acetylglucosamine, and in the third, 6′,6-sulfo-sialyl-Le^x, both positions are sulfated (Hemmerich et al. 1995). However, with the limited amounts of oligosaccharide material derived from GlyCAM-1, the potencies of these oligosaccharides relative to the non-sulfated sialyl-Le^x could not be determined. Chemical synthesis approaches have been extremely rewarding here, for they have enabled (1) the clear assignment of the position of sulfation on sialyl-Le^x that elicits the strongest L-selectin binding signals, (2) the identification of novel, more potent ligands, and (3) the discovery of a processing pathway of the sialyl-ligands that may be important for regulation of the expression of selectin ligand activity in vivo. These several developments are highlighted below.

3.3 Chemically Synthesized Sulfated Forms of the Le^x Pentasaccharide, and Their Interactions with L-Selectin

In the first of the synthetic approaches aimed at assessing the L-selectin binding signals elicited by 6′-, or the 6-sulfated sialyl-Le^x, the three sulfated variants of the sialyl-Le^x sequence (Fig. 3) were synthesized as pentaglycosylceramides (Komba et al. 1996). In initial experiments, the ability of these three sulfated compounds to support L-selectin binding appeared comparable to that of the non-sulfated sialyl-Le^x (Yoshino et al. 1997). However, this was soon found to be a reflection of the presence of two minor impurities in each of the sulfated compounds. Once these were removed, the 6-sulfo-sialyl-Le^x pentasaccharide clearly stood out as the preferred ligand for L-selectin, whereas the 6′-sulfo-sialyl-Le^x analogue was inactive, and the 6′,6-disulfo-sialyl-Le^x had intermediate activity (Galustian et al. 1997b). The 6-sulfo-sialyl-Le^x sequence is immunochemically detectable on high endothelial venules (Mitsuoka et al. 1997, 1998). Cell lines that lack L-selectin ligands, when transfected with α1–3 fucosyltransferase (Fuc-TVII) and *N*-acetylglucosamineb:6 sulfotransferase, have now been shown to acquire the ability to support L-selectin-mediated cell adhesion (Kimura et al. 1999; Bistrup et al. 1999; Hiraoka et al. 1999; Tangemann et al. 1999). Collectively, these results establish the 6-sulfo-sialyl-Le^x sequence as a functional ligand for L-selectin.

The second chemical synthesis approach (Auge et al. 1997) was undertaken with the knowledge that among epithelium-derived oligosaccharides that support E- and L-selectin binding, there are Le^x (and also Le^a) sequences which have a sulfate at position 3 of the terminal galactose instead of a 3-linked sialic acid (Green et al. 1992; Yuen et al. 1992). As 3′-sulfo-Le^x and 3′-sulfo-Le^a are, overall, more potent ligands for L-selectin than the sialyl analogues (Green et al. 1992, 1995; Galustian et al. 1997a, 1999), the pentasaccharides 6′,3-sulfo-Le^x, and 6,3-sulfo-Le^x, and also the monosulfated 6′-sulfo-Le^x (Fig. 3) were synthesized (Auge et al. 1997). In the form of neoglycolipids, these were compared with 3′-sulfo-Le^x pentasaccharide and also with the sulfated sialyl-Le^x analogues for their ability to support L-selectin binding (Galustian et al. 1999). As

SIALYL-LE^X SERIES

Galβ1-4GlcNAcβ1-3Galβ1-4Glc
 |2,3 |1,3
NeuAcα Fucα
Sialyl-Le^x

HSO$_3$
 |6
Galβ1-4GlcNAcβ1-3Galβ1-4Glc
 |2,3 |1,3
NeuAcα Fucα
6'-sulfo-sialyl-Le^x

       HSO$_3$
        |6
Galβ1-4GlcNAcβ1-3Galβ1-4Glc
 |2,3 |1,3
NeuAcα Fucα
6-sulfo-sialyl-Le^x

HSO$_3$ HSO$_3$
 |6 |6
Galβ1-4GlcNAcβ1-3Galβ1-4Glc
 |2,3 |1,3
NeuAcα Fucα
6',6-sulfo-sialyl-Le^x

SULFO-LE^X SERIES

Galβ1-4GlcNAcβ1-3Galβ1-4Glc
 |3 |1,3
HSO$_3$ Fucα
3'-sulfo-Le^x

HSO$_3$
 |6
Galβ1-4GlcNAcβ1-3Galβ1-4Glc
        |1,3
        Fucα
6'-sulfo-Le^x

HSO$_3$
 |6
Galβ1-4GlcNAcβ1-3Galβ1-4Glc
 |3 |1,3
HSO$_3$ Fucα
6',3'-sulfo-Le^x

       HSO$_3$
        |6
Galβ1-4GlcNAcβ1-3Galβ1-4Glc
 |3 |1,3
HSO$_3$ Fucα
6,3'-sulfo-Le^x

Fig. 3. Chemically synthesized sequences based on the Le^x pentasaccharide. The term pentasaccharide is used here to denote the fucosylated tetrasaccharide backbone that these sequences share. The sialyl-Le^x-containing sequences were synthesized as glycosylceramides (Kameyama et al. 1991; Komba et al. 1996) and the non-sialylated sulfo-Le^x series as the free oligosaccharides (Auge et al. 1997)

with the sialyl-Le^x series, the 6-sulfo form of the 3'-sulfo-Le^x pentasaccharide gave the greatest binding signal, the monosulfated, 6'-sulfo-Le^x analogue gave no detectable binding signal with L-selectin and 6',3-sulfo-Le^x was consistently less potent than the 3'-sulfo-Le^x. Thus on oligosaccharides of this series, sulfate at position 6 of the N-acetylglucosamine enhances the L-selectin binding signal elicited both by the 3'-sialyl- and the 3'-sulfo-Le^x sequences, whereas sulfate at position 6 of the terminal galactose hinders the binding. There is, however, an apparently conflicting, and as yet unexplained, finding in structural terms. This is the observation that cells transfected with a 6'-sulfotransferase (which transfers sulfate to the galactose) in addition to two other glycosyltransferases (Fuc-TVII and O-glycan core 2 enzymes) acquire the ability to support L-selectin binding (Bistrup et al. 1999; Tangemann et al. 1999); and in one cell line, the occurrence of enhanced L-selectin binding was observed to cells which were additionally transfected with the 6-sulfotransferases. This issue is considered below (see Sect. 3.6).

3.4 Clues to the Existence of Novel Biosynthetic Pathways for Selectin Ligands

Work with the impurities (Galustian et al. 1997b), in the synthetic pentaglyco-sylceramides which initially concealed the striking differences between the 6'- and the 6-sulfated forms of sialyl-Le^x, has opened up some exciting research directions. Using mass spectrometry, it was revealed that the impurities are analogues of the synthetic compounds in which the sialic acid is modified. In the 'superactive' component, the sialic acid is de-N-acetylated, whereas in the inactive component there is, in addition, a modification of the carboxyl group (Galustian et al. 1997b). Clearly, these are by-products that arose at the final stages of preparation (deprotection) of the intended compounds. Strong alkaline conditions are used to stabilize the sulfate group during the deprotection step (Galustian et al. 1997b), and under these conditions, some loss can occur of the N-acetyl group of the sialic acid. Moreover, this can proceed to the modification of the carboxyl group of the sialic acid by the reaction of the amino group at C-5 with the carboxyl group at C-2 of the six-membered ring to give an intermolecular amide bond. The stronger L-selectin binding signal with the de-N-acetylated compound was corroborated by examining the intentionally synthesized, de-N-acetyl sialyl compound (Komba et al. 1999). The striking differences in bioactivity relative to that of the N-acetylated compound raised the possibility (Galustian et al. 1997b) that such a modification of the selectin ligand may occur in vivo. The occurrence of de-N-acetyl sialic acid has been reported among sialoglycolipids (GM3 and GD3) in certain cell lines and tumor tissues (Hanai et al. 1988; Manzi et al. 1990; Sjoberg et al. 1995) and it has been proposed (Manzi et al. 1990) that a reversible de-N-acetylation and re-N-acetylation occurs in vivo. Evidence is now forthcoming that supports the concept of such a post-biosynthetic processing of N-acetyl neuraminic acid on selectin ligands. A monoclonal antibody (Mitsuoka et al. 1999) produced by immunizing mice with the unfractionated 6-sulfo-sialyl-Le^x pentaglycosyl-ceramide (Komba et al. 1996; Galustian et al. 1997b) has proven to be a powerful tool in this respect. This antibody happens to be directed to the minor impurity 6-sulfo-sialyl-Le^x with the modified sialic acid carboxyl group (Mitsuoka et al. 1999). Immunocytochemical findings with this antibody have shown that this form of the 6-sulfo-sialyl-Le^x exists on leukocytes (information is awaited regarding its occurrence on endothelium). Moreover, enzymatic studies using sonicates of leukocytes as the enzyme source, and the chemically synthesized de-N-acetyl form of 6-sulfo-Le^x pentasaccharide as the substrate, together with immunochemical studies with the above mentioned antibody (Mitsuoka et al. 1999), have provided evidence for the presence of an enzyme which converts the substrate to the analogue with the modified carboxyl group. This has been tentatively designated 'cyclic' sialic acid (Mitsuoka et al. 1999). It will be interesting to investigate if, as proposed (Mitsuoka et al. 1999), sialyl-Le^x analogues containing this inactive 'cyclic' sialic acid form act as a dormant pool of selectin ligands which, upon appropriate stimulation, are hydrolyzed by cellular enzymes and become active in selectin-dependent cell adhesion.

3.5 The Second Class of Sulfated L-Selectin Ligand

Much of the foregoing discussion has been concerned with the specificity of L-selectin binding to several immobilized acidic oligosaccharide analogues based on the Le^x pentasaccharide sequence. This is a class of oligosaccharide ligand to which L-selectin binds in a calcium-dependent manner. A second class of ligand, of which sulfatide (3′-sulfated galactosyl-cermide) is the prototype, encompasses various sulfated mono- and disaccharides in which the position of the sulfate on galactose is not critical, and the binding is partly or wholly calcium-independent (Suzuki et al. 1993; Green et al. 1995). The second class of L-selectin ligand includes ganglio series sequences with a terminal galactose that is variously sulfated 3-O, 4-O, 6-O, 3,4-O-di- or 3,6-O-di-sulfated (Feizi 1993 and references therein, and Bertozzi et al. 1995). There is increasing evidence that L-selectin binds also to glycosaminoglycans. When presented as neoglycolipids, glycosaminoglycan disaccharides of keratan sulfate, heparin and chondroitin sulfate types are bound (Green et al. 1995), leading to the conclusion that in contrast to the tri- and longer oligosaccharides, clustered short oligosaccharides with 6-O sulfation of N-acetylgalactosamine, N-acetylglucosamine or glucosamine, 4-O sulfation of N-acetylgalactosamine, 2-O-sulfation of uronic acid, N-sulfation of glucosamine and, to a lesser extent the non-sulfated uronic acid-containing disaccharides, can all support L- selectin binding. Endothelial cells contain heparan sulfate glycosaminoglycans with various sulfation patterns, also including non-sulfated glucosamine, to which immobilized L-selectin binds with high avidity (Norgard-Sumnicht et al. 1995). All this suggests that endothelial glycosaminoglycans may be among natural ligands for this selectin. Such interactions of L-selectin have been proposed (Green et al. 1995) to provide a link between the selectin-mediated and the integrin-mediated adhesion systems in leukocyte extravasation cascades, for glycosaminoglycans serve as reservoirs for inflammatory chemokines; these are short-range stimulators of lymphocyte migration which trigger integrin activation (Tanaka et al. 1993).

3.6 Possible Co-Operativity Between the Long and Short Ligands for L-Selectin

There is much to be learnt about the interplay and possible co-operativity of these two classes of L-selectin ligand in the natural setting. CHO cells transfected to express the recently cloned galactoseb: 6′-sulfotransferase concomitantly with CD34, fucosyltransferase VII and O-glycan core 2 enzyme were shown to be able to support L-selectin binding, as were the cells transfected with an *N*-acetylglucosamine-6-sulfotransferase (Bistrup et al. 1999). When co-transfected, the two sulfotransferases were found to synergize and elicit enhanced L-selectin binding (Bistrup et al. 1999). We have suggested (Feizi and Galustian 1999) that short saccharides 6′-sulfated at galactose may be among products of the galactoseb:6′-sulfotransferase. The sequences of the 6′-sulfated oligosaccharides in the transfected cell line have not yet been characterized. Definition of these will be crucial to understanding the biochemical basis of the synergistic property of the products of this enzyme.

4 Oligosaccharide Ligands for P-Selectin

4.1 P-Selectin Interactions with Defined Saccharide Sequences

In the initial studies of carbohydrate sequences recognized by P-selectin, the involvement of the Le^x and Le^a series was readily shown. Monoclonal antibodies designated anti-CD 15 (anti-Le^x) were found to be inhibitory in assay systems designed to measure the adhesion of natural P-selectin (on platelet membranes or incorporated into phospholipid vesicles) or recombinant P-selectin (expressed on COS-1 cells) to blood neutrophils, blood monocytes or to cultured myeloid (HL60) or monocytoid (U937) cells (Larsen et al. 1990). In the same study, the binding of neutrophils to activated platelets and the binding of HL60 to COS-1 cells transfected to express P-selectin were inhibited by a preparation of a Le^a pentasaccharide (structure 2, Fig. 2). Three other reports (Corral et al. 1990; Moore et al. 1991; Polley et al. 1991) described diminished adhesion to activated platelets when neutrophils were treated with sialidases; and a chemically synthesized methylglycoside of the 3′-sialyl-Le^x pentasaccharide was found to be approximately 30-fold more active than a non-sialylated Le^x analogue as an inhibitor of neutrophil binding to activated platelets (Polley et al. 1991). P-selectin at the surface of activated human platelets was shown to bind 3′-sialyl-Le^a pentasaccharide sequence rather more strongly than the 3′-sialyl-Le^x analogue, but did not bind their asialo-analogues (Handa et al. 1991). Binding of the recombinant soluble form of the human P-selectin-IgG chimera to 3′-sialyl Le^a trisaccharide linked to protein was also clearly documented (Nelson et al. 1993). As with L-selectin, a recombinant soluble P-selectin-IgG chimera was shown to bind to sulfatide (Aruffo et al. 1991) and to the HNK-1 glycolipid (Needham et al. 1993); moreover, sulfatide was found to inhibit the binding of the soluble P-selectin to U937 monocytic cells (Aruffo et al. 1991). The binding of P-selectin (purified from human platelets) to the myeloid cell line HL60 (Skinner et al. 1991) as well as the binding of activated platelets to 3′-sialyl-Le^a and -Le^x glycolipids (Handa et al. 1991) could be inhibited with various sulfated polysaccharides. Thus, considerable similarities were found between P- and L-selectins in their ability to bind to these structurally characterized compounds.

4.2 PSGL-1, the Counter-Receptor Displaying Two Classes of Ligand for P-Selectin

On the leukocyte surface, the counter-receptors for P-selectin are protease susceptible. Consequently, leukocyte glycoproteins have been a major focus of interest as the target molecules. Although multiple glycoproteins carry the sialyl-Le^x sequence on myeloid cells, attempts to identify the counter-receptors have so far revealed a single glycoprotein, PSGL-1, the assembly of which has been the subject of extensive investigations. Highlighted here are some of the interesting developments which clearly illustrate that the mode of presentation of the ligands to this carbohydrate-binding protein is crucial for recognition.

Human PSGL-1 is a 402-residue disulfide-linked homodimeric (type 1) membrane glycoprotein with an extracellular domain that contains multiple O-linked glycans and three potential sites for addition of N-glycans (Sako et al. 1993; McEver 1994). On human myeloid cells, this glycoprotein has features of a physiological counter-receptor for P-selectin, since a specific monoclonal antibody, PL1, directed to the NH$_2$-terminus of PSGL-1 is able to block rolling of neutrophils on P-selectin both in vitro (Moore et al. 1995) and in vivo (Norman et al. 1995). Furthermore, PSGL-1 has been shown to be localized by immunoelectron microscopy to the tips of neutrophil microvilli (Moore et al. 1995), and by rotary shadowing and electron microscopy, it can be visualized as an extended molecule of approx. 50 nm (Li et al. 1996a). Taken together, these findings indicate that PSGL-1 is optimally orientated at the leukocyte cell surface to mediate effective interactions with P-selectin expressed on endothelial cells.

With monoclonal antibodies that recognize protein determinants of PSGL-1, it has been shown (Moore et al. 1995; Vachino et al. 1995) that the protein is expressed on all blood leukocytes but not necessarily in a form that binds P-selectin. For example, T lymphocytes were shown to express high levels of PSGL-1 constitutively, but only following activation were they able to bind to P-selectin in a PSGL-1-dependent manner (Vachino et al. 1995). Where examined, activities of enzymes involved in biosynthesis of sialyl-Le^x have been demonstrated as essential in cells that produce a functional PSGL-1. Lymphocyte activation was accompanied by increased activity of β1–6 GlcNAc transferase (core 2 enzyme) and α1–3 fucosyltransferase, but no apparent increase in the levels of the PSGL-1 protein (Vachino et al. 1995). Using stably-transfected CHO cells which do not express α1–3 fucosyltransferase, and lack core 2 containing O-glycans, it was shown that the generation of high-avidity forms of PSGL-1 depended on the co-expression of the core 2 enzyme and an α1–3 fucosyltransferase (Li et al. 1996b). Also, human cancer cells that do not express PSGL-1, but do express the appropriate glycosyltransferases for synthesis of sialyl-Le^x, bind poorly to P-selectin (Handa et al. 1995). By transfecting these cells with PSGL-1 cDNA, it was possible to induce high levels of binding to P-selectin. However, in a construct where the O-glycosylation sites were provided wholly by another glycoprotein, CD43, the binding of P-selectin was not lost (Pouyani and Seed. 1995). This indicates that a unique protein moiety on PSGL-1 is not critical for P-selectin recognition.

At the NH_2-terminal region of the mature PSGL-1 protein there are three clustered tyrosine residues which are potential sites of sulfation. Three groups have demonstrated independently that sulfation of at least one of those tyrosines is an additional requirement for high avidity binding by P-selectin (Sako et al. 1993; Pouyani and Seed 1995; Wilkins et al. 1995). This contrasts with observations on the L-selectin counter-receptor GlyCAM-1 (Rosen et al. 1996), where sulfation is predominantly on oligosaccharide chains. By generating deletion mutants it was shown that constructs expressing only the first 19 or 20 amino acids of mature PSGL-1 that contain the three tyrosines and two potential O-glycosylation sites, when immobilized, were sufficient for P-selectin binding to P- but not E-selectin, although the binding was considerably lower than to the full-length construct (Sako et al. 1993; Pouyani and Seed 1995). When the three tyrosine residues were substituted with tryptophan, there was a marked reduction in avidity of P-selectin binding such that binding could only be detected in a multivalent cell binding assay (Sako et al. 1993). Treatment with sialidase, or excluding the presence of a fucosyltransferase resulted in loss of P-selectin binding. Moreover, the binding to the sulfated tyrosine-containing 19 amino acid fragment was abolished when Thr-16, a potential O-glycosylation site was replaced with alanine. Thus, O-glycosylation of the polypeptide with sialyl-Le^x (or related sequence) is essential to generate sufficient avidity for detecting binding. Collectively, these data suggest that for P-selectin, in contrast to E-selectin, there is co-operativity between oligosaccharides and sulfotyrosines. For a description of the sequences of the 3'-sialyl-Le^x type O-glycans on PSGL-1, and the advances being made in knowledge on the interplay of carbohydrate and sulfotyrosine on PSGL-1 in P-selectin recognition, the reader is referred to excellent reviews by McEver and Cummings 1997 and Cummings. 1999).

5 Perspectives

There is still much to be learnt about details of the oligosaccharide ligands for the selectins, and also about the degree of involvement of glycolipids as functional counter-receptors. Lipid-linked oligosaccharides immobilized in vitro, can certainly support rolling and tethering mediated by E- and L-selectins (Alon et al. 1995), but do they do so on the cell membranes of leukocytes and of endothelial cells?

It is clear that the blood group-related Le^x system is an integral part of physiologically relevant ligands for the human and murine selectins. Analogues based on the isomeric, Le^a system are overall more potent ligands when examined in vitro. Their involvement and relative importance in vivo in the pathobiology of human epithelial diseases warrants detailed study. Immunochemical evidence for the occurrence on murine leukocytes of the Le^a series sequences bound by E-selectin needs to be reconciled with evidence for the major contribution made by products of Fuc-TVII to the biosynthesis of functional ligands for the selectins. The lack of immuno-detection of the Le^x sequence is most likely a reflection of the limitation of the reagent used.

It is interesting to recall that the several 'exotic' capping sequences as ligands for the selectins, such as the 3'-sulfated analogues of 3'-sialyl-Le^x and -Le^a on epithelia, and the 6-sulfo and 6'-sulfo analogues of 3'-sialyl-Le^x on endothelia, and also the diverse backbone and the core sequences on glycolipids and glycoproteins of hematogenous cells, were discovered as a result of a relatively small number of biochemical studies of glycoconjugates from target tissues. With improved micro-techniques for carbohydrate ligand discovery (Stoll et al. 2000) the opportunities are enhanced for widening our knowledge of the repertoire of natural oligosaccharide sequences that can serve as ligands for these biologically important molecules.

The molecular basis of the mutual dependence of the two types of ligand, carbohydrate and sulfated amino acid protein, as presented on the counter-receptor for P-selectin, will be important to elucidate. This knowledge may have an important bearing on the recognition elements for other lectin-type proteins of the innate immune system, which recognize proteins, but also can bind to sulfated polysaccharides.

Acknowledgement. The author is supported by the Medical Research Council.

References

Alon R, Feizi T, Yuen C-T, Fuhlbrigge RC, Springer TA (1995) Glycolipid ligands for selectins support leukocyte tethering and rolling under physiologic flow conditions. J Immunol 154: 5356–5366

Aruffo A, Kolanus W, Walz G, Fredman F, Seed B (1991) CD62/P-Selectin recognition of myeloid and tumor cell sulfatides. Cell 67:35–44

Auge C, Dagron F, Lemoine R, Le Narvor C, Lubineau A (1997) Syntheses of sulfated derivatives as sialyl Lewis^a and sialyl Lewis^x analogues. In: Chapleur Y(ed) Carbohydrate mimics: concepts and methods. Verlag Chemie, Weinheim pp 365–383

Berg EL, Robinson MK, Mansson O, Butcher EC, Magnani JL (1991) A carbohydrate domain common to both sialyl Le^a and sialyl Le^x is recognized by the endothelial cell leukocyte adhesion molecule ELAM-1. J Biol Chem 266:14869–14872

Berg EL, Magnani J, Warnock RA, Robinson MK, Butcher EC (1992) Comparison of L-selectin and E-selectin ligand specificities:the L-selectin can bind the E-selectin ligands sialyl-Le^x and sialyl Le^a. Biochem Biophys Res Commun 184:1048–1055

Bertozzi CR, Fukuda S, Rosen SD (1995) Sulfated disaccharide inhibitors of L-selectin: deriving structural leads from a physiological selectin ligand. Biochemistry 34:14271–14278

Bevilacqua MP, Nelson RM (1993) Selectins. J Clin Invest 91:379–387

Bistrup A, Bhakta S, Lee JK, Belov YY, Gunn MD, Feng-Rong Z, Huang CC, Kannagi R, Rosen SD, Hemmerich S (1999) Sulfotransferases of two specificities function in the reconstitution of high endothelial cell ligands for L-selectin. J Cell Biol 145:899–910

Brandley BK, Swiedler SJ, Robbins PW (1990) Carbohydrate ligands of the LEC cell adhesion molecules. Cell 63:861–863

Corral L, Singer MS, Macher BA, Rosen SD (1990) Requirement for sialic acid on neutrophils in a GMP-140 (PADGEM) mediated adhesive interaction with activated platelets. Biochem Biophys Res Commun 172:1349–1356

Crocker PR, Feizi T (1996) Carbohydrate recognition systems: functional triads in cell-cell interactions. Curr Opin Struct Biol 6:679–691

Cummings RD (1999) Structure and function of the selectin ligand PSGL-1. Braz J Med Biol Res 32:519–528

Feizi T (1985) Demonstration by monoclonal antibodies that carbohydrate structures of glycoproteins and glycolipids are onco-developmental antigens. Nature 314:53–57

Feizi T (1992) Blood group-related oligosaccharides are ligands in cell-adhesion events. Biochem Soc Trans 20:274–278

Feizi T (1993) Oligosaccharides that mediate mammalian cell-cell adhesion. Curr Opin Struct Biol 3:701–710

Feizi T, Galustian C (1999) Novel oligosaccharide ligands and ligand-processing pathways for the selectins. Trends Biochem Sci 24:369–372

Feizi T, Stoll MS, Chai W, Lawson AM (1994) Neoglycolipids: probes of oligosaccharide structure, antigenicity and function. Methods Enzymol 230:484–519

Foxall C, Watson SR, Dowbenko D, Fennie C, Lasky LA, Kiso M, Hasegawa A, Asa D, Brandley BK (1992) The three members of the selectin receptor family recognize a common carbohydrate epitope, the sialyl Lewis[x] oligosaccharide. J Cell Biol 117:895–902

Fukuda M, Hiraoka N, Yeh JC (1999) C-type lectins and sialyl Lewis X oligosaccharides. Versatile roles in cell-cell interaction. J Cell Biol 147:467–470

Fukushima K, Hirota M, Terasaki PI, Wakisaka A, Togashi H, Chia D, Suyama N, Fukushi Y, Nudelman E, Hakomori S (1984) Characterisation of sialosylated Lewis[x] as a new tumor-associated antigen. Cancer Res 44:5279–5285

Galustian C, Childs RA, Yuen C-T, Hasegawa A, Kiso M, Lubineau A, Shaw G, Feizi T (1997a) Valency dependent patterns of reactivity of human L-selectin towards sialyl and sulfated oligosaccharides of Le[a] and Le[x] types: relevance to anti-adhesion therapeutics. Biochemistry 36:5260–5266

Galustian C, Lawson AM, Komba S, Ishida H, Kiso M, Feizi T (1997b) Sialyl-Lewis[x] sequence 6-O-sulfated at N-acetylglucosamine rather than at galactose is the preferred ligand for L-selectin and de-N-acetylation of the sialic acid enhances the binding strength. Biochem Biophys Res Comm 240:748–751

Galustian C, Lubineau A, le Narvor C, Kiso M, Brown G, Feizi T (1999) L-selectin interactions with novel mono- and multisulfated Lewis[x] sequences in comparison with the potent ligand 3'-sulfated Lewis[a]. J Biol Chem 274:18213–18217

Gooi HC, Feizi T, Kapadia A, Knowles BB, Solter D, Evans MJ (1981) Stage specific embryonic antigen SSEA-1 involves a1–3 fucosylated type 2 blood group chain. Nature 292:156–158

Gooi HC, Thorpe SJ, Hounsell EF, Rumpold H, Kraft D, Forster O, Feizi T (1983) Marker of peripheral blood granulocytes and monocytes of man recognized by two monoclonal antibodies VEP8 and VEP9 involves the trisaccharide 3-fucosyl-N-acetyllactosamine. Eur J Immunol 13:306–312

Green PJ, Tamatani T, Watanabe T, Miyasaka M, Hasegawa A, Kiso M, Stoll MS, Feizi T (1992) High affinity binding of the leucocyte adhesion molecule L-selectin to 3'-sulphated-Le[a] and -Le[x] oligosaccharides and the predominance of sulphate in this interaction demonstrated by binding studies with a series of lipid-linked oligosaccharides. Biochem Biophys Res Commun 188:244–251

Green PJ, Yuen C-T, Childs RA, Chai W, Miyasaka M, Lemoine R, Lubineau A, Smith B, Ueno H, Nicolaou KC, Feizi T (1995) Further studies of the binding specificity of the leukocyte adhesion molecule, L-selectin, towards sulphated oligosaccharides – suggestion of a link between the selectin- and the integrin-mediated lymphocyte adhesion systems. Glycobiology 5:29–38

Hanai N, Dohi T, Nores GA, Hakomori S (1988) A novel ganglioside, de-*N*-acetyl-GM3 (II3NeuNH2LacCer), acting as a strong promoter for epidermal growth factor receptor kinase and as a stimulator for cell growth. J Biol Chem 263:6296–6301

Handa K, Nudelman ED, Stroud MR, Shiozawa T, Hakomori S-I (1991) Selectin GMP-140 (CD62; PADGEM) binds to sialosyl-Le[a] and sialosyl-Le[x] and sulfated glycans modulate this binding. BBRC 181:1223–1230

Handa K, White T, Ito K, Fang H, Wang S-S, Hakomori S-I (1995) P-selectin-dependent adhesion of human cancer cells: requirement for co-expression of a 'PSGL-1-like' core protein and the glycosylation process for sialosyl-Le[x] or sialosyl-Le[a]. Int J Oncol 6:773–781

Harlan JM, Liu DY (1992) Adhesion: its role in inflammatory disease. W.H. Freeman & Co, New York

Hemmerich S, Leffler H, Rosen SD (1995) Structure of the *O*-glycans in GlyCAM-1, an endothelial-derived ligand for L-selectin. J Biol Chem 270:12035–12047

Hiraoka N, Petryniak B, Nakayama J, Tsuboi S, Suzuki M, Yeh JC, Izawa D, Tanaka T, Miyasaka M, Lowe JB, Fukuda M (1999) A novel, high endothelial venule-specific sulfotransferase expresses 6- sulfo sialyl Lewis(x), an L-selectin ligand displayed by CD34. Immunity 11:79–89

Kameyama A, Ishida H, Kiso M, Hasegawa A (1991) Synthetic studies on sialoglycoconjugates 22: total synthesis of tumor-associated ganglioside, sialyl Lewis X. J Carbohydr Chem 10: 549–560

Kimura N, Mitsuoka C, Kanamori A, Hiraiwa N, Uchimura K, Muramatsu T, Tamatani T, Kansas GS, Kannagi R (1999) Reconstruction of functional L-selectin ligands on a cultured human endothelial cell line by cotransfection of $\alpha 1 \times 3$ fucosyltransferase VII and newly cloned GlcNAcβ–6-sulfotransferase cDNA. Proc Natl Acad Sci USA 96:4530–4535

Kogelberg H, Rutherford T (1994) Studies on the three-dimensional behaviour of the selectin ligands Lewis[a] using NMR spectroscopy and molecular dynamics simulations. Glycobiology 4:49–57

Kogelberg H, Frenkiel TA, Homans SW, Lubineau A, Feizi T (1996) Conformational studies on the selectin and natural killer cell receptor ligands sulfo-and sialyl-lacto-*N*-fucopentaoses (SuL-NFPII and SLNFPII) using NMR spectroscopy and molecular dynamics simulations. Comparisons with the nonacidic parent molecule LNFPII. Biochemistry 35: 1954–1964

Kojima N, Handa K, Newman W, Hakomori S-I (1992) Multi-recognition capability of E-selectin in a dynamic flow system, as evidenced by differential effects of sialidases and anti-carbohydrate antibodies on selectin-mediated cell adhesion at low vs high wall shear stress: a preliminary note. Biochem Biophys Res Commun 189:1686–1694

Komba S, Ishida H, Kiso M, Hasegawa A (1996) Synthesis and biological activities of three sulfated sialyl Le[x] ganglioside analogues for clarifying the real carbohydrate ligand structure of L-selectin. Bioorg Med Chem 4:1833–1847

Komba S, Galustian C, Ishida H, Feizi T, Kannagi R, Kiso M (1999) The first total synthesis of 6-sulfo-de-*N*-acetylsialyl Lewis[x] ganglioside: a superior ligand for human L-selectin. Angew Chem Int Ed 38:1131–1133

Larkin M, Ahern TJ, Stoll MS, Shaffer M, Sako D, O'Brien J, Lawson AM, Childs RA, Barone KM, Langer-Safer PR, Hasegawa A, Kiso M, Larsen GR, Feizi T (1992) Spectrum of sialylated and non-sialylated fuco-oligosaccharides bound by the endothelial-leukocyte adhesion molecule E-selectin. Dependence of the carbohydrate binding activity on E-selectin density. J Biol Chem 267:13661–13668

Larsen E, Palabrica T, Sajer S, Gilbert GE, Wagner DD, Furie BC, Furie B (1990) PADGEM-dependent adhesion of platelets to monocytes and nuetrophils is mediated by a lineage-specific carbohydrate LNF III (CD15). Cell 63:467–474

Li F, Erickson HP, James JA, Moore KL, Cummings RD, McEver RP (1996a) Visualization of P-selectin glycoprotein ligand-1 as a highly extended molecule and mapping of protein epitopes for monoclonal antibodies. J Biol Chem 271:6342–6348

Li F, Wilkins PP, Crawley S, Weinstein J, Cummings RD, McEver RP (1996b) Post-translational modifications of recombinant P-selectin glycoprotein ligand-1 required for binding to P- and E-selectin. J Biol Chem 271:3255–3264

Liebecq J (1992) International Union of Biochemistry and Molecular Biology. Nomenclature and related documents. Portland Press, London

Lowe JB, Stoolman LM, Nair RP, Larsen RD, Berhend TL, Marks RM (1990) ELAM-1-dependent cell adhesion to vascular endothelium determined by a transfected human fucosyltransferase cDNA. Cell 63:475–484

Macher BA, Buehler J, Scudder P, Knapp W, Feizi T (1988) A novel carbohydrate differentiation antigen on fucogangliosides of human myeloid cells recognized by monoclonal antibody VIM-2. J Biol Chem 263:10186–10191

Maly P, Thall AD, Petryniak B, Rogers CE, Smith PL, Marks RM, Kelly RJ, Gersten KM, Cheng G, Saunders TL, Camper SA, Camphausen RT, Sullivan FX, Isogai Y, Hindsgaul O, von Andrian UH, Lowe JB (1996) The alpha(1,3)fucosyltransferase Fuc-TVII controls leukocyte trafficking through an essential role in L-, E-, and P-selectin ligand biosynthesis. Cell 86: 643–653

Manzi AE, Sjoberg ER, Diaz S, Varki A (1990) Biosynthesis and turnover of O-acetyl and N-acetyl groups in the gangliosides of human melanoma cells. J Biol Chem 265:13091–13103

McEver RP (1994) Selectins. Curr Opin Immunol 6:75–84

McEver RP, Cummings RD (1997) Role of PSGL-1 binding to selectins in leukocyte recruitment. J Clin Invest 100:S97–103

McEver RP, Moore KL, Cummings RD (1995) Leukocyte trafficking mediated by selectin-carbohydrate interactions. J Biol Chem 270:11025–11028

Mitsuoka C, Kawakami KN, Kasugai SM, Hiraiwa N, Toda K, Ishida H, Kiso M, Hasegawa A, Kannagi R (1997) Sulfated sialyl Lewis X, the putative L-selectin ligand, detected on endothelial cells of high endothelial venules by a distinct set of anti-sialyl Lewis X antibodies [published erratum appears in Biochem Biophys Res Commun 1997 Apr 17;233(2):576]. Biochem Biophys Res Commun 230:546–551

Mitsuoka C, Sawada KM, Ando FK, Izawa M, Nakanishi H, Nakamura S, Ishida H, Kiso M, Kannagi R (1998) Identification of a major carbohydrate capping group of the L-selectin ligand on high endothelial venules in human lymph nodes as 6-sulfo sialyl Lewis X. J Biol Chem 273: 11225–11233

Mitsuoka C, Ohmori K, Kimura N, Kanamori A, Komba S, Ishida H, Kiso M, Kannagi R (1999) Regulation of selectin binding activity by cyclization of sialic acid moiety of carbohydrate ligands on human leukocytes. Proc Natl Acad Sci USA 96:1597–1602

Moore KL, Varki A, McEver RP (1991) GMP-140 binds to a glycoprotein receptor in human nuetrophils: evidence for lectin-like interaction. J Cell Biol 112:491–499

Moore KL, Patel KD, Bruehl RE, Li F, Johnson DA, Lichenstein HS, Cummings RD, Bainton DF, McEver RP (1995) P-selectin glycoprotein ligand-1 mediates rolling of human neutrophils on P-selectin. J Cell Biol 128:661–671

Müthing J, Spanbroek R, Peter-Katalinic J, Hanisch FG, Hanski C, Hasegawa A, Unland F, Lehmann J, Tschesche H, Egge H (1996) Isolation and structural characterization of fucosylated gangliosides with linear poly-N-acetyllactosaminyl chains from human granulocytes. Glycobiology 6:147–156

Needham LK, Schnaar RL (1993) The HNK-1 reactivity sulfoglucuronyl glycolipids are ligands for L-selectin and P-selectin but not E-selectin. Proc Natl Acad Sci USA 90:1359–1363

Nelson RM, Dolich S, Aruffo A, Cecconi O, Bevilacqua MP (1993) Higher-affinity oligosaccharide ligands for E-selectin. J Clin Invest 91:1157–1166

Norgard KE, Han H, Powell L, Kriegler M, Varki A, Varki NM (1993) Enhanced interaction of L-selectin with the high endothelium venule ligand via selectively oxidized sialic acids. Proc Natl Acad Sci USA 90:1068–1072

Norgard-Sumnicht K, Varki A (1995) Endothelial heparan sulfate proteoglycans that bind to L-selectin have glucosamine residues with unsubstituted amino groups. J Biol Chem 270: 12012–12024

Norman KE, Moore KL, McEver RP, Ley K (1995) Leukocyte rolling in vivo is mediated by P-selectin glycoprotein ligand-1. Blood 86:4417–4421

Osanai T, Feizi T, Chai W, Lawson AM, Gustavsson ML, Sudo K, Araki M, Araki K, Yuen C-T (1996) Two families of murine carbohydrate ligands for E-selectin. Biochem Biophys Res Commun 218:610–615

Phillips ML, Nudelman E, Gaeta FCA, Perez M, Singhal AK, Hakomori S-I., Paulson JC (1990) ELAM-1 mediates cell adhesion by recognition of carbohydrate ligand, sialyl-Le^x. Science 250:1130–1132

Polley MJ, Phillips ML, Wayner E, Nudelman E, Singhal AK, Hakomori S-I., Paulson JC (1991) CD-62 and ELAM-1 recognize the same carbohydrate ligand, sialyl-Le x. Proc Natl Acad Sci USA 88:6224–6228

Pouyani T, Seed B (1995) PSGL-1 recognition of P-selectin is controlled by a tyrosine sulfation consensus at the PSGL-1 amino terminus. Cell 83:333–343

Rosen SD (1993) L-Selectin and its biological ligands. Histochemistry 100:185–191

Rosen SD, Bertozzi CR (1996) Leukocyte adhesion: two selectins converge on sulphate. Curr Biol 6:261–264

Sako D, Chang XJ, Barone KM, Vachino G, White HM, Shaw G, Veldman GM, Bean KM, Ahern TJ, Furie B (1993) Expression cloning of a functional glycoprotein ligand for P-selectin. Cell 75:1179–1186

Sekine M, Nakamura K, Suzuki M, Inagaki F, Yamakawa T, Suzuki A (1988) A single autosomal gene controlling the expression of the extended globoglycolipid carrying SSEA-1 determinant is responsible for the expression of two extended globogangliosides. J Biochem Tokyo 103:722–729

Simanek EE, Mc Garvey GJ, Jablonowski JA, Wong CH (1998) Selectin-carbohydrate interactions: from natural ligands to designed mimics. Chem Rev 98:833–862

Sjoberg ER, Chammas R, Ozawa H, Kawashima I, Khoo KH, Morris HR, Dell A, Tai T, Varki A (1995) Expression of de-N-acetyl-gangliosides in human melanoma cells is induced by genistein or nocodazole. J Biol Chem 270:2921–2930

Skinner MP, Lucas CM, Burns GF, Chesterman CN, Berndt MC (1991) GMP-140 binding to neutrophils is inhibited by sulfated glycans. J Biol Chem 266:5371–5374

Smith PL, Gersten KM, Petryniak B, Kelly RJ, Rogers C, Natsuka Y, Alford III JA, Scheidegger EP, Natsuka S, Lowe JB (1996) Expression of the a(1–3)fucosyltransferase Fuc-TVII in lymphoid aggregate high endothelial venules correlates with expression of L-selectin ligands. J Biol Chem 271:8250–8259

Stoll MS, Feizi T, Loveless RW, Chai W, Lawson AM, Yuen C-T (2000) Fluorescent neoglycolipids: improved probes for oligosaccharide ligand discovery. Eur J Biochem 267:1795–1804

Stroud MR, Handa K, Salyan MEK, Ito K, Levery SB, Hakomori S-I, Reinhold BB, Reinhold VN (1996a) Monosialogangliosides of human myelogenous leukemia HL60 cells and normal human leukocytes. 1. Separation of E-selectin binding from nonbinding gangliosides, and absence of sialosyl-Le^x having tetraosyl to octaosyl core. Biochemistry 35:758–769

Stroud MR, Handa K, Salyan MEK, Ito K, Levery SB, Hakomori S, Reinhold BB, Reinhold VN (1996b) Monosialogangliosides of human myelogenous leukemia HL60 cells and normal human leukocytes. 2. Characterization of E-selectin binding fractions, and structural requirements for physiological binding to E-selectin. Biochemistry 35:770–778

Suzuki Y, Toda Y, Tamatani T, Watanabe T, Suzuki T, Nakao T, Murase K, Kiso M, Hasegawa A, Tanado-Aritomi K, Ishizuka I, Miyasaka M (1993) Glycolipids are ligands for a lymphocyte homing receptor, L-selectin (LECAM-1), binding epitope in sulfated sugar chain. Biochem Biophys Res Commun 190:426–434

Takada A, Ohmori K, Takahashi N, Tsuyoka K, Yago A, Zenita K, Hasegawa A, Kannagi R (1991) Adhesion of human cancer cells to vascular endothelium mediated by a carbohydrate antigen, sialyl Lewis A. Biochem Biophys Res Commun 179:713–719

Tanaka Y, Adams DH, Shaw S (1993) Proteoglycans on endothelial cells present adhesion-inducing cytokines to leukocytes. Immunol Today 14:111–115

Tangemann K, Bistrup A, Hemmerich S, Rosen SD (1999) Sulfation of a high endothelial venule-expressed ligand for L-selectin. Effects on tethering and rolling of lymphocytes. J Exp Med 190:935–942

Thorpe SJ, Feizi T (1984) Species differences in the expression of carbohydrate differentiation antigens on mammalian blood cells revealed by immunofluorescence with monoclonal antibodies. Biosci Rep 4:673–685

Tiemeyer M, Swiedler SJ, Ishihara M, Moreland M, Schweingruber H, Hirtzer P, Brandley BK (1991) Carbohydrate ligands for endothelium leukocyte adhesion molecule-1. Proc Natl Acad Sci USA 88:1138–1142

Tyrrell D, James P, Rao N, Foxall C, Abbas S, Dasgupta F, Nashed M, Hasegawa A, Kiso M, Asa D, Brandley BK (1991) Structural requirements for the carbohydrate ligand of E-selectin. Proc Natl Acad Sci USA 88:10372–10376

Vachino G, Chang XJ, Veldman GM, Kumar R, Sako D, Fouser LA, Berndt MC, Cumming DA (1995) P-selectin glycoprotein ligand-1 is the major counter-receptor for P-selectin on stimulated T cells and is widely distributed in non-functional form on many lymphocytic cells. J Biol Chem 270:21966–21974

Wilkins PP, Moore KL, McEver RP, Cummings RD (1995) Tyrosine sulfation of P-selectin glycoprotein ligand-1 is required for high affinity binding to P-selectin. J Biol Chem 270:22677–22680

Yoshino Y, Ohmoto H, Kondo N, Tsujishita H, Hiramatsu Y, Inoue Y, Kondo H, Ishida H, Kiso M, Hasegawa A (1997) Studies on selectin blockers. 4. Structure-function relationships of sulfated sialyl Lewis X hexasaccharide ceramides toward E-P, P-, and L-selectin binding. J Med Chem 40:455–462

Yuen C-T, Lawson AM, Chai W, Larkin M, Stoll MS, Stuart AC, Sullivan FX, Ahern TJ, Feizi T (1992) Novel sulfated ligands for the cell adhesion molecule E-selectin revealed by the neoglycolipid technology among O-linked oligosaccharides on an ovarian cystadenoma glycoprotein. Biochemistry 31:9126–9131

Yuen C-T, Bezouska K, O'Brien J, Stoll MS, Lemoine R, Lubineau A, Kiso M, Hasegawa A, Bockovich NJ, Nicolaou KC, Feizi T (1994) Sulfated blood group Lewis[a]. A superior oligosaccharide ligand for human E-selectin. J Biol Chem 269:1595–1598

Structures and Functions of Mammalian Collectins

Uday Kishore[1] and Kenneth B.M. Reid[2]

1 Introduction

Protein–carbohydrate interactions serve multiple functions in the immune system. Many animal lectins (sugar binding proteins) mediate both pathogen recognition and cell–pathogen interactions using structurally related calcium-dependent carbohydrate recognition domains (C-type CRDs). The collectins are a group of mammalian lectins containing collagen regions. They include mannose-binding lectin (MBL), lung surfactant protein A (SP-A), lung surfactant protein D (SP-D), bovine conglutinin (BC), and collectin-43 (CL-43). Pathogen recognition by these collectins is mediated by binding of terminal monosaccharide residues characteristic of bacterial and fungal cell surfaces. The broad selectivity of the monosaccharide binding site and the geometrical arrangement of the multiple CRDs in the intact collectins explain the ability of these proteins to bind tightly to arrays of carbohydrate structures normally found on the surfaces of the micro-organisms and thus mediate discrimination between self and non-self.

The primary structure of each of the collectins is organised into four regions: an N-terminal region involved in the formation of inter-chain disulphide bonds, a collagenous region composed of Gly-Xaa-Yaa repeats, an α-helical neck peptide, and a C-terminal C-type CRD (Hoppe and Reid 1994). The collectins are large oligomeric structures, each assembled from multiple copies of a single polypeptide chain (with the exception of human SP-A, which has two closely related types of chains; Fig. 1). The C-type CRDs are spaced, in a trimeric orientation, at the end of triple-helical collagenous stalks. It is becoming increasingly clear that the CRD domains, by binding to carbohydrate ligands on the cell surface of the pathogens, fulfill a recognition function that can bring about effector functions, such as complement activation (by MBL) or phagocytosis (by SP-A and SP-D; Hoppe and Reid 1994; Kishore et al. 1997; Crouch 1998).

[1] Institute of Molecular Medicine, John Radcliffe Hospital, University of Oxford, Headington, Oxford OX3 9DS, UK
[2] Medical Research Council Immunochemistry Unit, Department of Biochemistry, University of Oxford, South Parks Road, Oxford OX1 3QU, UK

Results and Problems in Cell Differentiation, Vol. 33
Paul R. Crocker (Ed.): Mammalian Carbohydrate Recognition Systems
© Springer-Verlag Berlin Heidelberg 2001

A

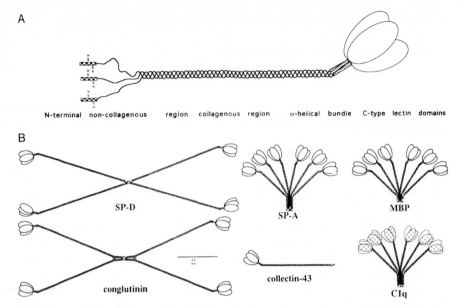

N-terminal non-collagenous   region collagenous region   α-helical bundle   C-type lectin domains

B

SP-D

conglutinin

SP-A

collectin-43

MBP

C1q

Fig. 1A,B. Domain organisation and assembly of the collectin molecules. **A.** The primary structure of each of the collectins is organised into four regions: an N-terminal, non collagenous region involved in the formation of interchain disulphide bonds, a collagenous region composed of Gly-Xaa-Yaa repeats, an α-helical neck peptide, and a C-terminal C-type CRD domain. The collectins are large oligomeric structures, each assembled from multiple copies of a single polypeptide chain (with the exception of SP-A, where two types of chain are formed). The C-type CRDs are spaced, in a trimeric orientation, at the end of triple helical collagenous stalks. It is becoming increasingly clear that the CRD domains, by binding to carbohydrate ligands on the cell surface of the pathogens, fulfill a recognition function that can bring about effector functions, such as complement activation (by MBL) or phagocytosis (by SP-A and SP-D; Hoppe and Reid 1994). **B.** SP-D and BC appear to be cruciform under the electron microscope, with four arms of equal length ending in globular heads. Oligomers of SP-D are also found. MBL and SP-A seem to resemble serum complement protein, C1q in their overall organisation. The bovine CL-43 exists as monomers (Hoppe and Reid 1994)

2 Mannose-Binding Lectin (MBL)

MBL, a Ca^{2+}-dependent lectin synthesized primarily in the liver, is an acute phase reactant which shows a modest (1.5- to 3-fold) rise in serum levels during stress or infection. The presence of glucocorticoid responsive elements and a cytokine responsive element homologue in the 5′ flanking region of MBL genes (Sastry et al. 1989) suggests that MBL is an acute phase reactant.

MBL binds to carbohydrate structures found on a wide range of pathogenic organisms. On binding to these carbohydrate structures, MBL can bring about activation of the serum complement system, thus allowing recruitment of a variety of inflammatory, killing and clearance mechanisms, in an antibody-independent manner. MBL is, therefore, considered to play an important role in innate immunity – especially in very young children or immunodeficient individuals. MBL is also found in amniotic fluid, nasal secretions, middle ear fluid, saliva and inflamed sites – such as rheumatic joint fluid (Turner 1996a).

2.1 Molecular Structure and Assembly of MBL

MBL is composed of multimers of identical polypeptide chains of 32 kDa. Three 32-kDa chains combine to make a structural subunit of 92 kDa and oligomeric forms of MBL are composed of two to six of the 92 kDa subunits. Each 32-kDa chain is composed of: an N-terminal region containing cysteine residues involved in inter-chain and inter-subunit disulphide bonding, a region of Gly-Xaa-Yaa repeating triplets which is involved in the formation of collagen-like triple-helical structure, an α-helical neck region of approximately 34 residues, and a C-terminal globular domain which contains the 14 invariant and 18 highly conserved amino acid residues characteristic of the 120-residue C-type carbohydrate recognition domain (CRD). In the electron microscope, the highest oligomeric form of the MBL, the hexamer of the 92-kDa structural unit, appears as a bouquet-like structure with six globular 'heads' each connected by collagen-like strands to a central core (Lu et al. 1990). The largest oligomeric form of serum MBL, which consists of 18 identical 32-kDa polypeptide chains arranged as a hexamer of timers of these chains, has a molecular weight of 576 kDa. MBL isolated from liver appears to be composed of six identical 32-kDa chains and has a molecular weight of 192 kDa. There are seven cysteine residues at positions 5, 12, 18, 135, 202, 216 and 224 (based on numbering of the 32-kDa chain of the mature protein). The disulphide bond arrangement has not been determined, but the residues at positions 5, 12, and 18 are probably involved in inter-chain and inter-subunit disulfide bonds. The remaining four residues are expected to form two intra-chain disulphide bonds (135 to 224 and 202 to 216) characteristic of that observed in other C-type lectin domains.

2.2 Biological Functions of MBL

MBL renders innate immunity by (1) recognition of carbohydrate structures on the pathogens via CRDs, and (2) activation of the serum complement system in order to recruit its inflammatory, opsonisation and killing mechanisms. Two possible mechanisms have been proposed for the MBL-mediated activation of the classical complement pathway. Initially it was shown that the higher oligomers (pentamers and hexamers) of MBL could interact with the proenzyme C1r2-C1s2 complex after binding to mannan-coated erythrocytes (Lu et al. 1990). The C1-esterase activity expressed by the bound complex results in C4 and C2 cleavage, leading to assembly of the C3 convertase of the classical pathway (C4b2a; Ohata et al. 1990). However, it is now considered that the MBL oligomers normally circulate with two serine proteases, designated MBL-associated serine proteases (MASP-1 and MASP-2), each of which show approximately 40% amino acid sequence identity to the complement enzymes C1r and C1s (Matsushita and Fujita 1992; Thiel et al. 1997). The MASP-2 shows functional similarity to C1s since it activates the C4 and C2 components of complement after interaction of MBL with targets, such as mannan. However, at present it is not clear what the stoichiometry is of the MBL-MASP-1/MASP-2 complexes and what the precise role of each of the MASP enzymes within such complexes is.

2.3 Interaction of MBL with Micro-organisms

The MBL CRDs bind glycans terminating with N-acetylglucosamine or mannose. Such terminal residues are relatively rare in mammalian tissues (Rademacher et al. 1988) but occur more commonly on microbial surfaces. MBL, via its CRDs, binds to a wide range of pathogens, which include Gram-negative and Gram-positive bacteria, yeasts, viruses, mycobacteria and parasites. Complement activation, brought about by the MBL-MASP-1/MASP-2 complexes interacting with the pathogen surface, results in coating of the target pathogen with large amounts of activated C4 and C3 which leads to opsonisation. One of MBL's main functions is the enhancement of the killing and clearance of pathogens by bringing about antibody-independent activation of the complement system, thus zero, or low, levels of MBL may greatly increase risk to certain infections in young children and immunodeficient individuals (Turner 1996b).

Native and recombinant MBL have been shown to bind wild-type virulent *Salmonella montevideo*, expressing a mannose-rich O-polysaccharide (Kuhlman et al. 1989), leading to uptake and killing by phagocytosis. It is likely that MBL-mediated opsonisation involves a complement amplification process and the deposition of C3b / iC3b on microbial surfaces. Non-encapsulated *Listeria monocytogenes*, non-encapsulated *Haemophilus influenzae* and non-encapsulated *Neisseria meningitidis*, *N. cirera* and *N. subflava* show avid binding to MBL. *Streptococci*, *Escherichia coli*, and *N. meningitidis* serogroup show intermediate binding, while encapsulated *N. meningitidis*, *H. influenzae* and *S. agalactive* show weak affinity towards MBL. In binding to different isogenic mutants of *N. meningitidis*, the structure of lipopolysaccharide (LPS) appears to be an important determinant (Jack et al. 1998). The sonicates of *Mycobacteria leprae* and *M. tuberculosis* bind strongly to MBL, probably because of the high D-mannose content found on the surface of these pathogens. MBL also binds to acapsular *Cryptococcus neoformans* and mediates agglutination (Schelenz et al. 1995).

MBL is known to inhibit both haemagglutinin (HA) activity and infectivity of several strains of viruses, in addition to acting as an opsonin and enhancing neutrophil reactivity against the virus. It has been shown that MBL, as a β inhibitor of influenza A virus (IAV) acts by binding to high mannose structures and masking the cell attachment site of HA (Hartshorn et al. 1993). MBL is also capable of initiating complement mediated neutralisation of IAV (Anders et al. 1990). It has been demonstrated that MBL inhibits HIV infection of CD4[+] lymphoblasts and binds to HIV-infected U937 cell lines (Ezekowitz et al. 1989). It can further activate the classical complement pathway by binding to *gp120* from HIV-1 and *gp110* from HIV-2 (Haurum et al. 1993).

2.4 Gene Organisation and Genetics of MBL

The 7 kb gene encoding human MBL is located on the long arm of chromosome 10, within a gene cluster, at 10q11.2-q23, which also includes genes for SP-A, SP-D and a pseudogene of SP-A. Four exons encode the four distinct regions seen in the mature 32-kDa polypeptide chain of MBL: the N-terminal cysteine rich region (exon 1), a collagenous region (exons 1 and 2) and an α-helical neck region (exon 3), followed by a CRD region (exon 4).

There are four allelic forms of the MBL gene which provide structural variants of the MBL polypeptide chain. These allelic forms are designated A, B, C and D. The allele A, being the most common, is taken as the normal, or wild-type, form. In the B and C alleles, one glycine residue is replaced by aspartic acid (allele B) or glutamic acid (allele C) within the collagenous region. In the D allele, an arginine residue in the collagenous region is replaced by a cysteine residue (Madsen et al. 1994). Each of these three substitutions probably affects the formation of a stable collagen-like triple-helix. The homozygotes with respect to the B, C, or D alleles have undetectable or trace amounts of MBL. The heterozygotes (A/B, A/C or A/D) have lower levels of MBL (approximately 15% of that of A/A homozygous individuals). The A/A homozygous individuals can also have low levels of MBL since there are variants within the promotor region of the gene which influence serum levels (Madsen et al. 1995; Turner 1996b). An average value of 1 µg/ml is found in the sera of Caucasians, but there is a wide variation in individual values (0–5 µg/ml), due to the B, C and D alleles and variants in the promoter region.

2.5 Crystal Structure of Trimeric CRDs of MBL

The crystal structure of the trimeric CRDs of MBL, together with the neck region, shows that each CRD is approximately $45 \times 25 \times 25$ Å, and contains two calcium ions, designated sites 1 and 2 (Weis et al. 1991, 1992, 1998; Sheriff et al. 1994). The CRD starts in a β-strand, followed by a ten residue α-helix, an extended stretch of ten residues, and a second α-helix. A short turn after this helix leads into a second β-strand that turns sharply at a conserved glycine (158 of MBL-A). The backbone then enters a region with no regular secondary

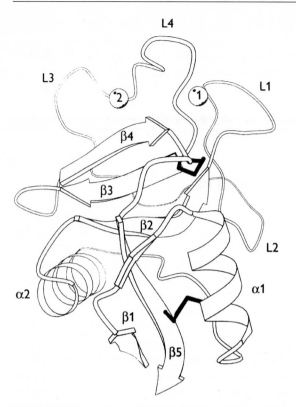

Fig. 2. Ribbon diagram of MBL-A CRD. The α-helices, β-strands, and loops are labelled as *α, β, and L*. The two Ca²⁺ are represented by *spheres*, and the disulphide bonds are shown in *black* (Weis et al. 1998)

structure, containing two loops, an extended stretch, and two more loops, for a total of about 45 residues (Fig. 2). After the fourth loop, the backbone forms two anti-parallel β-strands that end in a tight turn. The structure enters a short loop, and finishes with a β-strand that pairs in an anti-parallel orientation with the first N-terminal strand. The pairing of the first and last β-strands places the beginning and the end of the CRD next to each other. Structure-based sequence alignment of the CRDs reveals the essential sequence determinants of the fold (Weis et al. 1991). There are two disulphide bonds formed by four invariant cysteines: the outer disulphide links the first α-helix to the last β-strand, and the inner disulphide links the beginning of the third β-strand to the loop following the fourth β-strand. A second set of conserved residues form the calcium binding sites.

MBL has a broad carbohydrate specificity well-suited to recognise a variety of pathogenic surfaces. The common feature of these sugars is the presence of equatorial hydroxyl groups in the stereochemistry of the 3- and 4-OH groups of D-mannose. The structural basis of MBL and carbohydrate specificity has been investigated by high resolution X-ray crystallographic analysis of the two rat proteins, serum MBL-A and liver-associated MBL-C. The structure of MBL-A complexed with a high mannose oligosaccharide at 1.7 Å resolution (Weis et al. 1992), and a series of MBL-C structures complexed with methyl glycosides of mannose, N-acetylglucosamine, and fucose at 1.7–1.9 Å, have revealed that the binding occurs through direct co-ordination of the site 2 Ca^{2+} involving covalent bonding of the 3- and 4-OH groups of the ligand (Fig. 3 A). In the interaction of mannose with MBL, there are three non-polar van der Waals contacts: the C4 of mannose contacts the Cβ of residue 189 of MBL-A, the 2-OH contacts a carbon atom, and the exocyclic C6 contacts the terminal methyl group of Ile-207 (Fig. 3 B). Site-directed mutagenesis has revealed that only C4-Cβ contact is energetically significant (Iobst et al. 1994): replacement of His-189 with Ala, and that of Ile-207 with Val, has little effect on ligand binding. However, changing His-189 to Gly, which removes the Cβ contact, reduces the affinity for mannose significantly. Complexes of MBL-C with methyl glycosides of N-acetylglucosamine and fucose confirm that the contacts with the hydroxyl groups equivalent to the 3- and 4-OH groups are the principal determinants of the recognition. Di-, tri-, and higher oligomannose oligosaccharides complexed with MBL-A and MBL-C confirm that the site interacts only with a single sugar moiety of the ligand (Weis et al. 1992).

Despite their affinity for a range of monosaccharides, MBLs do not trigger complement-mediated lysis or opsonic reaction to host cells. This can partly be explained by the relative sparsity of terminal mannose, GlcNAc, and fucose residues on vertebrate glycoproteins and glycolipids. The crystal structures of the rat MBL-A (Weis and Drickamer 1994) and human MBL (Sheriff et al. 1994)

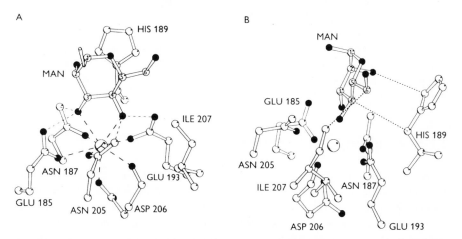

Fig. 3A,B. Binding of mannose at the rat MBP-A calcium site. **A.** Side chain oxygen atoms of residues 185, 187, 193 and 205, and the main chain carbonyl oxygen of residue 206, form the pentagonal, equatorial Ca^{2+} ligands (Weis et al. 1998). **B.** Van der Waals contact (*dashed lines*) between mannose and residues His 189 and Ile 207 (Weis et al. 1998)

trimers provide a likely explanation for the ability of MBL to distinguish non-self from self. A hydrophobic interface between the neck and the CRD maintains a fixed spatial relationship between the two, such that the sugar/calcium binding sites are 53 Å (rat) and 45 Å (human) apart in the trimer. The terminal mannose residues in vertebrate mannose-rich oligosaccharides are about 20–30 Å apart. The binding sites in the MBL trimer are therefore too far apart to interact multivalently with such oligosaccharides. In contrast, pathogenic cell surfaces present dense, repetitive arrays of ligands that can span the distance between binding sites in the MBL CRD trimers, resulting in highly avid multivalent interactions. MBL-A has proven to be an excellent system to probe the determinants of carbohydrate specificity in other C-type lectins. The interactions made by MBP-A and its ligands allow new specificities to be engineered onto the C-type CRD scaffold without affecting the overall structure of the protein. The pair of Glu-185 and Asn-187, when replaced with Gln and Asp, respectively, confers upon MBL-A preferential binding for galactosides, instead of mannose (Drickamer 1992).

3 Surfactant Protein A (SP-A)

SP-A accounts for 5% of the weight of the surface active mixture of phospholipids and proteins which are essential constituents of pulmonary surfactant. Nearly all the SP-A is tightly associated with lipids, such as dipalmitoylphosphatidylcholine (DPCC) and sphingomyelin. SP-A is considered to play an important role in surfactant secretion and uptake by type II alveolar cells, and also in the organisation of tubular myelin. Both SP-A and SP-D appear to provide innate immunity against lung pathogens by binding to carbohydrates on the pathogens, agglutinating them and triggering effector mechanisms which kill and opsonise the pathogens. SP-A can also interact with macrophages and increase their chemotactic, phagocytic and oxidative properties. It is therefore considered that SP-A may play an important role in the rapid recognition and clearance of pathogens, especially in immunodeficient individuals (Wright 1997).

3.1 SP-A Suprastructure and Assembly

SP-A has a hexameric structure in which six structural subunits of 105 kDa associate to yield a molecule of 630 kDa. Each subunit is composed of three 35-kDa polypeptide chains which are held together by disulphide bonds located in the N-terminal halves of the chains. The overall shape of SP-A is very similar to that of the complement protein C1q, both molecules appearing in the electron microscope as a bouquet-like structures with six globular heads linked by collagen-like strands to a fibril-like central core. The mature forms of the two SP-A polypeptide chains designated α_2 (product of the SP-A II gene) and α_3 (product of the SP-A I gene) are both composed of 248 residues which include: an N-terminal segment (7 residues), a collagen-like region (73 residues), the neck region (26 residues) and a CRD domain (123 residues). There is one N-linked glycosylation site at Asn-198 in each chain. There are seven cysteine residues in the mature α_3 polypeptide and six in the mature α_2 polypeptide. The cysteines at positions 6, 48 and 65 in the α_3 polypeptide are considered to form inter-chain disulphide bonds and are involved in oligomer formation (Elhalwagi et al. 1997). The remaining four cysteine residues in each chain are the conserved intra-chain cysteine residues found within the CRD domain. The collagen region contains 23–24 Gly-Xaa-Yaa repeats. Prolines are present in the X position of 4–5 triplets, and at the Y position of 15–16 triplets, most of which are hydroxylated. These hydroxyprolines greatly enhance the stability of collagen triple-helix. The stability of the triple-helix is also contributed to by the alternating charged residues present in the two halves of the collagen domain of SP-A, the two halves being separated by an interruption following the 12th triplet, called the 'hinge region'. As in the C1q, the hinge region introduces a bend in the collagen segment which tilts the CRD trimers away from the core in the whole SP-A molecule. The 38–40 residue long SP-A α-helical coiled-coil neck region is organised into 'heptad repeats'. The co-operative interactions between the neck and CRDs may contribute to SP-A interactions with type II cells (Sano et al. 1998). The CRDs of SP-A have been shown to mediate a variety of interactions including modulation of alveolar type II cell functions, binding and aggregation of phospholipids, and recognition of bacterial, viral and fungal organisms. Mutational analysis has suggested that the carbohydrate binding site of SP-A co-localises with a major Ca^{2+}-binding site (McCormack et al. 1994; Sano et al. 1998).

3.2 SP-A Gene and Genomic Organisation

Two transcribed SP-A genes (SP-A I and SP-A II) and one pseudogene have been localised to chromosome 10q21–24, within a cluster that includes the SP-D and MBL genes (Hoover and Floros 1998). Human SP-A contains two types of chain (α_2 and α_3) and it has been proposed that each trimeric subunit has one α_2 chain and two α_3 chains. The sequences and genomic organisations of the SP-A I and SP-A II genes are very similar, each being composed of seven exons. The expressed proteins are each encoded within four exons (I-IV): I, covering part of the 5′ untranslated region, the leader peptide, the N-terminal region and part of the collagen-like sequence; II, covering the remainder of the collagen-like sequence; III, the α-helical neck sequence; and IV, the CRD plus the 3′ untranslated sequences. Allelic variants of each gene are generated by splicing variability in the 5′ untranslated regions and by sequence variability in the 3′ untranslated regions (Floros and Hoover 1998).

The SP-A mRNA and protein are expressed in epithelial cells and the non-ciliated bronchiolar cells (Clara cells) of the terminal bronchioles and conducting airways (Khoor et al. 1993). It has also been detected in the serous glands of proximal human trachea, in the endocytic compartment of macrophages, rat intestinal epithelia, human and rat mesentery and human inner ear. SP-A is synthesised as a precursor with a 17- to 28-residue-long leader sequence. After synthesis, SP-A is hydroxylated in the endoplasmic reticulum (ER) by the enzyme prolyl hydroxylase, via the formation of 4-hydroxyproline. SP-A is also co-translationally glycosylated with mannose-rich carbohydrates in the ER, followed by post-translational modifications, such as sialylation and sulphation, contributing to the charge heterogeneity of the molecule (McCormack 1998).

3.3 SP-A-Carbohydrate Interaction

The ability of various monosaccharides to compete for binding of SP-A to mannan, a yeast-derived polymer of mannose, has the order: N-acetylmannosamine >L-fucose, maltose >glucose >mannose. Galactose, D-fucose, glucosamine, mannosamine, galactosamine, N-acetylglucosamine and N-acetylgalactosamine do not inhibit mannan binding. SP-A binds specifically to mannose-rich carbohydrates which are likely components of bacterial lipo-oligosaccharides or capsular oligosaccharides. The CRDs of SP-A bind avidly, in a Ca^{2+} - dependent manner, to LPS via the lipid A domain, whose structure closely resembles that of phosphatidylcholine. SP-A shows an acute phase response to LPS aerosolisation (van Helden et al. 1997), leading to upregulation of SP-A receptor expression on the macrophage surface. SP-A also binds, via its CRDs, to the major surface glycoprotein of *Pneumocystis carinii*, implicated in pneumonia in immunosuppressed subjects (Zimmerman et al. 1992). SP-A is also known to bind to glycolipids, such as lactosylceramide and galactosylceramide (Childs et al. 1992).

3.4 SP-A-Phospholipid Interactions

SP-A forms a part of the mixture of phospholipid and protein which lines the alveolar space and acts to reduce surface tension forces and prevent atelectasis during expiration. It can function as an inhibitor of phospholipid secretion by alveolar type II cells, via interaction with a high affinity receptor. SP-A binds tightly to dipalmitoylphosphatidylcholine (DPCC) and galactosylceramide (GalCer) and preferentially enhances DPCC uptake by type II cells, as well as the incorporation of this lipid into lamellar bodies. Since DPCC is the principal component responsible for the biophysical properties of pulmonary surfactant, SP-A may play an important role in phospholipid homeostasis in the alveolar space. Mutational studies have implicated the CRD as the major phospholipid interaction site (McCormack et al. 1994). The lipid binding site may overlap with the Ca^{2+}/carbohydrate binding site. SP-A is considered essential for the formation of tubular myelin and other surfactant aggregates, and in concert with surfactant protein B, facilitates rapid adsorption and spreading of surface active phospholipids at the air-liquid interface of the alveoli (Suzuki et al. 1989). SP-A gene-deficient mice have little or no tubular myelin (Korfhagen et al. 1996). SP-A augments the adsorption of phospholipids to an air-liquid interface and improves surface tension of cycled surfactant mixtures *in vitro*. A possible involvement of the N-terminal region, and not the collagenous region, in lipid aggregation, is suggested by the fact that disruption of the interchain disulphide bond at Cys-6 completely blocks lipid vesicle crosslinking by SP-A (McCormack et al. 1994).

3.5 SP-A-Type II Cell Interaction

The alveolar type II cells synthesise and secrete surfactant, and also internalise surfactant from the alveolar space. The interaction of SP-A with type II cells has been shown *in vitro* to inhibit lipid secretion and to promote the uptake of lipid by cells, suggesting a role of SP-A in the regulation of surfactant turnover and metabolism. SP-A interacts with the type II cells via its CRDs. This interaction involves a small disulphide loop containing Cys-204 and Cys-218 (Kuroki et al. 1988) and a region of Glu-202 to Met-207. Mutants of rat SP-A, in which Glu-195 was changed to Gln and Arg-197 to Asp, converted SP-A from a mannose-binding to a galactose-binding lectin (McCormack et al. 1994). These mutations also reduce SP-A binding to the type II cells, confirming the CRD as the interacting region. The CRDs have also been implicated in the ability of SP-A to increase lipid uptake (Kuroki et al. 1988). However, SP-A interaction with type II cells and regulation of type II cell function is complex in nature and needs further investigation.

3.6 Interaction of SP-A with Phagocytes

Binding of SP-A to alveolar macrophages has been found to be calcium-, temperature- and concentration-dependent, with an apparent K_d of 2–4 nm (Kuroki et al. 1988; Wintergerst et al,1989). It is inhibitable by mannosyl-BSA, the CRD of SP-A, and the collagen region of C1q, suggesting the involvement of both collagenous and CRD regions (Wintergerst et al. 1989). Studies on the functional implications of SP-A binding to alveolar macrophages, neutrophils, peripheral blood monocytes, monocyte-derived macrophages, and bone-marrow derived macrophages have formed the basis for the SP-A role in pulmonary defense (Table 1).

SP-A increases intracellular calcium and inositol triphosphate (IP3) concentrations in alveolar macrophages. The calcium response correlates with IP3 generation, and is necessary for SP-A stimulated phagocytosis. SP-A also stimulates chemotaxis via cell interaction involving its collagen region. It also stim-

Table 1. Biological functions proposed to be mediated by SP-A

Surfactant homeostasis
Receptor-mediated inhibition of surfactant secretion from type II cells
Receptor-mediated enhanced uptake of surfactant phospholipids by type II cells
Surfactant biophysical activity
Enhanced phospholipid adsorption to the monolayer
Prevention of protein inhibition by proteinaceous pulmonary oedema
Maintenance of tubular myelin
Phospholipase A_2 inhibition
Myosin clearance
LPS clearance
Host defense functions
Microbial binding and aggregation
Macrophage and neutrophil activation and chemotaxis
Enhanced microbial phagocytosis and killing (via superoxidative burst)
Antiproliferative and antiinflammatory effects on lymphocytes
Modulation of allergic reactions

---

ulates directional actin polymerisation (Tino and Wright 1996). These two effects are preceded by receptor binding and transmission of intracellular signals. SP-A also modulates the production of several mediators of inflammation, such as cytokines (TNF-α, CSF) which control inflammatory response and recruitment of other immune cells and reactive species, leading to tissue damage (Tino and Wright 1998).

3.7 Interaction of SP-A with Pathogens and Allergens

SP-A has been shown to act as an opsonin for herpes simplex virus type I via alveolar macrophages. Deglycosylated SP-A does not bind to infected HEp-2 cells expressing viral proteins over the cell surface (Van Iwaarden et al. 1992). SP-A has been shown to bind IAV via its sialic acid residues and thereby neutralises the virus (by inhibiting virus-mediated agglutination of RBCs). Deglycosylation of SP-A completely prevents binding to IAV (H3N2)-infected HEp-2 cells. SP-A, but not SP-D, has been shown to bring about phagocytosis of H3N2 by rat alveolar macrophages (Benne et al. 1995). Preincubation of IAV with SP-A also enhances the ability of the virus to stimulate the respiratory burst of neutrophils (Hartshorn et al. 1997), however, without protecting neutrophils against virus-induced deactivation.

Increased attachment of SP-A-coated *Staphylococcus aureus* to macrophages has been reported, but this interaction does not lead to phagocytosis (McNeely and Coonrod 1993). SP-A enhances the binding and opsonisation of *E. coli* J 5 (containing O-antigen deficient rough LPS), but not *E. coli* O 111 (with O-antigen containing smooth LPS), to macrophages, suggesting that SP-A binding to Gram-negative bacteria is dependent on LPS structure (Pikkar et al. 1995). SP-A has been reported to bind *Streptococcus pneumoniae* (Group A *Streptococcus* and Group B *Streptococcus*; Tino and Wright 1996). SP-A binds, aggregates, and promotes phagocytosis by macrophages of *H. influenzae*. Low binding to *H. influenzae* type b without agglutination and phagocytosis has also been reported (Tino and Wright 1996). SP-A binds Bacillus Calmette-Guerin (BCG) via CRDs and brings about phagocytosis by several types of phagocytic cells (Weikert et al. 1997). SP-A enhances phagocytosis, agglutination, and killing of *Klebsiella pneumoniae* K21a strain (capsule containing Man α1 Man sequences), either via opsonisation, or through activation of macrophages via the mannose receptor (Kabha et al. 1997). *Mycoplasma pulmonis* is considered to be involved in pneumonia and exacerbation of asthma and chronic obstructive pulmonary disease (COPD). SP-A binds, independent of calcium and sugar, and brings about phagocytosis and mycoplasmal killing by interferon-γ-activated murine alveolar macrophages. It also involves generation of peroxynitrite by alveolar macrophages (Hickman-Davis et al. 1999). SP-A also enhances phagocytosis of *M. tuberculosis* by a direct interaction with human monocyte-derived macrophages and human alveolar macrophages, possibly via upregulation of the macrophage mannose receptor activity. Carbohydrate moieties on SP-A seem important for this interaction. SP-A also enhances the attachment of *M. tuberculosis* to alveolar macrophages, and thus may contribute to the development of tuberculosis in patients with HIV infection (Downing et al. 1995).

SP-A contributes to the clustering of *P. carinii in vivo* by interacting with *gpA*, a mannose- and glucose-rich glycoprotein expressed on cysts and trophozoites of the pathogen (Crouch 1998). SP-A has recently been shown to bind and agglutinate, via CRDs, pathogenic unencapsulated *C. neoformans*, but not the capsulated forms (Schelenz et al. 1995). SP-A and SP-D have been shown to agglutinate *Aspergillus fumigatus* conidia and also to enhance binding, phagocytosis and killing of conidia by human alveolar macrophages and circulating neutrophils (Madan et al. 1997a). It appears that SP-A and SP-D may have an important immunological role in the early antifungal defense response in the lung, through inhibiting infectivity of the conidia by agglutination and by enhancing uptake and killing of conidia as well as hyphae of *A. fumigatus* by phagocytic cells.

SP-A has been reported to bind a variety of allergenic pollens such as Lombardy poplar, Kentucky blue grass, cultivated rye and short ragweed, via CRDs (Malhotra et al. 1993). SP-A and SP-D have been shown to bind whole mite extracts (*Dermatophagoides pteronyssinus*, *Derp*) and the purified glycoprotein allergens, in a carbohydrate-specific and calcium-dependent manner and inhibit specific IgE binding to allergens (Wang et al. 1996). SP-A and SP-D can also bind a range of allergens/antigens present in the 3-week culture filtrate of *A. fumigatus* and to purified allergens, *gp45* and *gp55*, and inhibit the ability of allergen specific IgE from aspergillosis patients to bind these allergens (Madan et al. 1997b). The blocking of IgE binding is probably mediated either by steric hindrance posed by the surfactant molecules already bound to the allergens' carbohydrate structures, or by the recognition of the same binding site of both IgE and surfactant proteins. The possible protective roles played by SP-A and SP-D against airborne allergens are further supported by their ability to block allergen-induced histamine release from basophils isolated from patients having allergic broncho pulmonary aspergillosis (ABPA) and asthmatic children (Madan et al. 1997b; Wang et al. 1998). SP-A and SP-D have also been shown to have antiproliferative effects on peripheral blood mononuclear cells, which were isolated from asthmatic children and challenged with *Derp* allergens (Wang et al. 1998). It therefore appears that SP-A and SP-D may modulate the development of asthmatic symptoms by both inhibiting histamine release in the early phase of allergen challenge and suppressing lymphocyte proliferation in the late phase of asthmatic attacks where there is bronchial inflammation. The inhibitory effects of SP-A and SP-D against *Derp* and *A. fumigatus* allergens suggest a general defense role in the allergen sensitisation/ allergic reactions. Recently, IgE has been shown to have a central role in the induction of lung eosinophil infiltration and Th2 cytokine production (Coyle et al. 1996). Antigen-IgE complexes, bound to CD23, allow B cells to facilitate antigen presentation to antigen specific T cells, resulting in a greatly amplified T cell response. Thus, through their ability to inhibit IgE binding to allergens, SP-A and SP-D may represent a novel approach for the treatment of asthma (1) by preventing degranulation of mast cells, and also (2) by possibly inhibiting CD23/IgE-enhanced antigen processing and presentation to $CD4^+$ T cells and subsequent activation of Th2 cytokine production.

4 Surfactant Protein D (SP-D)

SP-D is one of the surfactant proteins found in the air space lining material in the lungs. Although it does bind to specific phospholipids in the lung surfactant, it shows quite different properties to those of the hydrophobic peptides, SP-B and SP-C, which are strongly associated with lipids. Pulmonary SP-D is produced by alveolar type II cells and nonciliated bronchiolar alveolar cells. However, SP-D may not be a lung specific protein since low levels of material antigenically similar to SP-D are found in normal human serum and, animal studies indicate the presence of SP-D, or SP-D like proteins, in gastric mucosa, tracheobronchial, lacrymal and salivary glands (Crouch 1998).

4.1 Molecular Structure and Assembly of SP-D

SP-D is composed of oligomers of a 130-kDa subunit formed from three identical polypeptide chains (43 kDa each) which have an *N*-linked oligosaccharide structure at Asn-70. As judged by electron microscopy, human SP-D is assembled into a 520-kDa tetrameric structure with four of the 130-kDa, homotrimeric subunits linked via their N-terminal regions, but trimers, dimers and monomers of the 130-kDa subunit are also seen in SP-D preparations. The triple-helical arms in each 130-kDa subunit are approximately 46 nm in length and although appearing flexible, show no sharp bends or distribution which is consistent with there being no interruptions to the Gly-Xaa-Yaa repeat in the collagenous region of SP-D. Clusters of three CRDs are held together by the α-helical coiled-coil region found at the C-terminal end of the collagen-like triple helix present in each 130-kDa subunit. Up to eight of the 520-kDa tetrameric structures can undergo further oligomerization to give SP-D multimers having a large array, of up to 96 (8 × 12), CRDs (Crouch 1998). Each chain contains four distinct regions: a 25-residue-long N-terminal distinct region which contains cysteine residues involved in inter-chain disulphide bonding, a 177-residue-long collagen region, a 28-residue-long neck region, and a 125-residue-long CRD region. Unlike SP-A, the SP-D collagen region contains hydroxylysine and hydroxylysylglycosides. An important structural feature is the repeating heptad pattern of hydrophobic residues, in the 'a' and 'd' positions, seen in the α-helical coiled-coil neck region of the SP-D molecule. The alignment of these hydrophobic residues allows the formation of a self-associating triple-stranded parallel α-helical bundle which determines the trimeric orientation of the CRDs and possibly also acts as a nucleation point for triple-helix formation (Hoppe et al. 1994). The two cysteines at positions 15 and 20 are considered to be involved in the inter-chain disulphide bonding while the four within the CRD region are considered to be involved in intra-chain disulphide bonding.

4.2 Interaction of SP-D with Carbohydrate and Lipid Ligands

The order of preference of human SP-D in solid phase competition assays using maltosyl-BSA as the ligand is maltose >glucose, mannose, fucose >galactose, lactose, glucosamine >N-acetylglucosamine. SP-D is known to bind to the glucose-containing core oligosaccharides of LPS (Kuan et al. 1992), and mannose-rich N-linked oligosaccharides of the haemagglutinin of IAV and the *gpA* of *P. carinii*. SP-D shows high affinity binding to phosphatidylinositol (PI) and glucosyl-ceramide in a calcium- and sugar-dependent manner (Persson et al. 1992). Substituting Glu-321 to Gln and Asn-323 to Asp in the CRDs reverses the relative carbohydrate binding specificity from maltose >glucose >galactose to galactose >maltose, glucose. Phospholipid binding by SP-D appears to involve neck as well as CRD regions (Kishore et al. 1996).

4.3 Interaction of SP-D with Pathogens and Allergens

SP-D binds to carbohydrates/LPS on the surfaces of a variety of pathogens such as IAV, Gram-negative bacteria (*E. coli*, *Salmonella*, *Pseudomonas aeruginosa*,

K. pneumoniae) and fungal organisms (*C. neoformans, P. carinii, A. fumigatus*). The binding of SP-D to these organisms leads to a reduction in infectivity, agglutination and enhanced killing (Reid 1998).

SP-D CRDs bind the glycoconjugates expressed near the sialic acid binding site on the haemagglutinin (or neuraminidase) of specific strains of IAV (Hartshorn et al. 1996) and act as potent inhibitors of HA-mediated agglutination, thereby causing viral aggregation. SP-D is a more potent inhibitor of IAV infectivity than SP-A or MBL. The susceptibility of various IAV strains to neutralisation by SP-D directly correlates with specific differences in the number of glycoconjugates expressed on the HA (Reading et al. 1997). Inoculation of mice with IAV in the presence of mannan is known to increase viral replication, suggesting involvement of the CRDs.

SP-D binds to glycoconjugates (LPS) expressed by a variety of Gram-negative bacterial strains involved in lung pathogenesis, such as *K. pneumoniae, Ps. aeruginosa, H. influenzae,* and *E. coli.* In contrast to SP-A, SP-D binding to LPS and subsequent agglutination of bacteria is calcium-dependent and inhibitable by competing sugars, LPS and rough mutant forms of LPS (but not by lipid A; Kuan et al. 1992). SP-D does not recognise the capsular polysaccharides of *K. pneumoniae,* and the presence of a well-formed capsule limits interaction of SP-D with underlying LPS molecules. SP-D contributes to the clustering of *P. carinii in vivo* by interacting with *gpA,* a mannose- and glucose-rich glycoprotein expressed on cysts and trophozoits (Crouch 1998). SP-D has recently been shown to bind and agglutinate, via CRDs, pathogenic unencapsulated *C. neoformans* (but not the capsulated) forms (Schelenz et al. 1995).

4.4 SP-D Gene Organisation and Genetics

Human SP-D is encoded by a single gene with seven exons spanning >11 kb of DNA on the long arm of chromosome 10, at a locus on 10q22.2–23.1, which also includes the SP-A genes. The first protein-encoding exon (exon 2) includes sequences corresponding to the signal peptide, N-terminal region and first seven Gly-Xaa-Yaa triplets. The remainder of the collagen region (the total of 59 Gly-Xaa-Yaa repeats), is encoded by three exons of 117 bp each. The sixth exon encodes the coiled-coil neck region, whereas the seventh exon codes for the CRD.

4.5 SP-D Crystal Structure

The crystal structure of a trimeric fragment of human SP-D, composed of an α-helical coiled-coil 'neck' region and the three CRD domains, have recently been determined at 2.3 Å resolution (Hakansson et al. 1999). The fold of the CRD is similar to that of MBL. The novel central packing of one of the tyro-

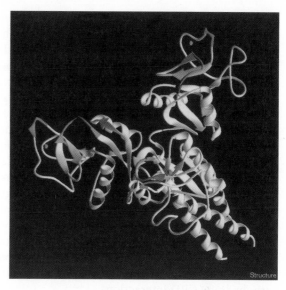

Fig. 4. Ribbon diagram showing the overall main chain structure of the trimeric α-helical coiled-coil and three lectin domains of human lung surfactant protein D. The positions of the three carbohydrate-binding calcium ions, one in each monomer, are shown as black spheres

sine side chains within the coiled-coil results in an asymmetric orientation of the CRDs. There are three calcium ions bound to each SP-D monomer (Fig. 4), one at the carbohydrate binding site and two in a second site previously described for MBL (Weis et al. 1992). The average distance between the two carbohydrate binding calcium ions is 51 Å. The position of the conserved calcium ligands in SP-D (Glu 321, Asn 323, Glu 329, Asn 341 and Asp 342) matches very closely with those of MBL-A. There is a central cavity between the three CRD domains which presents a positively charged surface, which has been suggested to be a putative interacting region for LPS, or cell surface receptors.

5 Cell Surface Receptors for Collectins

SP-A, MBL and conglutinin, but not SP-D, have been shown to bind to a molecule, designated as the cC1qR/collectin receptor, which is now generally regarded as being a membrane associated form of an intracellular, multifunctional Ca^{2+}-binding protein, called 'calreticulin' (Eggleton et al. 1997). However, given its primarily intracellular localisation and lack of membrane anchorage and signal transduction components, its candidature as a receptor molecule is debatable. The binding of C1q, SP-A and MBL, but not SP-D, to another cell surface molecule, defined as $C1qR_p$ has also been demonstrated (Nepomuceno et al. 1997). $C1qR_p$, a 126-kD glycoprotein, is a novel type I membrane protein composed of a C-type CRD, five epidermal growth factor-like domains, a transmembrane domain, and a short cytoplasmic tail. It is expressed on the surfaces of monocytes/macrophages, neutrophils, endothelial cells, and microglia. $C1qR_p$ is considered to mediate phagocytosis by binding to the collagen regions of C1q, MBL and SP-A. Since ligand–$C1qR_p$ ligation triggers phagocytosis without inducing release of proinflammatory cytokines, further dissection of the $C1qR_p$ system may provide a basis for antimicrobial therapy without an inflammatory response. It could be potentially useful in regulating the phagocytic capacity of myeloid cells (a prophylactic treatment for immunocompromised individuals at risk from infection). A putative SP-D receptor, $gp340$, appears to bind SP-D through its CRD regions in a calcium- dependent and sugar- independent manner (Holmskov et al. 1997). The $gp340$ has been described as a new member of a scavanger receptor cysteine-rich superfamily containing multiple scavanger receptor type B domains. Although the interaction between $gp340$ and SP-D involves the CRD region, the inability of maltose to block the interaction indicates a protein–protein interaction, rather than a carbohydrate–CRD interaction.

6 SP-A and SP-D Gene Knock-out Mice

Mice lacking SP-A mRNA and protein *in vivo* [designated SP-A (–/–)] have been generated using gene knock-out technology (Korfhagen et al. 1996). SP-A knock-out mice survive and breed normally, having normal levels of SP-B, SP-C and SP-D, phospholipid composition, secretion and clearance, and incorporation of phospholipid precursors. Lungs of SP-A (–/–) mice have markedly decreased tubular myelin figures. *P. aeruginosa* (Le Vine et al. 1998) and Group B *Streptococci* (Le Vine et al. 1999a) are cleared less efficiently than the wild-type mice. These mice are also more susceptible to respiratory syncitial virus (RSV) infection than the control mice (Le Vine et al. 1999b). These studies on SP-A knock-out mice demonstrate that SP-A has an important role in the innate immune system of the lung *in vivo*.

Mice bred after disruption of the SP-D gene have shown remarkable abnormalities in surfactant homeostasis and alveolar cell morphology. They also show a progressive accumulation of surfactant lipids and apoproteins in the alveolar space, hyperplasia of type II cells with massive enlargement of intracellular lamellar bodies, and an accumulation of alveolar macrophages (Botas et al. 1998). Thus, SP-D deficient mice resemble mice deficient in granulocyte-macrophage colony stimulating factor (GM-CSF; Dranoff et al. 1994). The SP-D receptor, $gp340$, which is a scavenger receptor, may serve as a route for internalisation of lipids in macrophages and may mediate the macrophage proliferation induced by exposure to oxidised low density lipoproteins. However, an association of SP-D and gp340, the GM-CSF pathway, and surfactant clearance remain to be investigated.

7 SP-A and SP-D in Human Diseases

The expression of SP-A and SP-D within the lung makes these collectins specific markers for lung diseases (Kuroki et al. 1998). The measurement of SP-A and SP-D in amniotic fluids and tracheal aspirates reflects lung maturity and the production levels of the lung surfactant in infants with respiratory distress syndrome (RDS). The SP-A concentrations in bronchoalveolar lavage (BAL) fluids are significantly decreased in patients with acute respiratory distress syndrome (ARDS) and also in patients at risk to develop ARDS. A significant increase of SP-A and SP-D in BAL fluids and sputum is diagnostic for pulmonary alveolar proteinosis (PAP). The BAL fluid from the alveolar proteinosis patients forms a very good source of human SP-A (Strong et al. 1998). The SP-A and SP-D concentrations in BAL fluids from patients with idiopathic pulmonary fibrosis (IPF; McCormack et al. 1995) and interstitial pneumonia with collagen vascular diseases (IPCD) are lower than those in healthy controls. SP-A and SP-D appear in the circulation in specific lung diseases, such as PAP, IPF, IPCD and ARDS (Kuroki et al. 1993). SP-A is also a marker for lung adenocarcinomas and can be used to differentiate lung adenocarcinomas from other types and metastatic cancers, and to detect metastasis of lung adenocarcinomas.

8 Bovine Collectins: Conglutinin (BC) and Collectin-43 (CL-43)

To date, conglutinin (BC) and collectin-43 (CL-43) have only been identified as being present in the serum of Bovidae. BC, the first mammalian lectin to be discovered, is known for its ability to agglutinate complement-coated erythrocytes – a reaction called 'conglutination'. Conglutination is brought about via binding of its CRDs to the complement component, iC3b, covalently attached to the erythrocytes (Hirani et al. 1985). BC is a cruciform-shaped tetramer of identical polypeptides, each being 44 kDa. It binds well to oligosaccharides containing nonreducing terminal GlcNAC. BC binds to the high mannose group on the α-chain of the complement degradation product iC3b – which contains the single high mannose oligosaccharide at Asn-917, covalently attached to another protein/carbohydrate via the activated thiol ester (Holmoskov and Jensenius 1996). This binding is inhibitable by GlcNAC and mannose. BC shows opsonising activity *in vitro* toward *Salmonella typhimurium* and *E. coli* (Friis-Christiansen et al. 1990). This activity depends on the presence of the complement system. It is likely that by binding to iC3b deposited on the bacterial surfaces, BC interacts with macrophages. Subcutaneous injection of BC has been shown to protect mice challenged with *Salmonella typhimurium* (Friis-Christiansen et al. 1990). BC also interacts with IAV, as shown by inhibition of virus agglutinating activity as well as by inhibition of infection *in vitro* (Hartley et al. 1992).

CL-43 (molecular mass of single polypeptide 31.5 kDa) is known to exist mostly as monomer, comprising a 28-residue-long N-terminal region, 114 residues of a collagen region (38 Gly-Xaa-Yaa triplets), a 31-residues-long neck region, and a 128-residue-long CRD. CL-43 shows selectivity for L-fucose and mannose, similar to BC. No data concerning the biological role for CL-43 has yet been reported (Holmoskov and Jensenius 1996).

Acknowledgements. We wish to thank Alison Marsland for secretarial help.

References

Anders EM, Hartley CA, Jackson DC (1990) Bovine and mouse serum β-inhibitors of influenza A viruses are mannose-binding lectins. Proc Natl Acad Sci USA 87:4485–4489

Benne CA, Kraaijeveld CA, van-Strijp JA, Brouwer E, Harmsen M, Verhoef J, van Golde LM, van Iwaarden JF (1995) Interactions of surfactant protein A with influenza A viruses: binding and neutralization. J Infect Dis 171:335–341

Botas C, Poulain F, Akiyama J, Brown C, Allen L, Goerke J, Clements J, Carlson E, Gillespie AM, Epstein C, Hawgood S (1998) Altered surfactant homeostasis and alveolar type II cell morphology in mice lacking surfactant protein D. Proc Natl Acad Sci USA 95:11869–11874

Childs RA, Wright JR, Ross GF, Yuen CT, Lawson AM, Chai W, Drickamer K, Feizi T (1992) Specificity of lung surfactant protein SP-A for both the carbohydrate and the lipid moieties of certain neutral glycolipids. J Biol Chem 267:9972–9979

Coyle AJ, Wagner K, Bertrand C, Tsuyuki S, Bews J, Heusser C (1996) Central role of immunoglobulin (Ig) E in the induction of lung eosinophil infiltration and T helper 2 cell cytokine production: inhibition by a non-anaphylactogenic anti-IgE antibody. J Exp Med 183:1303–1310

Crouch EC (1998) Collectins and pulmonary host defense. Am J Respir Cell Mol Biol 19:177–201

Downing JF, Pasula R, Wright JR, Twigg H, Martin W (1995) Surfactant protein A promotes attachment of *Mycobacterium tuberculosis* to alveolar macrophages during infection with human immunodeficiency virus. Proc Natl Acad Sci USA 92:4848–4852

Dranoff G, Crawford AD, Sadelain M, Ream B, Rashid A, Bronson RT, Dickersin GR, Bachurski CJ, Mark EL, Whitsett JA (1994) Involvement of granulocyte-macrophage colony-stimulating factor in pulmonary homeostasis. Science 264:713–716

Drickamer K (1992) Engineering galactose-binding activity into a C-type mannose-binding protein. Nature 360:183–186

Eggleton P, Reid KBM, Kishore U, Sontheimer RD (1997) Clinical relevance of calreticulin in systemic lupus erythematosus. Lupus 6:564–571

Elhalwagi BM, Damodarasamy M, McCormack FX (1997) Alternate amino terminal processing of surfactant protein A results in cysteinyl isoforms required for multimer formation. Biochemistry 36:7018–7025

Ezekowitz RAB, Kuhlman M, Groopman JE, Byrn RA (1989) A human serum mannose-binding protein inhibits *in vitro* infection by the human immunodeficiency virus. J Exp Med 169:185–196

Floros J, Hoover RR (1998) Genetics of the hydrophilic surfactant proteins A and D. Biochem Biophys Acta 1408:312–322

Friis-Christiansen P, Thiel S, Svehag SE, Dessau R, Svendsen P, Andersen O, Laursen SB, Jensenius JC (1990) *In vivo* and *in vitro* antibacterial activity of conglutinin, a mammalian plasma lectin. Scand J Immunol 31:453–460

Hakansson K, Lim NK, Hoppe HJ, Reid KBM (1999) Crystal structure of the trimeric α-helical coiled-coil and the three lectin domains of human lung surfactant protein D. Structure 7:255–264

Hartley CA, Jackson DC, Anders EM (1992) Two distinct serum mannose-binding lectins function as β-inhibitors of influenza virus: identification of bovine serum β-inhibitor as conglutinin. J Virol 66:4358–4363

Hartshorn KL, Sastry KN, White MR, Anders EM, Super M, Ezekowitz RAB, Tauber AI (1993) Human mannose-binding protein functions as an opsonin for influenza A viruses. J Clin Invest 91:1414–1420

Hartshorn KL, Reid KBM, White MR, Jensenius JC, Morris SM, Tauber AI, Crouch E (1996) Neutrophil deactivation by influenza A viruses: mechanisms of protection after viral opsonization with collectins and hemagglutination-inhibiting antibodies. Blood 87:3450–3061

Hartshorn KL, White MR, Shepherd V, Reid KBM, Jensenius JC, Crouch EC (1997) Mechanisms of anti-influenza activity of surfactant proteins A and D: comparison with serum collectins. Am J Physiol 273:L1156–L1166

Haurum JS, Thiel S, Jones IM, Fischer PB, Laursen SB, Jensenius JC (1993) Complement activation upon binding of mannan-binding protein to HIV envelope glycoproteins. AIDS 7: 1307–1313

Hickman-Davis J, Gibbs-Erwin J, Lindsey JR, Matalon S (1999) Surfactant protein A mediates mycoplasmacidal activity of alveolar macrophages by production of peroxynitrite. Proc Natl Acad Sci USA 96:4953–4958

Hirani S, Lambris JD, Müller-Eberhard HJ (1985) Localization of the conglutinin binding site on the third component of human complement. J Immunol 134:1105–1109

Holmskov U, Jensenius JC (1996) Two bovine collectins: conglutinin and CL-43. In: Ezekowitz RAB, Sastry KN, Reid KBM (eds) Collectins and innate immunity. Landes Company, Chapman and Hall, Austin, USA

Holmskov U, Lawson P, Teisner B, TornØe I, Willis AC, Morgan C, Koch C, Reid KBM (1997) Isolation and characterization of a new member of the scavenger receptor superfamily, glycoprotein-340 (gp-340), as a lung surfactant protein-D binding molecule. J Biol Chem 272: 13743–13749

Hoover RR, Floros J (1998) Organisation of the human SP-A and SP-D loci at 10q22-q23. Physical and radiation hybrid mapping reveal gene order and orientation. Am J Respir Cell Mol Biol 18:353–362

Hoppe HJ, Reid KBM (1994) Collectins – soluble proteins containing collagenous regions and lectin domains – and their roles in innate immunity. Protein Sci 3:1143–1158

Hoppe HJ, Barlow PN, Reid KBM (1994) A parallel three-stranded α-helical bundle at the nucleation site of collagen triple-helix formation. FEBS Lett 344:191–195

Iobst ST, Wormald MR, Weis WI, Dwek RA, Drickamer K (1994) Binding of sugar ligands to Ca^{2+}-dependent animal lectins. I. Analysis of mannose binding by site-directed mutagenesis and NMR. J Biol Chem 269:15505–15511

Jack DL, Dodds AW, Anwar N, Ison CA, Law A, Frosch M, Turner MW, Klein NJ (1998) Activation of complement by mannose-binding lectin on isogenic mutants of Neisseria meningitidis serogroup B. J Immunol 160:1346–1353

Kabha K, Schmegner J, Keisari Y, Parolis H, Schlepper-Schaeffer J, Ofek I (1997) SP-A enhances phagocytosis of Klebsiella by interaction with capsular polysaccharides and alveolar macrophages. Am J Physiol 272:L344–L352

Khoor A, Gray ME, Hull WM, Whitsett JA, Stahlman MT (1993) Developmental expression of SP-A mRNA in the proximal and distal respiratory epithelium in the human fetus and newborn. J Histochem Cytochem 41:1311–1319

Kishore U, Wang JY, Hoppe HJ, Reid KBM (1996) The α-helical neck region of human lung surfactant protein D is essential for the binding of the carbohydrate recognition domains to lipopolysaccharides and phospholipids. Biochem J 318:505–511

Kishore U, Eggleton P, Reid KBM (1997) Modular organisation of carbohydrate recognition domains in animal lectins. Matrix Biol 15:583–592

Korfhagen TR, Bruno MD, Ross GF, Huelsman KM, Ikegami M, Jobe AH, Wert SE, Stripp BR, Morris RE, Glasser SW, Bachurski CJ, Iwamoto HS, Whitsett JA (1996) Altered surfactant function and structure in SP-A gene targeted mice. Proc Natl Acad Sci USA 93:9594–9599

Kuan SF, Rust K, Crouch E (1992) Interactions of surfactant protein D with bacterial lipopolysaccharides. Surfactant protein D in an *E. coli*-binding protein in bronocho-alveolar lavage. J Clin Invest 90:97–106

Kuhlman M, Joiner K, Ezekowitz RAB (1989) The human mannose-binding protein functions as an opsonin. J Exp Med 169:1733–1745

Kuroki Y, Mason RJ, Voelker DR (1988) Alveolar type II cells express a high-affinity receptor for pulmonary surfactant protein A. Proc Natl Acad Sci USA 85:5566–5570

Kuroki Y, Tsutahara S, Shijubo N, Takahashi H, Shiratori M, Hattori A, Honda Y, Abe S, Akino T (1993) Elevated levels of lung surfactant protein A in sera from patients with idiopathic pulmonary fibrosis and pulmonary alveolar proteinosis. Am Rev Respir Dis 147:723–729

Kuroki Y, Takahashi H, Chiba H, Akino T (1998) Surfactant proteins A and D: disease markers. Biochem Biophys Acta 1408:334–345

LeVine AM, Kurak KE, Bruno MD, Stark JM, Whitsett JA, Korfhagen TR (1998) Surfactant protein-A-deficient mice are susceptible to *Pseudomonas aeruginosa* infection. Am J Respir Cell Mol Biol 19:700–708

LeVine AM, Kurak KE, Wright JR, Watford WT, Bruno MD, Ross GF, Whitsett JA, Korfhagen TR (1999a) Surfactant protein A binds group B *Streptococcus* enhancing phagocytosis and clearance from lungs of surfactant protein-A-deficient mice. Am J Respir Cell Mol Biol 20:279–286

LeVine AM, Gwozdz J, Stark J, Bruno M, Whitsett J, Korfhagen T (1999b) Surfactant protein-A enhances respiratory syncytial virus clearance in vivo. J Clin Invest 103:1015–1021

Lu J, Thiel S, Wiedemann H, Timpl R, Reid KBM (1990) Binding of the pentamer / hexamer forms of mannan-binding protein to zymosan activates the proenzyme $C1r_2$ $C1s_2$ complex, of the classical pathway of complement, without involvement of C1q. J Immunol 144:2287–2294

Madan T, Eggleton P, Kishore U, Strong P, Aggrawal SS, Sarma PU, Reid KBM (1997a) Binding of pulmonary surfactant proteins A and D to *Aspergillus fumigatus* conidia enhances phagocytosis and killing by human neutrophils and alveolar macrophages. Infect Immun 65:3171–3179

Madan T, Kishore U, Shah A, Eggleton P, Strong P, Wang JY, Aggrawal SS, Sarma PU, Reid KBM (1997b) Lung surfactant proteins A and D can inhibit specific IgE binding to the allergens of *Aspergillus fumigatus* and block allergen-induced histamine release from human basophils. Clin Exp Immunol 110:241–249

Madsen HO, Garred P, Kurtzhals J, Lamm LU, Ryder LP, Thiel S, Svejgaard A (1994) A new frequent allele is the missing link in the structural polymorphism of the human mannan-binding protein. Immunogenetics 40:37–44

Madsen HO, Garred P, Thiel S, Kurtzhals JA, Lamm LU, Ryder LP, Svejgaard A (1995) Interplay between promoter and structural gene variants control basal serum level of mannan-binding protein. J Immunol 155:3013–3020

Malhotra R, Haurum J, Thiel S, Jensenius JC, Sim RB (1993) Pollen grains bind to lung alveolar type II cells (A549) via lung surfactant protein A (SP-A). Biosci Rep 13:79–90

Matsushita M, Fujita T (1992) Activation of the classical complement pathway by mannose-binding protein in association with a novel C1s-like serine protease. J Exp Med 176:1497–1502

McCormack FX (1998) Structure, processing and properties of surfactant protein A. Biochem Biophys Acta 1408:109–131

McCormack FX, Kuroki Y, Stewart JJ, Mason RJ, Voelker DR (1994) Surfactant protein A amino acids Glu195 and Arg197 are essential for receptor binding, phospholipid aggregation, regulation of secretion, and the facilitated uptake of phospholipid by type II cells. J Biol Chem 269:29801–29807

McCormack FX, King T, Bucher BL, Nielsen L, Mason RJ (1995) Surfactant protein A predicts survival in idiopathic pulmonary fibrosis. Am J Respir Crit Care Med 152:751–759

McNeely TB, Coonrod JD (1993) Comparison of the opsonic activity of human SP-A for *Staphylococcus aureus* and *Streptococcus pneumoniae* with rabbit and human macrophages. J Infect Dis 167:91–97

Nepomuceno RR, Henschen-Edan AH, Burgess WH, Tenner AJ (1997) cDNA cloning and primary structure analysis of $C1qR_p$, the human C1q/MBL/SPA receptor that mediates enhanced phagocytosis *in vitro*. Immunity 6:119–129

Ohata M, Okada M, Yamashina I, Kawasaki T (1990) The mechanism of carbohydrate-mediated complement activation by the serum mannan-binding protein. J Biol Chem 265:1980–1984

Persson AV, Gibbons BJ, Shoemaker JD, Moxley MA, Longmore WJ (1992) The major glycolipid recognised by SP-D in surfactant is phosphatidylinositol. Biochemistry 31:12183–12189

Pikkar JC, Voorhout WF, van Golde LMG, Verhoof J, Vanstrijp JAG, Van Iwaarden JF (1995) Opsonic activities of surfactant proteins A and D in the phagocytosis of Gram-negative bacteria by alveolar macrophages. J Infect Dis 172:481–489

Rademacher TW, Parekh RB, Dwek RA (1988) Glycobiology. Annu Rev Biochem 57:785–838

Reading PC, Morey LS, Crouch EC, Anders EM (1997) Collectin-mediated anti-viral host defense of the lung: evidence from influenza virus infection of mice. J Virol 71:8204–8212

Reid KBM (1998) Interactions of surfactant protein D with pathogens, allergens and phagocytes. Biochem Biophys Acta 1408:290–295

Sano H, Kuroki Y, Honma T, Ogasawara Y, Sohma H, Voelker DR, Akino T (1998) Analysis of chimeric proteins identifies the regions in the carbohydrate recognition domains of rat lung collectins that are essential for interactions with phospholipids, glycolipids, and alveolar type II cells. J Biol Chem 273:4783–4789

Sastry KN, Herman GA, Day L, Deignan E, Bruns G, Morton CC, Ezekowitz RA (1989) The human mannose-binding protein gene. Exon structure reveals its evolutionary relationship to a human pulmonary surfactant gene and localisation to chromosome 10. J Exp Med 170:1175–1189

Schelenz S, Malhotra R, Sim RB, Holmskov U, Bancroft GJ (1995) Binding of host collectins to the pathogenic yeast *Cryptococcus neoformans*: human surfactant protein D acts as an agglutinin for acapsular yeast cells. Infect Immun 63:3360–3366

Sherrif S, Chang CY, Ezekowitz RAB (1994) Human mannose-binding protein carbohydrate recognition domain trimerises through a triple α-helical coiled-coil. Nature Struct Biol 1:789–794

Strong P, Kishore U, Morgan C, Lopez Bernal A, Singh M, Reid KBM (1998) A novel method of purifying lung surfactant proteins A and D from the lung lavage of alveolar proteinosis patients and from pooled amniotic fluid. J Immunol Methods 220:139–149

Suzuki Y, Fujita Y, Kogishi K (1989) Reconstitution of tubular myelin from synthetic lipids and proteins associated with pig pulmonary surfactant. Am Rev Respir Dis 140:75–81

Thiel S, Vorup-Jensen T, Stover CM, Schwaeble W, Laursen SB, Poulsen K, Willis AC, Eggleton P, Hansen S, Holmskov U, Reid KBM, Jensenius JC (1997) A second serine protease associated with mannan-binding lectin that activates complement. Nature 386:506–510

Tino MJ, Wright JR (1996) Surfactant protein A stimulates phagocytosis of specific pulmonary pathogens by alveolar macrophages. Am J Physiol 270:L677–L688

Tino MJ, Wright JR (1998) Interaction of surfactant protein A with epithelial cells and phagocytes. Biochem Biophys Acta 1408:241–263

Turner MW (1996a) Mannose-binding lectin: the pleuripotent molecule of the innate immune system. Immunol Today 17:532–540

Turner MW (1996b) Functional aspects of mannose-binding protein. In: Ezekowitz RAB, Sastry KN, Reid KBM (eds) Collectins and innate immunity. Landes Company, Austin, USA

Van Helden HP, Kuijpers C, Steenvoorden D, Go C, Bruijnzeel PL, van Eijk M, Haagsman HP (1997) Intratracheal aerosolisation of endotoxin (LPS) in the rat: a comprehensive animal model to study adult (acute) respiratory distress syndrome. Exp Lung Res 23:297–316

Van Iwaarden JF, van Strijp JA, Visser H, Haagsman HP, Verhoef J, van Golde LM (1992) Binding of surfactant protein A (SP-A) to herpes simplex virus type 1-infected cells is mediated by the carbohydrate moiety of SP-A. J Biol Chem 267:25039–25043

Wang JY, Kishore U, Lim BL, Strong P, Reid KBM (1996) Interaction of human lung surfactant proteins A and D with mite (*Dermatophagoides pteronyssinus*) allergens. Clin Exp Immunol 106:367–373

Wang JY, Shieh CC, You PF, Lei HY, Reid KBM (1998) Inhibitory effect of pulmonary surfactant proteins A and D on allergen-induced lymphocyte proliferation and histamine release in children with asthma. Am J Respir Crit Care Med 158:510–518

Weikert LF, Edwards K, Chroneos ZC, Hager C, Hoffman L, Shepherd VL (1997) SP-A enhances uptake of bacillus Calmette-Guerin by macrophages through a specific SP-A receptor. Am J Physiol 272:L989–L995

Weis WI, Drickamer K (1994) Trimeric structure of a C-type mannose-binding protein. Structure 2:1227–1240

Weis WI, Kahn R, Fourme R, Drickamer K, Hendrickson WA (1991) Structure of the calcium-dependent lectin domain from a rat mannose-binding protein determined by MAD phasing. Science 254:1608–1615

Weis WI, Drickamer K, Hendrickson WA (1992) Structure of a C-type mannose-binding protein complexed with an oligosaccharide. Nature 360:127–134

Weis WI, Taylor ME, Drickamer K (1998) The C-type lectin superfamily in the immune system. Immunol Rev 163:19–34

Wintergerst E, Manz-Keinke H, Plattner H, Schlepper-Schafer J (1989) The interaction of a lung surfactant protein (SP-A) with macrophages is mannose dependent. Eur J Cell Biol 50:291–298

Wright JR (1997) Immunomodulatory functions of surfactant. Physiol Rev 77:931–962

Zimmerman PE, Voelker DR, McCormack FX, Paulsrud JR, Martin WJD (1992) The 120 kDa surface glycoprotein of *Pneumocystis carinii* is a ligand for surfactant protein A. J Clin Invest 89:143–149

Index